Peter Ax

The Phylogenetic System

WITHDRAWN

The Phylogenetic System

The Systematization of Organisms on the Basis of their Phylogenesis

Peter Ax

Georg-August University of Göttingen, German Federal Republic

Translated by Dr. R. P. S. Jefferies
British Museum (Natural History), London

with 90 illustrations

A WILEY–INTERSCIENCE PUBLICATION

JOHN WILEY & SONS

Chichester · New York · Brisbane · Toronto · Singapore

Original edition: Peter Ax, Das phylogenetische System (Systematisierung der lebenden Natur auf-grund ihrer Phylogenese)

© Gustav Fischer Verlag Stuttgart/Akademie der Wissenschaften und der Literatur 1984

Copyright of English edition © 1987 by John Wiley & Sons Ltd.

Library of Congress Cataloging in Publication Data:

Ax, Peter, 1927–
    The phylogenetic system.
    Translation of: Das phylogenetische System.
    'A Wiley–Interscience publication.'
    Includes indexes.
    1. Phylogeny.   2. Biology—Classification.   I. Title.
QH367.5.A913   1987   574'012   86–16002
ISBN 0 471 90754 5

British Library Cataloguing in Publication Data:

Ax, Peter
    The phylogenetic system : the systematization of living organisms on
    the basis of their phylogenesis.
    1. Evolution.
    I. Title   II. Das phylogenetische system
    English
575   QH366.2
ISBN 0 471 90754 5

Printed and bound in Great Britain

Contents

Preface

"Phylogenetic systematics claims to supply, or at least to endeavour to supply, a general system of reference which, on theoretically indisputable grounds, must demand primacy over all other conceivable attempts at classification. This claim is based on the fact that the genealogical connections of individuals, and thus the phylogenetic connections of species, are the only real link that binds all organisms within time. For this link is the only one created by the real process of reproduction."

W. Hennig (1913–1976) in: Aufgaben und Probleme stammesgeschichtlicher Forschung (posthumous, 1984, p.17).

"Hennig's approach to systematics was, in a word, rational. In a field in which prominent authorities have customarily dismissed theoretical problems by characterizing systematics as part science, part art, Hennig achieved preeminence simply by adopting the attitude that systematics is, and ought to be, a logical and scientific endeavor. It is this quality of Hennig's approach that made it possible for him to overcome opposition based on ideas that were only seemingly rational, and it was Hennig's consistent application of scientific methods to systematic problems that made the virtue of the phylogenetic system clear to so many other workers."

J. S. Farris in: The Willi Hennig Memorial Symposium (1979b, p.415)

The study of life developed into evolutionary biology during the last century. But its oldest constituent discipline—systematics—broke through to a strictly rational research strategy only in the middle of the present century—a strategy fully compatible with the basic axioms of evolutionary theory and satisfying the demands that the theory of knowledge makes on a science. For a long time attempts at classification cherished the use of intuition and the art of evaluation as legitimate tools in "dominating the diversity of living Nature". Now, however, systematics in its phylogenetic form can throw these tools aside.

Willi Hennig created the foundations of an objectively testable phylogenetic system of organisms. He did this by basing systematic work consistently upon our insights into the process of phylogenesis. His "Grundzüge einer Theorie der phylogenetischen Systematik" (1950) had little effect in the years after the War. But the publication of "Phylogenetic Systematics" (1966) in the United States "marks a turning point in the history of Systematic Biology" (Rosen, Nelson & Patterson 1979, p.vii), and "Die Stammesgeschichte der Insekten" (1969) was received as "one of the most important milestones in the study of the phylogeny of insects and their allies" (Kristensen 1975, p.1).

"Systematic Zoology"—the journal of the American Society of Systematic Zoology—developed in the seventies into a forum for stormy argument as to the aims of systematics and for embittered debate for and against a consistently phylogenetic system. The rapid development of this revolutionary movement in the English-speaking world is shown by several books (Eldredge & Cracraft 1980, Wiley 1981, Nelson & Platnick 1981), by the published results of various symposia (Cracraft & Eldredge 1979, Nelson & Rosen 1981, Joysey & Friday 1982) and by the "Proceedings of the Meetings of the Willi Hennig Society" (Funk & Brooks 1981, Platnick & Funk 1983). A historical survey is given by Dupuis (1978).

In Germany, likewise, there is a continual increase in contributions to the theory, methodology and terminology of phylogenetic systematics, as well as to its application in practice (Schlee 1969–81; D. S. Peters 1970, 1972; D. S. Peters & Gutmann 1971; Lauterbach 1972–83; Königsmann 1975; Kraus 1976; G. Peters 1978; G. Peters & Klausnitzer 1978; Weygoldt & Paulus 1979; Weygoldt 1979; Dohle 1980; Lorenzen 1981 etc.). Except for the presentation of the "invertebrates" by Hennig in the "Taschenbuch der Speziellen Zoologie" (1972, 1979–1980), however, a consistent phylogenetic systematization has found no appreciable place in the textbooks. Even the newly published parts of Kaestner's "Lehrbuch der Speziellen Zoologie" (1980, 1982) follow conventional lines, regardless of the introduction by the editor Gruner.

The dilemma for academic instruction is therefore obvious. I was motivated to write this book when I discovered that teaching a phylogenetic system of animals was not compatible with the traditionally based classifications in our textbooks of zoology. While taking account of recent advances which seem to me essential, I have sought in this book to present the logical presuppositions and methodological approaches by which the order in living Nature can be discovered—an order which arose in the real historical process of phylogenesis.

This book takes the arguments step by step. In it I intend to give the biology student a critical account of the construction and content of an order-structure that—in a manner consistent with the philosophical theory of knowledge—makes it possible to show the relationships which exist between species and closed communities of descent. And thus it becomes the basis of comparative biology.

The book may also be of interest to university lecturers teaching courses in the diversity of organisms. A clear understanding of the stem lineages of closed descent communities may make it easier to break away from the conventional framework of teaching the "bauplans" or "types" of animals and plants. I have therefore paid particular attention to clarifying two concepts: that of the stem lineage, within which particular agreements shared between present-day representatives of a descent community evolved as novelties; and that of the ground pattern, being the complex of features which each descent community possessed at the end of its stem lineage.

Since nobody can look everywhere at once, I base my exposition as a zoologist on animals—especially on groups such as the arthropods and vertebrates which receive special attention, for other reasons, in teaching. In the chapter on Plathelminthomorpha, however, I concentrate intentionally on the place where I myself can see most clearly. I try to establish what now seem the most likely hypotheses concerning the phylogenetic relationships between the Gnathostomulida and Plathelminthes and link this validation to an extended discussion of the ground pattern of the Bilateria and the basal systematization of this latter taxon.

Phylogenetic systematics can obviously claim to be valid, in principle and method, beyond and outside the animals. It should apply at least to all Eucaryonta which, with more than 20 features common to them alone (Cavalier–Smith 1981), are convincingly established as a great closed descent community in Nature. The advance of phylogenetic systematics into botany is exemplified in the surveys of Bremer & Wanntorp (1981) or Hill & Crane (1982). Its use in explaining the evolution of the "Procaryonta" (Stackebrandt & Woese 1981) should be examined by those competent to judge.

Dr. U. Ehlers critically read the text and made many suggestions for its improvement. Dr. B. Sopott-Ehlers helped in examining the proofs; she and Mrs. D. Bürger compiled the index of animal names. In obtaining literature I was able to rely on Dr. J. Gottwald. Mrs. I.-Chr. Höttermann patiently typed the manuscript. Mr. R. v. Sivers and Mr. B. Baumgart were responsible for draughting the relationship diagrams and other illustrations. I cordially thank all these collaborators.

I also wish to thank Dr. G. Brenner, General Secretary of the Academy of Science and Literature of Mainz, and Mr. B. von Breitenbuch and Dr. W.-D. von Lucius of the publishers Gustav Fischer for their unfailing understanding and willingness to oblige. I am grateful to Mr. L. Henn, of the Mainz Academy, for taking charge of the production.

October 1983, Göttingen.

Preface to the English edition

As against the German edition (1984) an important change in the English edition concerns the problem of the possible "identity" between stem species and particular daughter species—commonly circumscribed with the slogan "survival of stem species". With regard to this question, I adopt the new convincing arguments of Willmann (1985). Each single stem species dissolves in the splitting process into new daughter species, regardless of the possibility that one of the daughter species retains unchanged the genetic constitution and phenotype of the stem species (Chapter B, section I; F, section IV).

I am extremely grateful to Dr. R. P. S. Jefferies for his careful translation of this book and for a constructive exchange of views during the process. I also thank Dr. E. O. Wiley for valuable suggestions made during a visit to Göttingen.

Peter Ax,
February 1986, Göttingen.

A. Introduction

All of Man's attempts to understand the order of living Nature and to express it in an equivalent logical structure begin from shared identical features, similarities and differences between organisms. Nobody will deny that the blackbird and the mute swan, on the one hand, and the honey bee and the peacock butterfly, on the other, would have to be grouped together in any account of the order in living Nature. This is done on the basis of undeniable similarities in the structure and functioning of the wings. At the same time, the one pair of organisms must be placed far apart from the other because of profound differences between their flight mechanisms. Likewise it would be understandable to place the great man-like apes—the gorilla, chimpanzee and orang—together in a unity, on the basis of their mode of life as tree-dwelling brachiators, and to separate this unity sharply from that of Man on account of his upright gait and many characteristics connected with it.

On the other hand, nobody could deny that judgements and procedures of this sort are mere unobjectifiable expressions of opinion until there is some causal understanding of the resemblances and differences between various organisms.

This, however, was the logical stance of essentialist or typological classification from the zoology of Aristotle (384–322 BC) through the Systema Naturae of Linnaeus (1707–1778) up to the nineteenth century. Under this doctrine, organisms were placed together in unities on the basis of the shared possession of particular features which were chosen subjectively and set up as essential characteristics. Such unities, being groups of elements with one or more characteristics in common, corresponded to the classes of logic.

Even after the victory of evolutionary theory, the comparison of organisms remained the empirical basis for the subdivision of the diversity of living Nature. This diversity could now be interpreted causally as resulting from a real historical developmental process, and thus the decisive shift occurred from the subjective procedures of essentialist classification to the objective basis of a unique, rationally justifiable systematic structure. All living organisms of the earth are connected, by reproduction, as ancestor and descendant in an uninterrupted nexus of relationship. Through this insight, descent from common ancestors could become the **principium divisionis** of biological systematics. "Our classifications will come to be, as far as they can be made, genealogies" (Darwin 1859, p.486)[1]. Surprisingly enough, another century passed before the requirements for satisfying this demand were established. This was achieved by W. Hennig (1950, 1966, 1969)[2], through the theory and methodology of phylogenetic systematics.

I wish to argue here that only a consistently phylogenetic system can adequately represent the order that exists in Nature. The logical first step, therefore, is to ask: how does the reference system of phylogenetic systematics differ, not only from that of traditional, essentialist classifications, but also from those of its present-day rivals—"evolutionary classification" (Simpson 1961; Mayr 1969, 1974, 1975, 1981;

---

[1] "Genealogy" in its widest sense as meaning relationship through descent. The distinction between genealogical and phylogenetic relationship, which is obligatory for present purposes, is explained later (p.31)

[2] Also posthumous publications: (1979/80, 1982, 1983, 1984).

1

Bock 1976; Ashlock 1979 etc.) and "numerical taxonomy" (Sokal & Sneath 1963, Sneath & Sokal 1973)? The answer requires a precise **account of the relations between theory and practice in phylogenetic systematics**.

I shall begin the argument by way of the **taxon concept** (plural: taxa). In biology, every systematic attempt operates with groups of organisms, each of which carries its own name. The species taxon consists of a group of particular individuals and is given a binomen. The supraspecific taxon is a bringing together of particular species and is given a uninomial designation.

A single mute swan is an individual of the taxon *Cygnus olor*, while a single blackbird represents the species *Turdus merula*. In a group of species with feathers and wings both are parts of the supraspecific taxon Aves. A single peacock butterfly is a representative of the species *Inachis io*, while the individuals of a hive of honey bees are representatives of the taxon *Apis mellifera*. Together with about 750,000 other species of winged insects, these two species are placed in a supraspecific taxon with the name Pterygota.

Individuals and species can therefore be placed together in taxa because the living matter of this earth is subject to evolution. On the basis of the theory of evolution[1] the following hypotheses have the character of axioms in the study of phylogenetic relationships:

1) Features arise as evolutionary novelties by way of alterations of the genetic information of single individuals in the populations of species.
2) Species arise by the splitting of species or fusion between species (hybridization).
3) Every closed community of descent arises by the splitting of a stem species common to it alone.
4) The step-wise sequence of species splittings in the course of time leads to kinship-connections, hierarchical in structure, between species and closed descent communities.

For its system of reference, phylogenetic systematics in principle accepts only the equivalents of real unities in Nature. As a logical consequence of the postulated axioms, a phylogenetic system of organisms comprises only two precisely definable forms of taxa.

1) The species taxon, being equivalent to a closed reproductive community in Nature in which evolution can occur and which, by splitting, can enter into the process of phylogenesis.
2) The supraspecific taxon, being equivalent to a closed descent community in Nature. The union of a stem species with all its descendent species makes the closed descent community into a historical product of phylogenesis. A group of organisms so constituted is called a monophyletic taxon (or monophylum for short).

---

[1] The term "theory of evolution" is used here and throughout this book to denote the theory which, in a very general way, postulates the evolution of living matter in time. This formulation asserts nothing about the causal mechanisms of the evolutionary process.

2

Phylogenetic systematics requires absolute **congruence** between the **descent communities** which have originated in Nature, on the one hand, and the humanly conceived **supraspecific taxa** of the phylogenetic system, on the other. Supraspecific taxa range from a minimum of two species and their stem species up to unities of unlimited size. But they are valid reflections of the order of living Nature only in so far as the species included within them are all, and solely, the descendants of a single stem species. The two chimpanzee species— *Pan troglodytes* and *Pan paniscus*—form the taxon *Pan*, the various primate species form the taxon Primates, the species of mammals form the taxon Mammalia, and the millions of species of recent and fossil multicellular animals form the taxon Metazoa. But each of these taxa is a valid unity of the phylogenetic system only if its parts have descended, respectively, from a stem species shared by them alone. If that is so, then these chosen taxa exemplify a number of levels of hierarchical order in living Nature. The stem species of the closed descent community *Pan* existed more recently than the stem species of the closed descent community Primates; the stem species of Primates is younger than that of the closed descent community Mammalia; and this, again, is younger than the stem species of the Metazoa. In this sequence, each of the descent communities that has arisen later is a component part of older, and increasingly more inclusive, descent communities in Nature.

The representatives of a monophyletic taxon must always go back to a stem species common to them alone. I must insist on this unequivocal formulation because the widespread mode of expression which speaks of deriving monophyletic groups from common stem species is absolutely useless. It is self-evident that any pair of species whatever, as for instance the mute swan and the peacock butterfly, had a shared stem species at some time in the phylogenesis of organisms. There is, however, no closed descent community in Nature which includes the mute swan and the peacock butterfly alone, since there never has existed a stem species shared by these two species only.

I now turn from theory to practice. To discover descent communities in Nature, the only available source of experience is comparison between organisms. To draw adequate data from this comparison, I start from a perfectly simple consideration— this is that, because of heritable alterations in the course of time, "evolutionary novelties" happen, one after the other, in the evolution of organisms. They show themselves as transformations of the phenotypes of species. Each single species is able to acquire in the course of its existence an indefinite number (0–n) of evolutionary novelties. But if a species with particular evolutionary novelties becomes, by the process of splitting, the stem species of a closed descent community, then the new features acquired by it will, as a rule, pass to the descendent species. And in the course of later splittings of the descendent species these features will be passed on to a new set of descendants. In this way a descent community arises, in which the evolutionary novelties of the stem species are reflected in the shared possession of identical features among the descendants. Each evolutionary novelty of the stem species is called an autapomorphy. And the corresponding identical feature shared between the adelphotaxa (sister species, sister groups) that arise from the split is called a synapomorphy, as proposed by Hennig.

This brings me back to the comparison of species. What I have already said shows

3

that, among the whole potential of similarities and agreements between different species, identical features are of interest only if they can be seen as the transmitted evolutionary novelties of a stem species that was common to these species alone. Solely by proving this specific quality in agreements that have come into existence once only can different species be traced back to a common origin. Solely by proving synapomorphies, therefore, can we claim to find closed descent communities in Nature and to set up supra-specific taxa as their equivalents in a phylogenetic system.

It is simple and logical in theory to demand the proof of synapomorphous agreements as a basis for setting up monophyletic taxa, but the satisfaction of this demand seems full of problems in practice. If the stem species of closed descent communities were known, the evolutionary novelties could be read off directly from them as autapomorphies. But as a rule such species are not known. Equally the recent and fossil representatives of closed descent communities do not have particular common features openly labelled as synapomorphies. We therefore need an objective methodology by which particular agreements between different species can be hypothesized as identicalities which first appeared in phylogenesis as evolutionary novelties in the stem species that was shared by those species alone. The exposition of such methods is a central aim of this book. Here, however, I must first conclude the argument. Objective hypotheses about the existence of synapomorphous agreements between species, and associated hypotheses about the phylogenetic relationships between species, are the pre-requisites for setting up a phylogenetic system.

On the basis of these remarks, the specific characteristics of the phylogenetic system can now be confronted, step by step, with the **concepts behind alternative classificatory endeavours**.

In the first place, phylogenetic systematics logically accepts all traditional groups as supraspecific taxa, from whatever classification they come, if they fulfil the conditions for being recognized as equivalents of closed descent communities in Nature. Thus the species group of the birds, which was first introduced into zoology by Aristotle, is accepted into phylogenetic systematics as the taxon Aves from the Systema Naturae of Linnaeus (1758). This is done because detailed agreements in the flight organs of the most various kinds of birds justify the following hypothesis: in the phylogenesis of birds flight mechanisms with feathers were evolved once only; they were present as an evolutionary novelty in a particular species in the past; and from this they were taken over into the pattern of features seen in the blackbird, mute swan, and all other wing-bearing species of bird.

Similarly, from the studies of Brauer (1885) phylogenetic systematics takes over the name Pterygota for an extensive monophylum in the system of the Insecta. It does so because it maintains the hypothesis that two pairs of wings, in the form of dermal excrescences in the meso- and metathorax, arose once only in the phylogenesis of the insects and represent, along with other features, an autapomorphy of the winged insects. According to this hypothesis, the honey bee, the peacock butterfly, and all other representatives of the taxon Pterygota go back to a single stem species common to them alone, which, as an evolutionary novelty, possessed two pairs of wings on the thorax.

4

Supraspecific taxa of the phylogenetic system must include all descendants of a particular stem species. This makes it necessary to justify why flightless, and even wingless, species are placed in the taxa mentioned. So as not to interrupt the flow of the argument I wish to keep this justification brief. The great running birds of the southern continents belong in the taxon Aves because they share with flighted birds a series of other identical features—for example, the construction of the leg with the intertarsal joint between the tibiotarsus and the tarso-metatarsus and with an opposable hallux. These features, in addition to the wings, can be interpreted as having arisen as evolutionary novelties once only in the phylogenesis of the Aves. Likewise, the flightless fleas can be placed in the taxon Pterygota because, in common with particular other groups of winged insects such as the beetles, the butterflies or the Hymenoptera, they have a totally specific pupal stage in the ontogeny. This can be regarded as an autapomorphy of a narrower taxon Holometabola within the Pterygota and thus postulated as an evolutionary novelty for the common stem species of all holometabolous insects. In consequence of these probability decisions, the lack of flying ability and of wings in a series of birds and particular insect groups must be seen as the results of evolutionary reduction.

This brings me to the other side of the penny. Phylogenetic systematics decisively rejects, without compromise, all groupings from traditional and from other competing classifications which do not represent equivalent descent communities in Nature.

I shall focus on the example of the great man-like apes. I have already said how understandable it was that, in a widely accepted classification, the apes *Pongo*, *Pan* and *Gorilla* are united in a taxon Pongidae which was contrasted with a second taxon Hominidae. Phylogenetic systematics does not deny the striking agreements between the three man-like apes, nor overlook the profound differences in comparison with Man. Nevertheless it rejects the conventional subdivision and this for two reasons: on the one hand, no agreements can be presented between *Pongo*, *Pan* and *Gorilla* which can be interpreted with sufficient certainty, and without contradictions, as evolutionary novelties of a stem species shared by these taxa alone; and on the other hand, there are a series of anatomical and biomolecular correspondences between *Pan*, *Gorilla* and *Homo* which can be hypothezised as autapomorphies of a stem species shared by these three taxa only. As a logical consequence, the taxon "Pongidae" must be eliminated from the phylogenetic system of Primates as an artificial combination of the great apes. At the same time, a supraspecific taxon consisting of gorilla, chimpanzee and Man must be constituted as a valid portrayal of a closed descent community in Nature.

Because of its **principium divisionis**, phylogenetic systematics is bound to dissolve even the most traditional groupings, known to everybody, if these can be recognized as artificial unities whose members can be separated into monophyletic taxa. Thus a class "Reptilia" has no place in the system of the vertebrates since, firstly, not a single common feature of the "reptiles" can be interpreted as an evolutionary novelty of a stem species shared only by rhynchocephalians, lizards, snakes, turtles and crocodiles. And, secondly, a phylogenetic systematization can be proposed to replace the traditional one. Thus, for example, the crocodiles can be united with the birds in a taxon Archosauria, because they possess a series of unique agreements which can be interpreted, without inconsistencies, as synapomorphies of the two taxa Aves and Crocodilia.

These examples, which I shall discuss in detail in the course of the book, lead to the following essential statement: Neither the degree of agreement between taxa, nor the breadth of the "gaps" between them, count in a phylogenetic system of organisms. However great the differences between *Gorilla gorilla* and *Homo sapiens* or between a crocodile and a bird, this has not the slightest importance for setting up a phylogenetic system. For this system, nothing counts but features identical between different taxa when these features can be seen as the evolutionary novelties of a stem species that was shared by these taxa alone.

These statements differ irreconcilably from the basic concepts of "**evolutionary classification**". Phylogenetic systematics has a "single-minded concern" with the course of phylogenesis, whereas the partisans of evolutionary classification believe that a maximum of information, resulting from various evolutionary analyses, should be included in a classification. In addition to phylogenetic relationship, they believe that genetic similarity and the extent of evolutionary change between different species and species groups should not merely be considered but, in some cases, should take precedence over such relationship. Evolutionary classification therefore expressly defends groups like the Pongidae and the Reptilia as legitimate supraspecific taxa, and this for two reasons. In the first place, the advocates of evolutionary classification believe that the agreements between the great man-like apes or the sub-groups of the traditional reptiles should be expressed in a classification as reflecting a supposedly high degree of genetic similarity and showing a particular unified grade of evolution. They believe that they should be so expressed whether or not the representatives of the Pongidae and the Reptilia go back, respectively, to a stem species common to them alone. In the second place, the profound evolutionary change caused by entry into a new "adaptive zone" — such as the invasion of open savanna in the stem lineage of the Hominini or the conquest of the air in that of the birds — demand that Man be separated in classification from the man-like apes, and the birds from the reptiles, in taxa of equal value with identical rank. They therefore support a family Hominidae versus the family Pongidae and a class Aves versus a class Reptilia.

Why not accept this "plausible argument" if, as it seems, a truly biological system can only be achieved by including a broad spectrum of evolutionary phenomena? The answer is simple. Even if there were an objective measure for recognizing the degree of genetic similarity and evolutionary transformation — which is not the case — it is simply impossible to serve two masters or to be in two places at once. In these sayings, popular wisdom expresses nothing else than the logical postulate, going back to Aristotle, of the unity of the **principium** or **fundamentum divisionis** (Günther 1971; Hennig 1975, 1984). It is asking the impossible if it is required that various aspects, incompatible with each other, should be expressed together in any particular systematic structure. For this opens the door to the logical error of wilfully jumping from one frame of reference to another. In fact, all the procedures of evolutionary classification are mere subjective evaluations. We are supposed to decide, in any given case, how far "satisfying" unities of classification can be achieved on the basis of hypothesized phylogenetic relationship, or how far they can more elegantly be reached by judging the degree of genetic similarity or of evolutionary transformation. The demand for a syncretically or eclectically oriented evolutionary classification that considers divergent parameters of evolution is logically indefensible. It can never supply a universally binding reference system for biology.

In the plurality of classificatory endeavours, "**numerical taxonomy**" or "**phenetics**" advocates a third approach. It sets up supra-specific taxa, neither on the basis of phylogenetic relationship, nor on that of any other evolutionary parameter. Instead, the unities of numerical taxonomy are based solely on the degree of phenetic (phenotypic) similarity between different organisms. This is calculated as the degree of overall similarity on the basis of all available features, with the aid of particular algorithms, whereby individual features receive the same value, without any weighting.

The claim made by numerical taxonomy, to have reached a highest possible measure of objectivity and stability in classification, has been refuted from various sides (Schlee 1971, 1975; Farris 1977; Mickevitch 1978; Wiley 1981 etc). For present purposes, however, the following established fact is more important. The connections worked out by phenetics between different species and groups of species, remain pure similarity connections (phenetic relationships)—they have no necessary link with the closed descent communities that exist in Nature. A correspondence between the unities of phenetics and the supraspecific taxa of a phylogenetic system can only result when, out of the overall resemblances between different organisms, those identical features which represent evolutionary novelties of the common stem species outweigh the disturbing noise of all other, irrelevant similarities (Wiley 1981). This, however, is not enough to bring the estimates of phenetics into a system of reference which expressly aims to reflect the order which has arisen in the process of phylogenesis.

These comments make it necessary to adopt a position on the use of the terms "**systematics**", "**classification**" and "**taxonomy**" in biology. The subject-matter of systematics is the theory and practice of discovering and presenting the order of living Nature. This order is based on the unbroken connection of all organisms within time and shows itself in the genealogical relationships between the individuals of species as well as the phylogenetic relationships between different species.

The Greek word "**systema**" signifies "structure" as well as "complex". The term "system" has many different meanings in popular speech and in the various sciences (Diemer 1968) but the following definition seems to indicate a consensus: "A system signifies a structure whose individual parts, by a particular ordering of their composition, form a whole" (Oeser, 1976, quoted from Wuketits 1978, p.147). The living order of Nature has the character of such a system. The particular ordering of the parts to form a whole depends on the kinship-relations of the parts.

A **class**, on the other hand, is a unification of single elements on the basis of agreements in freely chosen characteristics. These are established subjectively in the definition of the class.

On the basis of what we know of the existence of phylogenetic relationships, there can only be one objective way to present the order of living Nature. This unique prospect can be contrasted with an indefinite number of possibilities for placing elements of living Nature subjectively into classes on the basis of freely choosable aspects. Thus I intentionally refer to the structure erected on the basis of hypotheses of the phylogenetic relationships of organisms as a system, and not as a classification. And the methodological procedure of placing organisms in order when setting up this structure is **systematization** and not classification. The terms "classification" and "classifying" are limited to subordinate formal procedures such as the problem of applying the conventional categories of the Linnaean hierarchy (species, genus, family, order, class) to the taxa of various hierarchical levels of the phylogenetic system (G. C. D. Griffiths 1974).

As a result of this argument, I shall, with a quiet conscience, treat the words "systematics" and "taxonomy" as synonyms. Widespread definitions in which systematics is defined as the science of the diversity of organisms, while taxonomy shall be the theory and practice of classification, may sound reasonable (Simpson 1961, Mayr 1975). They mark no real difference, however, and are therefore empty formulas.

The status of a science depends on the maturity of its theoretical foundations and methodological procedures. It is also affected, however, by the **precision of the relevant terminological apparatus**. Different terms for one and the same thing, or, above all, terms with several meanings (equivocations), can, as fertile sources of continual misunderstanding, totally corrupt a scientific discipline. I therefore consider it a pressing duty, consistently to oppose the numerous terminological inadequacies in the study of phylogenetic relationships, and the confusions that result from these inadequacies. And it is also a duty strictly to define the vocabulary needed in building up a phylogenetic system.

Definitions apply to words, not to the things denoted by the words (Ghiselin 1966a). The definition establishes the meaning of a word, and the defined word thus becomes a term. Expressed the other way round, a term is a short and pregnant expression for the meaning of a word—a meaning which the definition circumscribes in detail.

In this connection, two basic forms of definition must be distinguished: 1) **operational definitions**, which include definite criteria for the recognition of the meaning-content; and 2) **theoretical (non-operational) definitions** which are completely independent of how the defined meaning of a word is to be discovered in practice.

Thus, if the word "fertilization" is defined as "the union of sexually differentiated gametes to form a zygote", this is without doubt an operational definition. For it includes a clear criterion by which the meaning of the defined word can be recognized in reality. It is possible to observe directly the entry of the sperm into the egg in suitable specimens, and it is also possible to repeat the fusion of gametes experimentally.

The study of phylogenetic relationships, on the other hand, concerns itself with the product of a process which has happened uniquely in time, which could not be directly traced and cannot be repeated. Logically the definitions of the meaning of the central terms of phylogenetic systematics are, and remain, theoretical definitions. But this is not all. In my view, there can be, in the **methodology of phylogenetic research, no actual criteria of recognition**—whether the word 'criterion' is used in the sense of a decisive mark or of a totally reliable touchstone.

Take, for example, the phrase already discussed of the "closed descent community". It can be defined theoretically as "a unity of Nature consisting of a stem species and all its descendants", but in no way tells us how a closed descent community in Nature is to be recognized. And even the source of our experience—the comparison of organisms—yields no single criterion by which we can read off or prove existence of a closed descent community in Nature. The documents left by evolution in the feature-patterns of species supply the methodological procedure of phylogenetic systematics with nothing more than **indications**. As with all indirect evidence, they alone give the reference point for formulating and justifying the hypotheses of phylogenetic systematics—hypotheses by means of which we seek to find the facts circumscribed by the theoretical definitions.

In seeking truth, in the form of the agreement of its own predicates with the facts created by Nature, phylogenetic systematics can in principle never aim at "provable" results. Being the insights of a historical science, its statements must always have the character of hypotheses. In actual fact, phylogenetic systematics decisively rejects all forms of unjustified and untestable speculation. It is based exclusively on empirically founded hypotheses concerning the kinship relations between species. And these hypotheses, in the framework of a logically consistent methodology, are always open to objective examination, by which they may be confirmed or may be rejected.

B. Evolutionary Species and Closed Communities of Descent as Real Unities in Living Nature

A basic condition for working out a phylogenetic system of living organisms is proof that the taxa placed in the systematization are real reflections of living Nature and not, perhaps, arbitrary creations of the human spirit. Before going further, therefore, I must justify the statement just made, that species and monophyletic supraspecific taxa are equivalents of unities in Nature, having reality and individuality. For this purpose I must add precision to the already mentioned term "class" and contrast it, in a new connection, with the meaning of the word "individual" (Hull 1976, 1979; Wiley 1980, 1981; Ax 1985c).

Classes possess members and have definitions. A single class is a construct for all those unities (elements) which agree in having the characteristics given by the class definition. Every unity to which the definition of a class applies is a member of this class, independent of the unity's origin or of its position in time and space. Classes have no limits in space or time and take no part in natural processes. The designations of classes are universals in the nominalist sense (universalia). They are words for putting together certain things by means of subjectively selected similarities—like poisonous snakes or toxic wastes.

Individuals are unities limited in time and space and having a particular unique history. As unities with coherence and continuity they are subject, unlike classes, to the processes that occur in Nature. A single individual can be characterized by the presence or absence of features when compared with other individuals. Unlike a class, however, an individual cannot be defined. The names of individuals are not general concepts but rather are proper names (nomina propria). Each individual can in principle receive a unique name, assigned only to itself.

In biology, the **single organism** is the **paradigm** of an **individual**. For each organism has an existence limited in time and space and represents a unique, unrepeatable manifestation. The phenomenon of the uniqueness of an individual is valid, in this connection, for twins having the same origin in time as well as for organisms that follow each other temporally in a line of descent, even if they are identical in all conceivable aspects. Each single organism passes through the process of individual development (ontogenesis) and is subject in its life-span to the influences of Nature, including the possibility of mutational change in its genetic information.

I. The evolutionary species—the working unit of evolution

Individuals as single organisms are the smallest units, and species as groups of individuals are the largest units, in which the process of evolution can take effect. This process shows itself by the origin of evolutionary novelties in single individuals as well as by the spread of these novelties in successive populations of species, by the origin of new species from existing species, and by the extinction of species.

If the species is taken to be the largest aggregate of individuals which evolves as a unit (Wiley 1981), this assumes a positive answer to the question: Among the various definitions of the word "species", is there a particular meaning of the term which, without contradiction, can be identified with the species viewed as the "unit of evolution"?

1. The biological species concept

The biological species concept, whose leading protagonist is E. Mayr, is now well known. He gives his latest **definition of the word "species"** in this form (1969, 1975):
"Species are groups of interbreeding natural populations that are reproductively isolated from other such groups."[1]

In this characterization the maintenance of a common gene pool by crossing and by reproductive barriers is made the central component of the cohesion between the individuals of a species. Nevertheless the biological species concept does not suffice for the interpretation of the species as the basic unit of evolution.

An essential objection concerns the lack of any objectifiable delimitation of the species in space and time. As a consequence of the definition, the biological species concept can only extend to populations that co-exist in time and space, and thus to synchronous and sympatric populations of various species within which crossings are at present possible, but between which reproductive isolation mechanisms in fact operate. Convinced advocates of the definition themselves stress the non-dimensional character of the biological species concept: "As one progresses geographically and chronologically further and further away from a single point, the species distinctiveness becomes more and more vague" (Bock 1979, p.28).

In the space dimension, use of the biological species concept will not decide whether phenetically identical, allopatric populations are members of one and the same species if these populations are reproductively separated from each other by ecogeographical barriers and if the potential for crossing and for producing fertile offspring between them cannot be proved.

But the question of delimiting the species in time is however even more important. From the astounding assertion that a species, consisting of populations mutually crossing with each other, is today simply not the same as its ancestor of one hundred generations ago, Bock (1979, p.29) deduces that: "Because the biological species concept is a non-dimensional one, it is not possible to speak of the age of a species or the origin of a species, or of the life and death of a species."

The inappropriateness of a non-dimensional species concept for the characterization of species as real units of Nature could scarcely be expressed more clearly than in this sentence. If a species in the course of one hundred, or however many more, generations has not split into at least two reproductively isolated units, it is self-evident that the populations that existed at the corresponding time in the past, and those that still exist today, belong to one and the same species. And this merely because, as members of an unbroken line of descent, they are in continuous

---

[1] Mayr (1982c, p.273): "A more descriptive definition is: A species is a reproductive community of populations (reproductively isolated from others) that occupies a specific niche in nature."

11

genealogical connection with each other. Probably nobody could seriously deny that the populations of present-day Man, like the human populations which inhabited the earth one hundred generations or a few thousand years ago, belong equally to the species *Homo sapiens*. For the problem of delimitation in time it does not matter in the slightest whether evolutionary novelties have, or have not, arisen in the lineages of a species if those lineages are held together by reproduction in time. Or whether, in other words, the look of successive populations of a species has suffered change or stayed constant.

There is a second essential criticism of the biological species concept—that it is limited to units with bisexual reproduction. For such is required if crossing and panmixia are to occur between the individuals of populations of a species. Units with solely unisexual (parthenogenetic) reproduction, or with asexual (vegetative) reproduction, cannot be covered by the species definition of the biological species concept.

2. The evolutionary species concept

The evolutionary species concept solves these problems by considering the dimensions of time and space, and also by considering all the various modes of reproduction and aiming at compatibility with the different models of species formation. According to this concept the **word "species" is defined** as follows, in the revised version of Wiley (1978, 1981):

"An evolutionary species is a single lineage of ancestor-descendant populations which maintains its identity from other such lineages and which has its own evolutionary tendencies and historical fate."

By "lineage", Wiley means a series of populations with a common history of descent which is not shared with other populations. By the "identity" of such a lineage he refers to the quality which a unity acquires as a by-product of its origin and which, in the totality of characteristics specific to the lineage, includes the ability to keep the species separate from other comparable unities.

Cohesion and continuity within time

I shall therefore consider the previously discussed problems again, using the definition of the evolutionary species, focusing first on the phenomena of the **coherence and continuity of species within time**—phenomena that are based on reproduction.

Species with bisexual reproduction have two chains binding them together—a vertical linkage through the relationship of parents and offspring and a horizontal linkage through the pairing of partners in the population. Species with exclusively parthenogenetic or vegetative reproduction—commonly but incorrectly referred to as "asexual" species—lack, on the other hand, an essential link in the chain of cohesion, because they have no pairing. Since, however, they form unbroken lines of descent, they fall under the definition of the evolutionary species, just like units that reproduce bisexually.

There are, of course, great differences between bisexually reproducing species,

12

on the one hand, and those with completely parthenogenetic reproduction, on the other, as concerns the significance of these species in the process of evolution. In parthenogenetic species, because gene flow is broken and recombination impossible, genetic and thus phenetic unity can only be maintained in successive generations by evolutionary stasis or by parallel answers to uniform environmental factors. Evolutionary novelties can spread over all the populations that co-exist in space only if there is horizontal linkage through pairing. If reproduction is parthenogenetic, therefore, evolutionary changes can show themselves only in the descendants of the female individuals hit by the mutation. Populations of parthenogenetically reproducing species may be able to diversify genetically in the short term and may have a corresponding adaptive capacity (White 1978). But judged by the long-term effects of the evolutionary process, they are cul-de-sacs in evolutionary history.

Delimitation of species in space

As to the **delimitation of evolutionary species in space**, I come back to the problem of how to place allopatric populations that are completely separate — a problem insoluble with the biological species concept. Wiley (1978, 1981) has emphasized the analogy of this widespread situation with the clones of parthenogenetic ("asexual") species. In both cases horizontal linkage through pairing is lacking. Since the only coherence is through the linkage of parent with offspring, phenetic continuity must be interpreted in the first place as the expression of evolutionary homoeostasis. What conclusion can be drawn, therefore, if the variation between allopatric populations is no greater than the variation within each single population? With the evolutionary species concept the answer is simple. All allopatric populations belong to one single species, because the hypothesis can be defended that, in the absence of evolutionary differentiation, the evolutionary tendencies of single populations remain the same. In this connection it does not matter whether cross-pairings between individuals of different allopatric populations are still possible or whether the original ability to cross has gone.

Life span of a species

This leads me to the next question — **the chronological delimitation of species**. What processes in Nature decide the "life span" of a species?

Assuming that the evolution of living matter has reached the level of separate sexual reproductive communities, then new species can in principle arise only from species that exist already. Their origin may happen in two different ways — through the **splitting** of the populations of a species by the erection of reproductive barriers, or through **hybridization** with the sexual union of members of populations from two distinct species. Logically these processes mark the instant of origin of new species, i.e. the beginning of their existence. Splitting is the normal case, whereas hybridization, at least among animals, is a rare exception.

I shall treat the first alternative by way of a schematic example (Fig. 1). The species A and w arose at geological time T2 by the splitting of the species v into two reproductively separate lineages. Through this process the species v thus became the

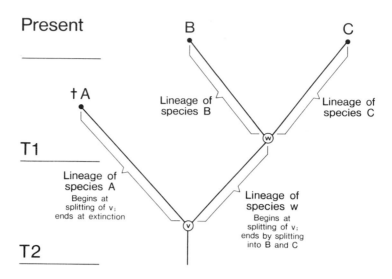

Fig. 1. Diagram of the phylogenetic relationship between the extinct species A and the recent species B and C, showing the possible existence of evolutionary species in time.

stem species of the two new species A and w. Suppose further that the species A died out at time T1, while a new splitting happened in the species w, producing species B and C which persist to the present day. This example gives the two basic possibilities for the ending of the life span of a species. Translating the terminology of Willmann (1985), the ending of the life span can be called the expiry of a species, and this process can consist either in extinction without bodily issue (species A) or in the dissolution of a species as it passes into its daughter species (species v into A and w or species w into B and C).

Leaving on one side for the moment the phenomenon of accidental extinction without descendants, there in principle results, for **every species, a regularly fixed existence in time**. This existence begins with the origin of the species by the splitting of a stem species (or from the fusion of two species) and lasts until the species itself splits into reproductively isolated daughter species. In our example, the lineage of the evolutionary species w began with speciation from species v and expired at its own splitting into the species B and C. This is a logically unassailable position, adopted by Hennig and other phylogenetic systematists, for the chronological limits of species. Seeing that two daughter species B and C have arisen from one stem species w, there can never exist after the splitting process a species completely identical with the former species w of before the split. This remains true, whether both or only one of the daughter species B and C have suffered evolutionary change with respect to the shared stem species.

Willmann (1985) has recently justified this interpretation in detail. He argues that the crucial initiating process in speciation is the evolution of reproductive isolating mechanisms in one of the two species that arose at the split. In this connection one fact is of special importance—namely that the isolating mechanisms evolve with respect, not to the stem species, but to the other daughter species. In other words, the

14

origin of the one daughter species is necessarily bound up with the origin of the other daughter species. They both arise from the split as new species. An inevitable result of these considerations is that the shared stem species expires as an independent unit at the instant of speciation.

Dissolution of a stem species

The postulate that a stem species necessarily dissolves in the splitting process has entered as a methodological rule into the practice of phylogenetic systematics. Opposition to this assumption and to its logical consequences has become an essential point in the opposition to the whole idea of setting up a consistently phylogenetic system (compare Hull 1979, Wiley 1981). It is therefore necessary to examine whether the postulated dissolution of a species at speciation is in fact consistent with what is now known of the various ways in which new species may arise. From this viewpoint, I wish briefly to consider the basic **models of species formation** that are today under discussion (Figs. 2 & 3); among the extensive new literature, I refer to the works of Mayr (1963, 1967, 1982a), Grant (1971), Bush (1975), Endler (1977), White (1978) and Wiley (1981).

Transformation

The widespread **view that species transform themselves in time into new species (species transformation, phyletic evolution, phyletic speciation)** starts from the thought that a species is able, during a long interval of time, to change into a second, third or fourth species without any splitting process. If, for example, the species A from the time T2 acquired an altered appearance by the development of evolutionary novelties in the time T1, then according to this view we should have to consider it as a new species B. A and B would be two arbitrarily delimited chronospecies in a lineage of descent (Fig. 2a).

It only needs a few words to show that this conception of "successional species" is incompatible with the evolutionary species concept. In a direct lineage of populations connected together continuously by the process of reproduction, an indefinitely large number of evolutionary changes may affect the genetic information, and thus the pattern of features, in successive generations. Without a splitting process, however, the single individuals of this lineage must remain parts of a single evolutionary unity. The transformation of the genome in time does not transform one species into a new and different species but changes the look of temporally successive generations of one and the same species. Thus, in our example, the species A from time T2 remains the species A in time T1, even after acquiring one, two or several evolutionary novelties (Fig. 2b). The model of species transformation does not represent a mode of speciation in Nature; rather it is the artificial product of a mistaken train of thought.

Hybridization

As concerns **the origin of new species through hybridization** there are two conceivable possibilities: 1)Two species completely fuse to form a third species and, as

15

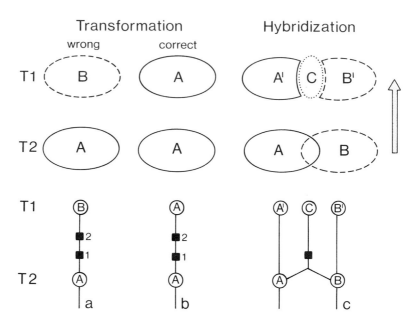

Fig. 2. Models of speciation—transformation and hybridization.
a) Artificial model of the transformation of species A into a new species B in time.
b) Transformation of the phenotype of one and the same species A through the acquisition of evolutionary novelties in time.
c) Hybridization—the origin of a new species C through the fusion of parts of two stem species A and B. T = time; ■ = evolutionary novelty.

In Figs. 2 and 3, the agreement in genetic constitution and phenotype between stem species and particular daughter species is intentionally expressed by choosing similar symbols. Thus, compared with the stem species A and B, the corresponding new daughter species are represented only by the symbols A' and B'.

its stem species, expire; no definite cases of such a reductive speciation are known (Wiley 1981); or 2) A new hybrid species arises from portions of two parental species. This latter alternative can be presented as a schematic example (Fig. 2c). Crosses between individuals of sympatric populations of species A and B are here supposed to produce viable hybrids which in turn produce fertile offspring. The hybrid populations acquire evolutionary novelties during the speciation process or afterwards and become a new evolutionary unity as the species C. The species A and B lose, in forming the hybrid species, only insignificant proportions of their total populations so that their gene pools remain more or less unaltered. Among plants, this mode of speciation is widespread when a new species originates by the allopolyploidy of parts of two stem species. Among the rare instances in animals, an obvious case is the origin of the edible frog *Rana esculenta* as a hybrid of the species *Rana ridibunda* and *Rana lessonae* (White 1978).

Splitting

This brings me to the crucial process in speciation—the **origin of new species by the splitting of an existing species**. In the space dimension the modes of fragmentation of a stem species, as discussed in the literature, can be subsumed basically in three models (Bush 1975, Mayr 1982a, Cracraft 1984). These are:

1) **allopatric speciation** (= dichopatric and peripatric speciation), where there is complete geographical separation of populations of the stem species.
2) **parapatric (and stasipatric) speciation**, where spatial contact is maintained during the splitting process.
3) **sympatric speciation**, which occurs within the area of distribution of the stem species.

In considering the **relation between stem species and daughter species**, however, a different subdivision becomes important, as follows (Fig. 3).

1) According to the **dichopatric model of speciation** (Cracraft 1984; allopatric speciation of mode a of Bush, 1975) a species with wide distribution becomes separated by certain geographical events into two extensive daughter units. Gene flow between these is interrupted by total reproductive isolation. The spatially separated populations, in a long period of time, grow gradually distinct from each other in that evolutionary novelties may appear in one fragment or, of course, in both fragments. If both fragments alter, two new species will arise which, in their genetic constitution and their phenotype, are clearly marked off from the common stem species.

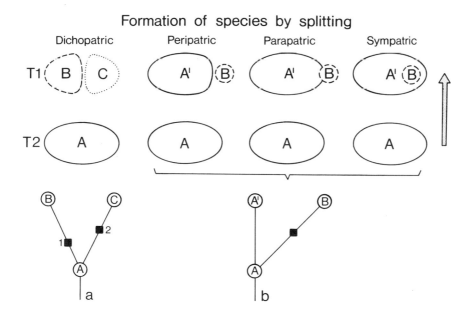

Formation of species by splitting

Dichopatric   Peripatric   Parapatric   Sympatric

Fig. 3. Models of speciation (continued)—splitting of stem species.
a) Splitting into two descendent species (B and C) of approximately equal size.
b) Separation of small populations (B). T = time, ■ = evolutionary novelty. Further explanation in text.

2) The **peripatric model of speciation** (Mayr 1982a; allopatric speciation of mode b of Bush, 1975) is based on a phenomenon known as the founder effect. In this case, a new evolutionary unity is formed by the separation and geographical isolation of a small founder population at the periphery of the area of distribution of the stem species. The peripheral isolate may evolve quickly because of the small number of individuals. The large population, however, remains unaltered.

This account holds in principle for the separation of small populations according to the **parapatric and sympatric models of speciation**, whatever the mechanism may be by which reproductive isolation is achieved.

From the situation just outlined, the following theoretical statements can be made concerning the relationship between stem species and daughter species.

As concerns the dichopatric model of speciation, it can be assumed that both of the species that arise from the split will, as a rule, acquire evolutionary novelties. For this case, therefore, the theory of speciation causes no controversies. The postulate that the stem species expires at the instant when it splits into two new daughter species causes no conflict.

As concerns the peripatric, parapatric and sympatric models of speciation, as listed under point 2), a different outcome must as a rule be assumed. Here it will normally be only the smaller populations, separated in various ways, which will develop particular evolutionary novelties. The large populations, on the other hand, will continue the genotype and phenotype of the stem species. These considerations lead to the widespread view that the stem species "survives" in one of the two daughter species (Schlee 1971, 1981; Eldredge & Cracraft 1980; Wiley 1981)—a view that was also presented in the original German version of this book.

The concept of the "surviving stem species", however, can be countered with the following argument. In the theoretical interpretation of the results of the splitting process, it is in principle unimportant whether only one species acquired evolutionary novelties, or whether both units that arise from the split undergo evolutionary changes. In this connection, I repeat Willmann's central assertion (1985): namely that those new daughter species which acquired reproductive isolating mechanisms can only develop those changes with respect to a second species which arose at the same time and which lives simultaneously. It cannot develop them with respect to a stem species which existed in time past. In other words, it does not matter in the slightest whether the second unity has retained the genotype and phenotype of the stem species, or not. For, in every case, this second unity must also be seen as a new species that arose in the splitting process. Of course, the results of the transmission, unchanged, of the feature pattern of the stem species to one daughter species need to be considered in detail as regards the methodology of phylogenetic systematics. I shall discuss this later (p.63).

Willmann (1985) even extends his interpretation to cover the origin of hybrids by allopolyploidy. Thus he maintains that, in the process of hybridization, the portions of two parental species do not give rise to one new hybrid species only. For the parental species likewise expire in the process and are superseded by two new daughter species—irrespective of the fact that their phenotypes agree in each case with the phenotype of one of the parental species.

3.   Unity in Nature—taxon of the system—category of the classification

The word "species" has been used in many different ways in biology. On the basis of the arguments just given, I now wish to clarify the logical and factual connections between the various meanings.

Unity in Nature

In Nature there are groups of individuals which form closed communities of reproduction. From the point of view of the evolutionary species concept they are called "evolutionary species".

The definition of the term "evolutionary species" covers all the corresponding real unities in Nature. The individual evolutionary species, on the other hand, is not definable—no more so than the single individual. It can and must be characterized, in comparison with other evolutionary species, on the basis of the feature-pattern which it alone possesses.

Every evolutionary species is a unity in Nature, linked to time and to space. Thus it has a clearly marked beginning, a limited span of existence in time, and a definite end. Every evolutionary species is potentially subject to the causal mechanisms of evolution. Every evolutionary species has a unique history in which it can evolve as a unity.

The evolutionary species is a really existing unity in Nature, ontologically having the characteristics of an individual (Mayr 1963, 1982c; Ghiselin 1966a, 1974, 1981; Hennig 1966, 1984; Löther 1972; Hull 1976, 1979; White 1978; Wiley 1978, 1980, 1981; Eldredge & Cracraft 1980). Accordingly each evolutionary species can be assigned a unique name which by international agreement must be a binomen.

The taxon in the system

Each group of individuals which can be hypothesized as being parts of an evolutionary species in Nature becomes a taxon in systematics. The species taxon is the basic unit of reference for the phylogenetic system of organisms. Each such taxon takes the name of the corresponding evolutionary species.

Category in the classification

Ever since Linnaeus the word "species" is used to indicate a particular category of the formal classification. In the Linnaean hierarchy of categorial designations, the species category has the quality of a logical "class". It represents a class among other classes, such as the Genus, Family, Order, Class or Phylum (p.234).

Since the species category is obligatory under the international rules of nomenclature (p.237), we have to live in systematics with the many different meanings of the word "species". This ambiguity was and is a regrettable source of misunderstanding although the logical connection between the various meanings is simple. Single individuals are hypothesized as members of an evolutionary species in Nature. And correspondingly a species taxon is created in the phylogenetic system of organisms. This taxon, given a binomen in accordance with the laws of biological nomenclature, is obligatorily placed in the species category.

4.   The recognition of evolutionary species

The theoretical definition of a species as a real unity in Nature is independent of the question whether, in the established meaning of the word "species", evolutionary species can in practice be recognized or not.

Nevertheless, this definition is exceptional among the theoretical definitions of phylogenetic systematics, because, in the case of extant bisexual species, an operational approach can immediately be deduced — that of the crossing experiment. However, this approach can be applied only to species the individuals of which can be kept under laboratory conditions or, speaking more generally, in captivity. If pairing attempts between individuals of spatially separated populations give a positive result, and fertile offspring arise from the crosses, then this doubtless suggests that the populations belong to only one evolutionary species. The failure of pairing between individuals of allopatric populations, or the production of sterile offspring, is more difficult to evaluate. Under the evolutionary species concept, as already discussed, negative results do not necessarily show the existence of separate evolutionary species. Furthermore the reproductive behaviour of individuals may be disturbed by the experimental conditions and pairing experiments fail in consequence. On the other hand, reproductive barriers which act in the natural environment between different species, may break down in captivity. The results of crossing experiments, therefore, need not always agree with the relationships in Nature. Their value for separating evolutionary species is very limited.

In separating species, therefore, it is necessary, even under the evolutionary species concept, to fall back on some well known arguments. The shared gene pool of an evolutionary species expresses a vast number of common characteristics in the phenotypes of members of the species. On the other hand, obstinate analysis of various aspects of the biology will usually reveal sufficient differences between the populations of different evolutionary species, even when, as with sibling species, the distinctions are extremely subtle. Wider discussion of this theme lies beyond the scope of the present book. In any case, agreements and differences revealed by structural, ethological, genetic and ecological analyses, together with information from biogeography, all provide data for our hypotheses concerning the evolutionary species of which living Nature is composed. Mistakes caused by over- or underestimating the number of species that really exist are not due to defects in the species concept. They stem from inadequacy in our systems of observation and method and not from inadequacy of the concept (Wiley 1978, p.23).

II. The closed community of descent—a historical product of phylogenesis

Having discussed the evolutionary species concept, I now wish once again to **define** precisely the term "**descent community**", as follows:

"A closed descent community in Nature is a unity comprising two or more evolutionary species together with the stem species that is common to them alone."

A supraspecific taxon of the phylogenetic system has already been termed a **monophyletic group or monophylum**. By Hennig the term monophylum was only suggested once, in passing (1953). Because it is short and meaningful I prefer it to the longer-winded "monophyletic group" or "monophyletic taxon".

In any case the use of these terms sharply separates the valid supraspecific taxa of the phylogenetic system from the various forms of non-monophyletic group of other classifications whatsoever. The need to eliminate such artificial groupings (paraphyla, polyphyla) from the phylogenetic system is dealt with in a separate chapter (p.179).

Since phylogenetic systematics requires equivalence between the closed descent communities in Nature and the monophyla of its system, it follows logically that the meaning in the definitions of these two terms must be identical in principle. I shall now give **two definitions of the word monophylum** from Hennig (1966) and a third, shorter version which emphasizes the inclusion of the stem species (Farris 1974, Bonde 1977, Wiley 1981):

"A monophyletic group is a group of species descended from a single ('stem') species, and which includes all species descended from this stem species" (Hennig 1966, p.73).

"A monophyletic group is a group of species in which every species is more closely related to every other species than to any species that is classified outside this group" (Hennig 1966, p.73).

"A monophyletic group is a group of species that includes an ancestral species (known or hypothesized) and all of its descendants" (Wiley 1981, p.76).

In accordance with these definitions we can also speak of the "**monophyly**" of a species group when its members, without exception, descend from a single stem species shared by them alone.

Before discussing the historical derivation of the word monophylum (p.26), I shall address the central problem of this chapter, which is to establish **the real existence of "closed descent communities in Nature = monophyla of the system**". I shall look at this from four aspects.

1) Each closed descent community has a particular **existence in space** which is determined by the geographical and ecological distribution of all its species. Existence in space may be very limited as, for example, the famous Darwin finches which, as the monophylum Geospizini, are represented only in the Galapagos Archipelago (13 species) and the Cocos Islands (one species).

2) Each closed descent community has an exactly determinable **origin in time**. This is given by the splitting of a particular evolutionary species into at least two de-

scendent species. At the finish of this process, a single species has retrospectively become the stem species of a supra-specific taxon. Thus the taxon Mammalia must, at some time in geological history, have existed as a closed descent community with a minimum of two species—just as today the monophylum *Pan* exists with two species, being the two chimpanzees *Pan troglodytes* and *Pan paniscus*. The finish of the temporal existence of a closed descent community presents a more difficult problem. A "proper" extinction, such as I have already discussed for the fate of evolutionary species in the process of splitting, cannot occur with a closed descent community. However, there are many monophyletic species groups—such as the Trilobita, Ammonoidea, Pterosauria etc.—which have died out in the course of the Earth's history, for whatever reason (Erben 1981).

3) In what does the continuity and cohesion of a closed descent community lie? Each descent community is a **historical continuum** as implied by the connection of its species through the chain of speciation-events in time. Since, however, the parts of a monophylum are broken up into reproductively separated units, there are no effective cohesive links—neither by the vertical connection between parents and offspring which exists in the populations of evolutionary species, nor as horizontal connections by way of pairing such as happens in populations of evolutionary species with bisexual reproduction.

4) Because of this situation **the closed descent community**, being a species group, is—unlike a single species—**not subject to the process of evolution**. Once a descent community has been established in Nature, there is no evolutionary mechanism which can affect it as a whole and transform it as a closed unity. Evolution acts in a closed descent community only by altering its mutually independent parts, and these are the individual species as the largest evolving units.

This statement may be illustrated by the already mentioned examples of Mammalia and *Pan*. At some time in the Carboniferous, the Mammalia were, with only two species, in principle one and the same closed descent community which they now are with thousands of living species. And the taxon *Pan* would remain in 100 million years time one and the same monophylum, even if a unity with thousands of species had by then developed from the total of two species which now exist. Irrespective of how many new species arise in time, and regardless of how profound the evolutionary changes of individual species may be, none of these alterations affect the monophylum as a unity. Rather they happen within the limits of a closed descent community which was definitively established in the history of Nature by the splitting of its stem species.

Because of these facts and the resulting conclusions no doubt can remain that closed descent communities really exist in Nature. Moreover, the connection in space and time which has just been outlined is the basis for the widespread view that **each closed descent community represents a unity having, ontologically, the essential characters of an individual** (Hennig 1966, Brundin 1972, Griffiths 1974, Bonde 1977, Patterson 1978a, Ghiselin 1980b, etc.). Wiley (1980, 1981) likewise stresses that closed descent communities (= natural supraspecific taxa of his terminology) have the character of individuals, but rightly emphasizes what has already been pointed out—that they are not subject to the effects of evolution and thus do not take part as unities in any natural process. This latter peculiarity, however, is a characteristic of classes. According to Wiley, therefore, closed descent communities, since they show a mixture of several of the characteristics of individuals and one of the characteristics of classes, are neither individuals nor classes. Rather they are historical groups which descend from single ontological individuals (evolutionary species). This subtly differentiated argument, however, seems in Wiley's own words "more likely to be of interest to philosophers than to systematists" (1981 p.75). In any case, closed descent communities are not classes, and thus not the artificial product of Man's imagination. Instead they are real, individual-like unities of Nature with a historical continuity. Correspondingly we give them their own names (nomina propria). Unlike the names of species, these are uninomials.

Descent communities (= monophyla) are not definable—any more than a concrete evolutionary species or a biological individual are definable. A single closed descent community can be characterized as a unity in Nature only by showing the existence of particular shared features among its representatives the species. These are identical features which were developed in the latest common stem species of the monophylum as evolutionary novelties (autapomorphies) and from this were taken over into the descendent species.

In interpreting the factual and logical **relationships between species and monophyletic species groups** I wish to discuss two further aspects.

Firstly Lorenzen (1976, 1981) argues that sometimes a monophyletic species group may consist of a single species. This contradicts the definition given above for a closed descent community or monophylum since this definition requires a minimum of two species, together with the stem species common to the group alone. But obviously the definition of a term can and must be changed, if there are new insights into the meaning covered by the definition. However Lorenzen's theoretical equation of a single species with a monophyletic species group, which latter must always be traceable back to a single stem species, is in fact mistaken. It is simply impossible that a single species can give rise to a single descendent species. As I explained above, in dealing with the artificial concept of the transformation of species, without a splitting process a particular species remains one and the same species. This is still true however many evolutionary novelties it may have acquired in its life span. It is therefore theoretically impossible that a single species can be the sole descendant of another (stem) species.

Secondly the meaning of the word monophyly sometimes enters into the characterization of species, as recently once again in the work of Mishler & Donoghue (1982) who advocate a pluralism of species concepts. As concerns monophyly they make a grave mistake. The concepts of monophyly and non-monophyly have nothing to do with species as real unities in Nature, regardless of what concept is used in defining the word "species". Obviously every single species may possess apomorphous features (evolutionary novelties) and plesiomorphous features by comparison with its sister species. Between the individuals and populations of one and the same species,

23

however, there can be neither synapomorphies nor symplesiomorphies[1]. The shared possession of apomorphous and plesiomorphous features is possible, in principle, only between different species. In other words, a single species or one of its populations cannot be either monophyletic nor non-monophyletic[2]. The concept of the monophyletic taxon can be applied only to groups of species which are considered, on the basis of probable synapomorphic agreements, as the descendants of a stem species common to them alone.

This theoretical discussion of closed descent communities in Nature and their equivalents in the phylogenetic system brings me to a final problem. Before putting my later arguments on the relationship between taxa in the system and categories of the Linnaean hierarchy (p.234), I should like to consider here "the **monotypic classificatory units**" of traditional systematics. In traditional classifications all categories above the species category serve to indicate the relative ranks of supraspecific taxa in the hierarchy. Thus they normally cover groups of species rather than single species.

In the phylogenetic system a single evolutionary species must, under some circumstances, be given the same rank as a particular monophyletic group of species (p.41). In the formal classification such single species are placed in the categories of supraspecific taxa. Thus the duck-billed platypus *Ornithorhynchus anatinus* is put, as the only recent species, in the monotypic monotreme genus *Ornithorhynchus* and furthermore in the monotypic family Ornithorhynchidae. The aardvark *Orycteropus afer* is the only recent species of the placental genus *Orycteropus*, of the monotypic family Orycteropodidae and even of the monotypic order Tubulidentata.

The problem of giving categorial designations to taxa of the phylogenetic system is examined in detail below. But if we drew the conclusion, on the basis of formal classificatory procedures, that a monotypic genus, family or order constituted a supraspecific, monophyletic taxon of the system , then that would be a mistake. *Ornithorhynchus anatinus* and *Orycteropus afer* remain obvious evolutionary species. They cannot, by some classificatory backstair, become closed descent communities in Nature.

---

[1] Definitions on pages 52 and 53.
[2] Added in press; basically the same arguments can be found in Willmann (1983).

C. Phylogenesis as the Process of Origin of Closed Descent Communities

I now wish to focus on the word which has become, quite simply, the characteristic term in all studies of descent relationships. The word is "phylogenesis". Nowadays it is often applied thoughtlessly and confused, without distinction, with the word "evolution". I wish to free it from these vague misconceptions and shall do so by considering what Haeckel originally meant by it and thus establishing its exact meaning.

In biology there are more than enough definitions of the word "**evolution**". These can together be subsumed as all conceivable processes and phenomena in the development of organisms which involve hereditary alterations in time. If "evolution" is defined, in Darwin's manner, as "descent with modification" (Eldredge & Cracraft 1980, Wiley 1981), then this meaning of the word covers two basic processes: the transformation of genetic information with the consequent phenetic alterations in the populations of evolutionary species; and the origin of new species through speciation.

As opposed to this, Haeckel in his work "Generelle Morphologie der Organismen" (1866) created the words "**Phylogenese**" and "**Phylogenie**" for a clearly circumscribed aspect of evolution, whereby "Phylogenese" ("**phylogenesis**" in English) means a particular process while "Phylogenie" ("**phylogeny**" in English) is the science of this process and its results.[1] Leaving aside any inadequacies due to the time when he wrote, there is in Haeckel the following completely unequivocal formulation:

> "Unter einem Stamm oder Phylon verstehen wir ... die Summe aller derjenigen Organismen-Formen, welche, wie z.B. alle Wirbelthiere oder alle Coelenteraten, von einer und derselben Stammform ihren gemeinsamen Ursprung ableiten" (volume 1, p.57).

> ["By the word 'Stamm'(stem) or 'Phylon' we understand the totality of all those forms of organisms, as for example all vertebrates or all coelenterates, which derive their common origin from one and the same stem-form"] (volume 1, p.57).

Haeckel repeats the definition of the word "**Phylon**" ("phylum") with variations and always indicates "Phylogenie" as the science of the developmental history of the organic "phyla". Already for Haeckel, however, the individual species make up "the higher individuality of the phylum". Consequently "the real content of this discipline is the concrete developmental history of species" (vol.2, p.304).

Thus under the word "**Phylon**", Haeckel indubitably understood what is now called a **closed descent community in Nature** and thus a group of species "as for example all vertebrates" which go back to one and the same stem species shared by them alone.

It would at first seem legitimate and desirable to take over unchanged from

---

[1] Although unusual in English, (as well as in the German literature of today), the discrimination between "phylogenesis" and "phylogeny" appears to me advisable and valuable. In any case, I intentionally use the term "phylogenesis" when referring explicitly to the process of formation of closed descent communities in Nature.

Haeckel the short and meaningful term "phylum" for a closed descent community in Nature, as recently proposed by Bock (1977b). This, however, is made impossible by the firmly accepted usage of the term "phylum" for a high-ranking category in the Linnaean classification. It would be intolerable to adopt into our terminology a double meaning for the word "phylum" as a second source of misunderstanding alongside the unavoidable ambiguity of the word "species".

Haeckel is himself responsible for the deplorable mix-up of different meanings for the word phylum. Although he powerfully campaigned against the artificiality and arbitrary application of categories like family, class and order, he used the word "phylum" for a category and thus placed the "phylum Vertebrata" at the top of a "ladder of subordinate categories" (volume 2, p.400). He insisted, however, that the phylum was "the only real category of the zoological and botanical system" and this for a simple reason. In his "Generelle Morphologie" Haeckel was undecided whether the five phyla which he accepted for the animal kingdom (= Metazoa) (i.e. Coelenterata, Echinodermata, Articulata, Mollusca, Vertebrata) were derived, separate to each other, from "autogenous stem forms", or whether all animals, at least, or perhaps even all organisms, had their common origin in a single stem form. "Either all organisms are members of a single basic stem (phylum) i.e. descendants of one and the same shared and autogenous stem form; or there exist different phyla, alongside each other, which have developed independent of each other from separate autogenous stem forms." In the latter case, the autogenous stem forms of all phyla would have arisen alongside each other "by direct transition of inorganic into organic matter." (Vol.2, p.419).

This alternative is today of historical interest only, but it clearly illuminates Haeckel's view of a phylum. It shows that the phylum was not, for him, one of many categories in the hierarchy of classification. Rather he saw it as the most inclusive real unity, comprising all those organisms which had arisen from a single shared stem species. In this connection, his often reproduced tree of descent is an unmistakable visual expression of the conceivability that all organisms are connected by descent (Fig. 4). And this classic diagram carries the significant caption: "A monophyletic tree of descent of organisms."

Thus Haeckel in the previous century not only introduced the words "phylogenesis" and "phylum", but also coined the adjective "**monophyletic**". The word "phylum" cannot be adopted unchanged but we can refer directly back to Haeckel by adopting the word "monophyletic", in the substantive form of "**monophylum**", for a closed descent community in Nature. Moreover this procedure does not involve unjustifiedly redefining a word which in his day would have had a significantly different meaning. There is only one, easily understood, difference from Haeckel's conception of the phylum. In his mind it was inextricably entangled with the categories of supraspecific taxa and he would therefore use the word "phylum" only for groups of organisms "above the Class category". For our purposes, on the other hand, all species groups down to a minimum of two species are valid monophyla in so far as they descend from a stem species shared by them alone, whatever category they may be assigned to. In this connection it does not matter that Haeckel knew nothing of the various possible non-monophyletic groupings (p.179) and consequently inserted a mass of artificial taxa into his tree of descent. Thus among the vertebrates alone he had "Pisces", "Anamnia" and "Reptilia".

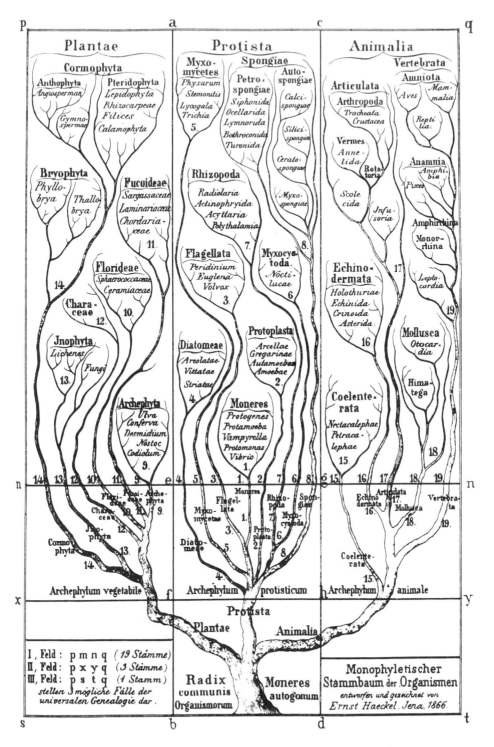

Fig. 4. The monophyletic tree of descent for organisms as given by Ernst Haeckel in his "Generelle Morphologie der Organismen" (1866).

After this analysis there is no reason to exchange the term "monophyly" for any other designation, such as "holophyly" for example (Ashlock 1971, 1972, 1974). Equally there is not the slightest reason to accept other definitions of the term "monophyly" (Simpson 1961; Mayr 1969, 1975; Ashlock 1971, 1972, 1974; Holmes 1980; Lorenzen 1981). Only the definition which I have given above for the word "monophylum" is logically consistent with the insight that closed descent communities exist in Nature.

Closed descent communities, however, are nothing other than the product of successive splittings of evolutionary species in time and space. Consequently there can only be one consistent theoretical **definition of the word** "phylogenesis":

"Phylogenesis is the process of origin of closed descent communities in Nature by the splitting of a stem species common respectively to each such community alone."

The goal of research in our science is to discover the products of phylogenesis and their ordering in relation to the sequence of speciations in time. Logically enough, it is called **phylogenetic systematics** and the aim of its endeavours is called a **phylogenetic system**.

Despite the fact that the meaning of the word phylogenesis was unequivocally established by Haeckel in 1866, the following "definition" of "phylogenesis" and its adjective "phylogenetic" has been recommended, more than a hundred years later, by Holmes (1980 p.45): "Phylogenetic pertains to evolutionary history." This definition of the word, or comparable vague formulations, enjoy a wide currency today, but nevertheless they are nothing but empty platitudes. For the purposes of phylogenetic systematics I reject them and all definitions of systematic terms based upon them.

In this connection I cannot avoid discussing the pair of words "**cladogenesis**" and "**anagenesis**" (Rensch 1947). Now that the term "anagenesis" has been stripped of the philosophically loaded concept of "higher development", widespread agreement exists in the modern literature as to the meaning of both these words:

"cladogenesis" is the process of origin of new lineages (branches) as the result of the splitting of species;

"anagenesis" is the process of evolutionary change in lineages i.e. the origin of evolutionary novelties in the populations of species (Dobzhansky, Ayala, Stebbins & Valentine 1977; White 1978; De Jong 1980; Wiley 1981).

Thus the two terms cladogenesis and anagenesis cover approximately the basic processes that I subsumed above under the term of evolution.

Is there, according to these definitions, a real difference in meaning between the words phylogenesis and cladogenesis? The origin of new lineages by the splitting of evolutionary species is inseparably connected with the origin of new descent communities. If a minimum of two lineages of evolutionary species has arisen from a particular species, then equally there exists in Nature a closed descent community that includes both lineages. In considering the phenomenon, more stress may be laid on the separation of the lineages or, alternatively, on the unity in Nature formed by those lineages. In point of fact, however, the words phylogenesis and cladogenesis are co-extensive in meaning and no terminological differentiation is justified. I treat them as synonyms, and the older word phylogenesis has priority.

The Greek word "**clados**" (= branch) has been applied in evolutionary research in yet another connection. Opponents of new scientific movements are quick to apply derogatory labels to them (Hull 1979). The term "cladism" thus arose as a reaction to Hennig's development of phylogenetic systematics. Accordingly, every student of phylogeny who orientated his work, with alleged narrow-mindedness, on the process of phylogenesis (= cladogenesis) could be branded as a "cladist". But curiously enough, the negatively loaded terms "cladism", "cladist" and "cladistics" seemed to take the fancy of some phylogenetic systematists, or, in any case, spread with astonishing speed in English-speaking countries. There the advances of phylogenetic systematics are now presented as "Advances in cladistics" (Willi Hennig Society: Funk & Brooks 1981, Platnick & Funk 1983); its frontiers are discussed by a philosopher under the title "The limits of cladism" (Hull 1979); the question is asked "Why to be or not to be a cladist?" (van Valen 1978, Rieppel 1980a); and beyond that there is now a stormy altercation about the "transformation of cladistics" (Platnick 1979), for cladists are separating into the camp of the "pattern cladists" and that of the "phylogenetic cladists" (Beatty 1982, Platnick 1982, Patterson 1982b, Brady 1982).

"Cladism, like all scientific movements, is neither immutable nor monolithic" (Hull 1979, p.417). If, however, the development or transformation of cladistics is bound up with abandoning the basic positions of phylogenetic systematics in Hennig's sense, then a strict terminological separation is imperative. **"Phylogenetic systematics" is, and remains, the appropriate name for a science whose goal is the discovery and representation of the products of phylogenesis.**

For the rest, the accusation that a logically consistent phylogenetic system neglects the "anagenetic component" of evolution, strikes into thin air. Rather is the opposite true. Phylogenetic systematics operates with the whole arsenal of evolutionary transformations of features in the lineages of evolutionary species. Out of this arsenal, it consistently selects those anagenetic elements which can be interpreted as synapomorphies between particular species and species groups. Only these allow monophyletic taxa to be set up and validated as being the equivalents of closed descent communities in Nature.

By **definition, phylogenesis covers the genesis of entire and individual unities in Nature**. It does not extend to the anagenetic transformations of the physical parts of these unities. Correspondingly, it is factually and etymologically mistaken to refer to the alteration of features, or the origin of new features in populations of species, as "feature phylogeny". The correct expression for this is **"feature evolution"**. Likewise, those widespread expressions "the monophyletic origin of a feature" and "the monophyly of a feature" are gross etymological solecisms. The meaning intended can correctly be caught by speaking of **"the unique evolution of a feature"** or "the unique origin of an evolutionary novelty".

One final question should be discussed. Why do I avoid the expression "**natural system**" which is colloquial to every biologist? If Nature itself has created a particular unique order, then nothing could seem more natural than to use Linnaeus' term "Systema Naturae" for Man's picture of this order. I hesitate because the word "naturalness" is historically loaded and ambiguous in use (Ghiselin 1969, Crowson 1970, Wiley 1981). The demand for naturalness in classificatory grouping was raised even from the essentialist Aristotelean viewpoint. That aspect is no longer important, but the opinion has by no means been eradicated that "natural taxa" should be groups of species which share the greatest number of features with each other. Finally, even in modern textbooks classifications are on offer under the heading of "natural system" that contain large numbers of non-monophyletic species groups—and thus "natural taxa" whose elements cannot be traced back to a stem species common to them alone. Whether this has sometimes been done intentionally, or whether it results from defective ideas on the nature of closed descent communities, is of no importance. Fallacious opinions and mistaken classificatory attempts of this sort must in every case be firmly rejected, and it is only consistent that the seductive term "natural system" should be rejected too.

D. Relationship and the Ancestor–Descendant Connection

I. General survey

The word "relationship" is used in biology with several completely different meanings. In view of the existence, as real unities in Nature, of individuals, species and closed descent communities the following particular meanings are of interest:

— "Genealogical relationship" for the connections between individuals and generations within species.
— "Phylogenetic relationship" for the connections between species and species groups which arose by the splitting of stem species i.e. the connections between the species of a closed descent community as well as between different descent communities.

Since the word "relationship" here covers two fundamentally different phenomena I must insist on a consistent corresponding terminology. Genealogical relationship retains its original meaning as the connection between "blood relatives". The extension of the word genealogy to include the relationships between species and supraspecific taxa should be rejected, to avoid the confusions which this creates (Brundin 1972).

The word "ancestor", as in the word pair "ancestor–descendant", is applied with at least three different meanings:

— The individual as ancestor
— The species as ancestor
— The supraspecific taxon as ancestor.

I shall now follow the link between the words relationship and ancestor. This will reveal two logically unobjectionable connections and an artefact.

1) Genealogical relationship—individuals as ancestors

I return to the mechanisms of cohesion in evolutionary species so as to reveal the two possible alternatives in the structure of genealogical connections.

— In species with **uniparental reproduction** (whether parthenogenetic or vegetative) a single individual in each case becomes the ancestor of the next generation. Because there is no horizontal linkage by pairing, the connection by relationship has a strictly hierarchical structure.
— In species with **biparental reproduction** two individuals in each case become the ancestors of the following generation. In addition to the vertical connection of parents with offspring there is also horizontal connection through pairing.

The result is a reticulate structure of relationships in the populations of bisexual species.

2) Phylogenetic relationship—species as ancestors

In the speciational process of **splitting**, one species becomes the ancestor (stem species) of two or several descendants (descendent species). Stated more exactly, descendent species arise from partial populations of the stem species.[1] As a logical result, the phylogenetic connections have a hierarchical structure comparable with the genealogical connections between the generations of an evolutionary species that reproduces uniparentally (Hennig 1984).

Here, however, it is also necessary to consider speciation by **hybridization**, in which a single descendant arises from individual portions of two ancestral species. In form, the result corresponds to the reticulate structure of genealogical connection in biparental species. The formation of species by hybridization, however, is not part of the process of phylogenesis, since it does not produce a closed descent community but only a species (though this, in turn, could become the point of origin of a descent community). And even if we accept Willmann's viewpoint (1985), that three new species arise at a hybridization (p.18), then these three species together would not form a closed descent community having a stem species common to them alone.

This leads into the following difficulty. From the terminological point of view, it is only consistent that the factual content corresponding to phylogenetic relationship should be separated from the special connections between parents and offspring which result from horizontal fusions between species. But this factual content must itself be included in the system of reference. At bottom this system remains a phylogenetic system, for it is based on the fundamental process of the splitting of species as the sole way in which closed descent communities can arise. Hybridization is a subordinate speciation process which can only occur when a hierarchical structure of phylogenetic relationships has already been created by the splitting of species.

Finally, monophyletic species groups can never be connected by horizontal linkage. The relationships between different descent communities are, in principle, phylogenetic connections and have a hierarchical structure.

3) Supraspecific taxa as ancestors—an artefact

Can there perhaps be, however, a vertical connection between different species groups in the sense that a particular species group constitutes the ancestor (stem group) of one or more other species groups (descendent groups)?

Looking at the rich offering of such notions in the phylogenetic literature, this might actually seem self-evident. Among fossils, at any rate, there is a pullulation of "**stem groups**". Thus the "Rhipidistia" are considered as the stem group of the Tetrapoda, and the "Cotylosauria" are referred to as stem reptiles, from which not the "Reptilia" only, but also the other Amniota, are supposed to have arisen. The

---

[1] For a treatment of the population as ancestor within the limits of a species, I refer to Engelmann & Wiley (1977).

"Thecodontia" are pointed out as a species group from which the Crocodilia, the Aves and others can be derived. And the "Therapsida" are presented, under the impressive title of "mammal-like reptiles", as the stem group of the Mammalia.

However, it is not fossil taxa only, but also groups of animals known solely from the present day, which have come to be categorized as supra-specific ancestors. In the controversial debates about the origin of the bilaterally symmetrical Metazoa, the Acoela (taxon Plathelminthes) have often been claimed as the ancestral group of the Bilateria in general. And within the Plathelminthes, the "Turbellaria" are supposed to be the stem group from which the parasitic trematodes and cestodes have evolved (p.250).

All these interpretations are downright wrong. They result from logical mistakes in working out the course and consequences of phylogenesis. I must state once more, with emphasis: as supraspecific products of phylogenesis, only the closed descent communities are real unities in Nature. And, in the process of phylogenesis they have had, each and every one of them, a single stem species only. No group of species—whether solely fossil, Recent and fossil, or known only from the Recent— can, as a group, be the ancestor of a closed descent community in Nature.

If the taxon proposed as "stem group" is itself hypothesized as a monophylum, then it necessarily follows, moreover, that not one single representative of this "stem group" is the stem species or ancestor of any other descent community whatsoever. No single species of the "stem group" Trilobita can have been the stem species of any other arthropod taxon. And this is because the Trilobita have been established as a monophylum (Lauterbach 1980b; p.213 below) and because not one of its constituent species could overstep the limits of the closed descent community Trilobita, once that community had arisen in Nature.

Of course, if the alleged "stem group" is nothing other than an artificial collection of similar species, then among them there may obviously be the stem species of closed descent communities. But if so, it is likewise obvious that this collection, as a non-monophyletic taxon, ought to be broken up. And this will be the fate of most so-called stem groups under logically consistent phylogenetic analysis.

The **concept of the supraspecific taxon as ancestor** is in every case an **artefact without real equivalent in Nature**. This is true, whether the alleged stem group is itself a closed descent community or not. Correspondingly, there can be no place in a phylogenetic system of organisms for supraspecific ancestors or stem groups. I shall discuss this assertion in more detail in considering the systematization of fossils (p.223).

II. The degree of phylogenetic relationship

It is the succession of shared stem species in time which determines the hierarchical structure of phylogenetic relationship connections. From the hierarchy of these connections, the degree of phylogenetic relationship results as "**the relative recency of common ancestry**" (Bigelow 1958). The nearer the common stem species lies to the present day, the closer is the degree of phylogenetic relationship between its descendants.

This brings me to the problem of **determining the degree of phylogenetic rela-**

tionship. Widespread expressions such as "two species are closely related to each other" or "the species X and Y are only very distant phylogenetic relatives" can be forgotten as meaningless tricks of speech. The *conditio sine qua non* is the comparison of at least three species or species groups. For such a "three-taxon statement", the degree of phylogenetic relationship can be unequivocally defined as follows:

"A particular taxon B is more closely related to another taxon C than to a third taxon A if, and only if, it has at least one stem species in common with C that is not also a stem species of A." (Hennig 1966, p.74).

According to this definition, it must theoretically be possible to specify **the relative degree of phylogenetic relationship** between any three evolutionary species, however they are chosen. In every case, two of them will be closer related to each other than are either of these two species to the third.

Looking, for example, at the relationship between the earthworm *Lumbricus terrestris*, the cockchafer *Melolontha melolontha*, and Man *Homo sapiens*, we reach the following result: *Lumbricus terrestris* and *Melolontha melolontha* are closer related to each other than either of them is to *Homo sapiens* because the earthworm and the cockchafer had once in their phylogenesis a shared stem species which was not at the same time the stem species of Man. I give the arguments for this conclusion later (p.68).

This statement is legitimate and very well grounded in the particular instance. But for my purposes—the discovery of closed descent communities in Nature— there is no value whatever in determining the degree of phylogenetic relationship between three randomly chosen species or species groups . Why? Because neither earthworm + cockchafer, nor earthworm + cockchafer + Man, have ever had in their phylogenesis a stem species common to them alone. And thus, in other words, neither the two first-named species, nor the three species together, constitute a monophyletic species group.

If, in studying phylogenetic relationships, we wish to discover closed descent communities in Nature, we need to go beyond the relative degree of phylogenetic relationship and to obtain absolute results. We can do this in the same way as in deciding the degree of genealogical relationship between different individuals of *Homo sapiens*. [1]

---

[1] The degree of genealogical relationship is a quantitative expression of how closely different people are related to each other. In German law, this degree is determined by the number of births through which the relationship is transmitted. A distinction is made, in this connection, between relationship in the direct line and collateral relationship.

Thus people are related in the direct line if one is descended from the other. For example, with grandfather, father and son, the father and son are related in the direct line in the 1st. degree, while the grandfather and the son (his grandson) are related in the direct line in the 2nd. degree.

People are related collaterally if they descend in common from a third person. Thus, in canon law, in the case of siblings and first cousins, the siblings (brothers and sisters) are related collaterally in the 1st. degree, while the children of siblings (first cousins) are collaterally related in the 2nd. degree.

In this book, the degree of phylogenetic relationship is defined, in strict analogy to collateral genealogical relationship, by the number of species splits through which the relationship is transmitted. Sister species or sister groups (= adelphotaxa) stand in the 1st. degree of phylogenetic relationship to each other because of their descent from their common stem species ("the same third person"). Thus the connection between adelphotaxa, in phylogenetic relationship, corresponds exactly to the connection between siblings in genealogical relationship—it is collateral relationship in the 1st. degree.

34

To illustrate **the absolute degree of phylogenetic relationship** I shall give a schematic example with the following assumptions. The three recent species A, B and C are imagined to form a closed descent community with the stem species v common to them alone (Fig. 5). For this "three-taxon connection" the first valid statement is to say: species B and C are phylogenetically closer related to each other than either of them is to species A only if they are the descendants of a stem species which is not at the same time the stem species of A.

Beyond this relative statement, however, we can define, in this instance, the absolute degree of phylogenetic relationship. The species B and C are in the first degree of phylogenetic relationship with each other because, within a larger closed descent community (A + B + C) they are the sole descendants of a stem species w, common to them alone, from the time horizon T1. This is the closest possible phylogenetic relationship which can exist between two evolutionary species.

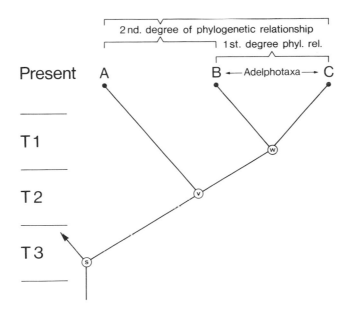

Fig. 5. Diagram to show a particular hypothesis of the relationships between the species A, B and C and to indicate the first two degrees of phylogenetic relationship.

The two species A and B, therefore, and likewise the two species A and C, stand in the second degree of phylogenetic relationship. They do this because each of these two pairs of species goes back to the next older stem species v from time T2. Their relationships are thus based on two successive species splits. The third degree of phylogenetic relationship, finally, is shared by species B and likewise species C with a particular species or species group symbolized by an arrow in the diagram. With this they have a stem species s in common, still farther back in the time T3. Three splittings, of the stem species s, v and w, lead here to the corresponding third degree of phylogenetic relationship.

The absolute degree of phylogenetic relationship between the species of a mono-phylum thus corresponds exactly to the number of splittings of species in time. If

35

so, then phylogenetic systematics should obviously give a maximum of insights if, within a monophyletic species group, it recognized the fifth, tenth or nth degree of phylogenetic relationship in which the individual species stand to each other. Such a goal, however, is unattainable in the practice of phylogenetic research, for two easily understood reasons. Firstly, the pattern of features of individual evolutionary species does not usually yield enough evidence, with current methods, for the sequence of successive species splits to be reconstructed step-by-step without gaps. Secondly, in the unfolding of a closed descent community in time, there may always have been further speciations whose products are incompletely documented as fossils (if at all) or which, in groups of organisms without fossilisable tissue, are inherently unknowable.

Are these statements a confession of surrender? In no way! In trying to discover closed descent communities in Nature, phylogenetic systematics is simply not interested in the fifth, tenth or nth degree of phylogenetic relationship. Its methodological approach follows an entirely different path.

The starting point for determining the degree of phylogenetic relationship is given, in principle, by the organisms of the present day (p.201,204). **Among the connections between recent species and species groups, phylogenetic systematics searches continuously and consistently for the first degree of phylogenetic relationship.** The search begins with three species of a supposedly monophyletic "three-taxon relationship". In the schematic example above, it finds, as already discussed, the first degree of phylogenetic relationship between the species B and C—these being sister species. On achieving this result, the first degree of phylogenetic relationship is sought again at the next higher hierarchical level. At this level, B + C together constitute the sister group of the species A. Stated the other way round, the taxon A is in the first degree of phylogenetic relationship to the species group B + C and is its sister species. The next step is to place the monophyletic taxon A + B + C in the framework of a new "three-taxon connection", so as to seek for the sister group of A + B + C.

For the terms "sister species" and "sister group", I introduce, on the basis of personal discussions with W. Hennig, the technical term **adelphotaxon**. This gives an internationally applicable expression for the first degree of phylogenetic relationship between species or species groups. It covers equally the meaning of the terms sister species and sister group. The **word adelphotaxon can be defined** as follows:

"Adelphotaxa are evolutionary species, or monophyletic species groups, of the first degree of phylogenetic relationship. They arise by the dichotomous splitting of a stem species common to them alone."

Phylogenetic research into relationships can thus be characterized in essence as a consistent and continuous search for the adelphotaxon connections between species and monophyletic species groups. Nevertheless it is here necessary to introduce a further essential consideration. Namely, we do not always need to start with Adam and Eve. The search for the first degree of phylogenetic relationship between taxa does not necessarily begin at the level of evolutionary species. Rather, instead of the three species A, B, and C of the above example, we could place in a three-taxon

36

connection three supra-specific taxa containing any number of species, though this is done under two presuppositions. Firstly, the three supra-specific taxa must themselves be thought to be monophyla. Secondly, the three taxa together must derive from a stem species common to them alone (p.70). Under these conditions the procedure is completely valid, for the supra-specific taxa are now represented by their stem species i.e. by evolutionary species of the past which must have possessed as autapomorphies exactly those evolutionary novelties which recur as synapomorphic agreements in their immediate descendants. And it is because of these synapomorphies that the groups are regarded as monophyla.

For reasons to be established later, I shall now assume the following: the three known mammalian taxa of the Monotremata, the Marsupialia and the Placentalia are each monophyletic species groups. Moreover, in the past they shared a stem species common to them alone. In that case, the Monotremata, Marsupialia and Placentalia can be placed in the position of the three recent species A, B and C of the schematic example.

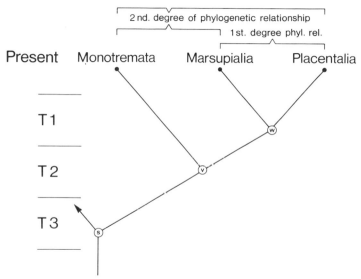

Fig. 6. A particular hypothesis of the relationships of the three mammalian taxa Monotremata, Marsupialia and Placentalia; also shown are the first two degrees of phylogenetic relationship.

Assuming that the connections shown in Fig. 6 can be established (p.86), I shall now present, for the sake of tidiness, the three first degrees of phylogenetic relationship in the case of the mammals. The Marsupialia and the Placentalia are adelphotaxa, or phylogenetic relatives of the first degree. This is because they are the sole descendants of the stem species w from time T1. The second degree of phylogenetic relationship exists between the Monotremata and the Marsupialia, and also between the Monotremata and Placentalia, because they go back to an older ancestor v from the time T2. The third degree of phylogenetic relationship, finally, is shared by the

37

Marsupialia and the Sauropsida[1], as well as the Placentalia and the Sauropsida, for these had a common stem species s, one stage farther back, at time T3.

In view of what I have said above, however, the establishment of the second and third degree of phylogenetic relationship between the individual mammalian taxa, and between each of them and the Sauropsida, is a side-issue. Rather, once again, it is crucial always consistently to establish the first degree of phylogenetic relationship. After deciding the adelphotaxon relationship between the Marsupialia and the Placentalia, we need to move up to the next higher hierarchical level. Here the Marsupialia and the Placentalia together constitute the adelphotaxon of the Monotremata. Thus it is the group Marsupialia + Placentalia, on the one hand, and the Monotremata, on the other, which are phylogenetic relatives of the first degree. And, after a further step upwards, we search for the first-degree phylogenetic relative of the Mammalia as a whole (Monotremata + Marsupialia + Placentalia). At this level, this is probably the taxon Sauropsida.

The fundamental requirement for the erection of a consistent phylogenetic system is the recognition of the first degree of phylogenetic relationship between recent species or monophyletic species groups with recent representatives.

When once established, the adelphotaxon relationships between species of the present day and supraspecific taxa with recent representatives remain unaltered if fossil taxa are integrated into the system using the methods of phylogenetic systematics (p.201).

---

[1] The hypothesis of an adelphotaxon relationship between the Mammalia and a taxon Sauropsida is disputed (p.91). The general considerations concerning the degree of phylogenetic relationship would, of course, still be completely valid if some differently constituted amniote taxon were placed in the position of the Sauropsida.

E. The Structure of the Phylogenetic System and the Equivalence between Relationship Diagram and Hierarchical Tabulation

I. Hierarchical structure

The hierarchical structure of phylogenetic relationships necessarily produces a phylogenetic system with a corresponding hierarchical structure.

I shall now complete my explanation of how phylogenetic relationships can be logically transcribed into an appropriate systematic structure by using the examples already discussed and suitably modify the relationship diagrams (Figs. 5–7 and 6–8).

In the relationships between three evolutionary species, the two sister species occupy the lowest level of hierarchy which is possible in fact. Thus, in the first example discussed, adelphotaxa B and C occupy level 3 (Fig. 7). On the next higher level they must be united to form a monophylum with its own name (taxon II in the diagram). The coordinate adelphotaxon on this level 2 is taxon I with the single species A. Then, on the first level of the hierarchy, taxa I and II, in their turn, are joined to form a new monophylum with its own name (taxon α).

Such a **diagram of phylogenetic relationships** can now be translated, without any additional information, directly into a **hierarchical tabulation**. I shall label the hierarchical sequence of taxa schematically by means of subsuming groups of numbers. In this way the adelphotaxon relationships can be clearly characterized as follows:

Taxon α (= I + II)
   1. Taxon I   (= A)
   2. Taxon II (= B + C)
      2.1 Taxon B
      2.2 Taxon C

In the mammalian example, I began with supraspecific taxa at an indeterminate hierarchical level, and recognized the Marsupialia and Placentalia as adelphotaxa (Fig. 8). These sister groups must be united to form a monophylum at the next higher level and this is given its own name of Theria. The coordinate adelphotaxon at this level was the species group Monotremata. Logically, therefore, the Monotremata and the Theria must be united to form a new taxon, with its own name, at the next higher level. This is the monophylum Mammalia.

The equivalent of this relationship diagram is the following hierarchical tabulation of the systematization. Here again, the adelphotaxa are indented to correspond with the sequence of subordination.

Taxon Mammalia
   1. Taxon Monotremata
   2. Taxon Theria
      2.1 Taxon Marsupialia
      2.2 Taxon Placentalia

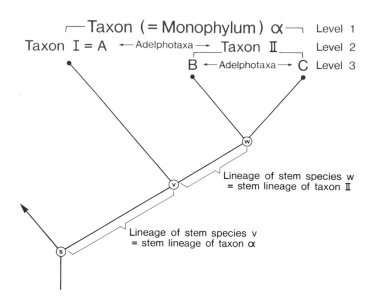

Fig. 7. A diagram of phylogenetic relationships for the three evolutionary species A, B and C with the arrangement of the adelphotaxa on identical levels. Further explanation in text.

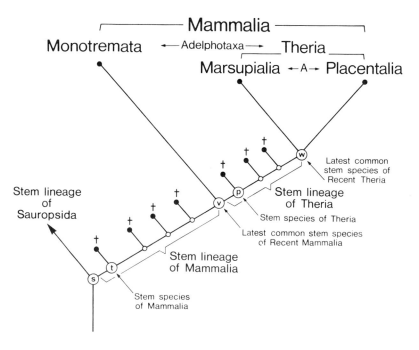

Fig. 8. The phylogenetic relationship of the three supraspecific taxa Monotremata, Marsupialia and Placentalia with the arrangement of adelphotaxa on identical levels. The stem species and stem lineages of closed descent communities are indicated.

40

From a complete analysis of every three-taxon connection there result three successive levels of hierarchy of the phylogenetic system. This statement can be exemplified by the following scheme:

Superordinate level   Mammalia

Coordinate level   Monotremata + Theria

Subordinate level   Marsupialia + Placentalia

This shows the structure of the phylogenetic system in a form that necessarily applies in general. Its hierarchical pattern is the result of constant sequences of superordination, coordination and subordination. If we begin on the lowest level with two evolutionary sister species (B and C in the example), then logically there is only the one way of building the system upwards in steps. If some higher level of coordination is selected, with two supraspecific adelphotaxa, then superordination follows upwards, and subordination necessarily follows downwards.

II.   Rank of taxa

In the structure of the phylogenetic system, as in every hierarchical ordering, a **sequence of ranks** between the taxa of the individual levels is necessarily linked with the sequence of hierarchical levels. Here the following law applies: the **adelphotaxa** of any given level of coordination have **identical rank**. This is easy to understand since they arose at exactly the same time by the splitting of the stem species common to them alone. The adelphotaxa of the superordinate level take the next higher rank, as products of an older stem species. And, logically enough, the adelphotaxa of the subordinate level take the next lower rank just because they together derive from a correspondingly younger stem species. Among supraspecific taxa, the smallest possible monophylum, comprising two sister species (and their common stem species) logically takes the absolutely lowest rank.

In assigning an identical rank to the adelphotaxa of any given level of coordination it does not matter in the slightest whether the adelphotaxa, after the splitting of their stem species, remain as two single species, or whether one or both of them develop in the course of phylogenesis, by further splittings of species, into closed descent communities. As I have already suggested in treating the formal aspects of classification (p.24), a single evolutionary species can thus take up any rank whatsoever in a consistent phylogenetic system if it is the sister species of a particular supraspecific taxon of the corresponding rank. In a correct phylogenetic analysis of the three recent monotreme taxa *Ornithorhynchus*, *Tachyglossus* and *Zaglossus*, the evolutionary species *Ornithorhynchus anatinus* (duck-billed platypus) must take the same rank as the supraspecific taxon of the echidnas. The Ornithorhynchidae and Tachyglossidae stand next to each other on one level of coordination, with an identical rank.

Here I must stress emphatically that these statements about the ranks of taxa in a phylogenetic system are, in fact and logic, absolutely independent of the subsidiary technical problem of how best to indicate the individual ranks in the hierarchy of the phylogenetic system. The knowledge contained in the phylogenetic system does not depend, in the slightest, on whether the single steps in rank are labelled with the categorial designations of the traditional Linnaean classification (by necessity or choice), with subsuming groups of numbers, or however else (p.234).

III. Graphical representation

Since Haeckel's classical conception of the "monophyletic tree of descent of organisms" (1866) the literature of the subject has been flooded with the most diverse representations of such trees, each claiming to show, in a suitable visual form, various aspects of the phenetic, genetic, evolutionary, genealogical, phylogenetic or cladistic relationships between organisms.

The range of such graphical designs extends from naturalistic trees with a trunk and branching limbs to simple line-diagrams. Sometimes they attempt to express supposed differences in the "grade of evolutionary organization" by spreading the sequence of organisms in a vertical direction. Sometimes they try to show the "different extent of evolutionary transformations" by arranging the taxa at varying distances in the horizontal dimension so that the connecting lines meet at correspondingly different angles. Sometimes the different numbers of species in the supraspecific taxa are shown by corresponding thickenings of the lines as they rise upwards. And sometimes, finally, the astonishing statement is made that phylogenetic relationships can only be adequately illustrated by using all three planes of space, with three-dimensional diagrams as a result.

The terms used for the graphical representations or relationships are no less varied. They range from the oldest designation "Stammbaum" (phylogenetic tree) to names like dendrogram, phenogram or phylogram to the newest word "cladogram" created by the opponents of phylogenetic systematics (Mayr 1965, Camin & Sokal 1965). However, there is absolutely no agreement on which term should be assigned to which particular form of diagram. On the contrary, the terms are used by different authors to cover the most varied aspects representing the relationships between taxa.

I expressly desist from all the terms mentioned. A graphical representation of phylogenetic relationship will be called exactly what it is—**a diagram of phylogenetic relationship**[1]. This may sound somewhat long-winded but in point of fact cannot be misunderstood and is totally necessary if we wish to escape an incurably confused terminology.

Phylogenetic relationships arise as the result of successive species splits in the time dimension. In this sense, they are of unidimensional nature and can be perfectly represented on the surface of a sheet of paper. Occasional attempts to express phylogenetic relationships in three dimensions are pointless.

---

[1] Or relationship diagram for short.

IV. Terminal taxa—stem species

The alternatives of the terminal taxon and the stem species are the basis for applying different symbols in the relationship diagram.

Recent evolutionary species, being the present-day end products of evolution, are provisionally terminal taxa. As unities which ontologically are individuals they may have a very long existence in time. Sooner or later they may die without issue and then become terminal taxa definitively. Or they may be involved in a splitting process and then become the stem species of a new closed descent community. **Fossil evolutionary species** may either be terminal taxa without descendants or stem species.

Closed descent communities are, as such, not subject to the action of evolution (p.22) and correspondingly are definitive terminal taxa. This is true not only for completely extinct descent communities like Pterosauria or Trilobita, but also for closed descent communities with recent species. Within the limits of their respective descent communities, these extant species may, of course, suffer evolutionary changes or split up into new species, but the descent communities themselves remain terminal taxa.

For these reasons, I use a uniform sign to symbolize recent evolutionary species, extinct evolutionary species without descendants and closed descent communities in Nature (= monophyla of the phylogenetic system), as terminal taxa. This sign is a large black dot.

Stem species, being alternatives to the terminal taxa, are shown by circles, in which letters can be placed if necessary. Stem species take this symbol: 1) whether they are in principle identifiable and also can be recognised as fossils, or not (p.225); or 2) whether they belong to a group of organisms without fossilisable tissues and thus can, in principle, never be discovered.

The evolutionary species concept is a lineal concept and, consequently, dots for terminal species and circles for stem species are great simplifications. Obviously they do not include the whole span of temporal existence of these closed working units of evolution. Rather, with a still extant species they mark the final part of its existence as being provisionally terminal (black dot); with a species which died without issue in a particular period of the past they signify the final part of its existence as definitively terminal (black dot); and with a stem species they show when it split into descendent species (open circle).

I now wish to discuss **the arrangement of terminal taxa and stem species** in diagrams of phylogenetic relationship. The symbols for stem species in the diagram follow one another vertically from below to above according to the sequence of speciations in time. Those terminal taxa which emerge simultaneously from a stem species common to them alone, being adelphotaxa, have an identical rank. In principle, they are placed in a single horizontal plane.[1] As an example I shall recapitulate the favoured hypotheses for the interconnections of the highest ranking mammalian taxa and consider how these hypotheses can be expressed as a correct phylogenetic

---

[1] For the treatment of fossil taxa see p.207.

relationship diagram (Figs. 6–8). The three taxa examined—Monotremata, Marsupialia and Placentalia—start out with equal rank, but this situation changes as soon as a sister-group relationship between two of them—Marsupialia + Placentalia—is shown to be probable.

When this has been done, these two alone remain next to each other on a single level while, at the next higher level, the Monotremata become coordinate with a new taxon Theria, being its adelphotaxon.

Terminal taxa and stem species are connected with each other in the diagram of phylogenetic relationship by straight lines, whether the terminal taxa are single evolutionary species or monophyletic species groups. Each forking of the lines in the diagram signifies a real process in Nature i.e. the splitting of a stem species into descendants.

V.   Stem lineages

Closed descent communities in Nature have stem lineages, within which their characteristic evolutionary novelties were evolved. I shall establish the existence of stem lineages and explain the two basic possibilities.

1)   The stem lineage of a closed descent community when the lineage comprises a single evolutionary species (Fig. 7)

In the lineages which connect the symbols of the species s and v as well as v and w, it is supposed that no speciations have occurred. In this case the stem lineage of taxon α corresponds to the lineage of stem species v, and the stem lineage of taxon II corresponds to the lineage of the stem species w. In other words, during the existence in time of the evolutionary species v, all those novelties (= autapomorphies) must have evolved which exist as synapomorphies between the subordinate adelphotaxa I and II. Then, in the lineage of the evolutionary species w, all the features of taxon II were additionally developed which the sister species B and C have in common as synapomorphies.

2)   The stem lineage of a closed descent community when the lineage comprises two or more evolutionary species (Fig. 8)

In the lineages connecting s and v and also v and w, it is supposed that several speciations have occurred. In this case the respective stem lineages represent complex segments of phylogenesis each made up of a number of evolutionary species corresponding in the minimum to the number of speciations.

In the mammalian example, the stem lineages of the Mammalia and the Theria form segments of this sort, made up of a large, but not precisely determinable, number of evolutionary species. Corresponding to the successive splittings of species, there results a whole chain of stem species, for which the beginning and end of this chain must be marked exactly.

At the beginning of the stem lineage of the taxon Mammalia there was the stem species t which arose from the splitting of the latest common stem species s of the recent Sauropsida and Mammalia. The evolutionary species t is the oldest stem species of the Mammalia, or simply **the stem species** from which the closed descent community Mammalia arose. The end of the stem lineage of the Mammalia is determined by the evolutionary species v—this is the **latest common stem species** of the highest ranking adelphotaxa (Monotremata and Theria) that have recent representatives.

Correspondingly, in the subsequent stem lineage the evolutionary species p is the stem species of the Theria. The upper end of the stem lineage of the Theria is formed by the evolutionary species w which is the latest common stem species of the highest ranking adelphotaxa with recent representatives—these are the Marsupialia and the Placentalia.

What justifies the claim that the stem lineages of the Mammalia and the Theria respectively consisted of a whole chain of evolutionary species? The ground pattern of the recent mammals includes more than 60 evolutionary novelties (p.87,149) which arose within the stem lineage of the Mammalia and which must all have been present in the latest common stem species of the Monotremata and the Theria. For the adelphotaxa Marsupialia and Placentalia about 20 synapomorphies are known (p.88–89) which must have evolved as autapomorphies in the stem lineage of the Theria. Even without empirical evidence, it seems *a priori* extremely unlikely that the evolutionary novelties of the Mammalia and of the Theria respectively arose within the limits of a single evolutionary species without any speciation processes. The rich fossil record of the phylogenesis of the mammals likewise indicates that this was not so (p.217). The stem lineages of the Mammalia and of the Theria comprised numerous species.

I shall now repeat the **essence** of what I have said **concerning the stem lineages** of closed descent communities in Nature. In a supraspecific unity with a small number of evolutionary novelties, the stem lineage may perhaps, theoretically, coincide with the temporal existence of a single evolutionary species—in other words, the unity has only one stem species. For a descent community with numerous autapomorphies, on the other hand, we must expect that the stem lineage will consist of a greater number of evolutionary species. To understand the composition of a closed descent community and to express it in a relationship diagram two speciations are of basic importance. With the splitting of the oldest stem species at the beginning of the stem lineage, a new descent community is introduced into Nature. And, at the end of the stem lineage, the splitting of the latest common stem species gives rise to the highest ranking sister groups with still living representatives within the descent community. These two speciations correspond exactly to the instants referred to by Hennig (1966, 1982) as the time of origin and the time of differentiation of a closed descent community.

The visual statements made by a phylogenetic relationship diagram can be summarized in a few sentences. The hierarchical structure of phylogenetic relationships can be unambiguously expressed by: arranging taxa of equal rank at identical levels in the horizontal dimension; by placing the stem species in a sequence corresponding to the levels of ordination in the vertical dimension; and by connecting the symbols with lines. The forkings of the lines show the minimum number of species splits for the unity shown in the diagram. The lines between the stem species represent the stem lineages, or portions of the stem lineages, of the closed descent communities in Nature.

VI.   Cladogram — phylogenetic tree — relationship diagram

In English-speaking countries there has been, in recent years, a controversy about two forms of diagram intended to show the connections between unities of living Nature — two forms that are alleged to be fundamentally distinct i.e. the cladogram and the phylogenetic tree (Tattersall & Eldredge 1977; Cracraft 1979; Eldredge 1979; Platnick 1979; Hull 1979; Wiley 1979b, 1981; Eldredge & Cracraft 1980; Patterson 1980; Nelson & Platnick 1981; Ball 1981).

The definitions vary between the individual authors. In so far as reference to evolutionary theory is acknowledged, the following difference in meaning between the words cladogram and phylogenetic tree can be extracted from the multiplicity of opinions. A "cladogram" is supposed to represent an abstract synapomorphy scheme — each branching (whose nature is not further specified) leads to unities characterized by evolutionary novelties common to them alone. A "phylogenetic tree", on the other hand, specifies that the branching points of the diagram are species splits, and it thus portrays the historic happenings which have caused the phylogenetic relationships between the unities displayed.

The distinction between cladogram and phylogenetic tree may, or may not, be useful. However that may be, I here assert with emphasis: if my viewpoint is correct that Nature has created phylogenetic relationships by successive speciations, then there can be only one correct form for displaying these relationships graphically. I claim that the diagram of phylogenetic relationship is this correct form. In other words, for the purposes of phylogenetic systematics, it is irrelevant what notions different authors may connect with the terms cladogram and phylogenetic tree. I shall remain consistent and use the expression "diagram of phylogenetic relationship" as an unambiguous term that cannot be misunderstood. The relationship diagram represents a graphical summary of those connections in living Nature that have arisen in the process of phylogenesis.

VII.   Formal conventions

It is desirable that diagrams of phylogenetic relationship should be uniform in construction, and this leads me to deal finally with two formal conventions.

The horizontal distances between the adelphotaxa, the vertical distances between the stem species, and the angles of furcation between the connecting lines convey no specific information. Therefore, within the spatial limits of a sheet of paper, I make them as equal as possible. From the start this avoids the illusion that varying distances or angles could perhaps hide some additional meaning. Whether, for the rest, the lines are drawn straight or curved, or made to run parallel upwards to save space, is finally a question of taste — about which, as is well known, there is no point in arguing. Nevertheless, stem lineages bent through rightangles, resembling the similarity diagrams commonly used in phenetics, should be avoided. For they produce the irritating visual impression that the course of phylogenesis can be divided into horizontal and vertical components.

A second, thoroughly accepted convention concerns the arrangement in the horizontal dimension for adelphotaxa of equal rank. With reference to the distinction treated below between plesiomorphy and apomorphy, the taxon with the relatively greater number of plesiomorphous (primitive) features is always placed on the left and the taxon with the correspondingly larger number of apomorphous (advanced) features is placed on the right. Thus, in my phylogenetic relationship diagrams of the Mammalia, the Monotremata are at the left and the Theria alongside them on the right. At the next level of subordination, again, the Marsupialia are at the left and the Placentalia at the right. Correspondingly, in the hierarchical tabulation of the phylogenetic system, the adelphotaxon with the greater number of plesiomorphous features is placed in the first position. Thus, to take the same example, 1. Monotremata, 2. Theria and 2.1 Marsupialia, 2.2 Placentalia. Obviously, nothing would change in the contents of the statement if the arrangement of the adelphotaxa were rotated through 180° in the relationship diagram (Theria left, Monotremata right) and the sequence in the tabulation correspondingly inverted (1. Theria, 2. Monotremata).

F. From Hypothesis of Relationship to Phylogenetic System

I. Phylogenetic systematics as an empirical science

Phylogenetic systematics claims to be an empirical science. It stakes this claim by asserting that, in solving its problems, it can obtain objective perceptions about the relationships between real unities in Nature i.e. between evolutionary species and closed descent communities. And these perceptions are based on experience.

The problem of whether this claim accords with K. Popper's philosophy of science (1974, 1976, 1979) has recently been intensively debated (Hull 1983). As is widely known, the decisive criterion for Popper in demarcating any form of science from non-science, metaphysics or speculation is the potential falsifiability of scientific statements. According to Popper, theories or hypotheses have a scientific character only if they can be refuted by experience. Thus a scientific theory (hypothesis) must have testable implications which make it, in principle, falsifiable on the basis of observationally or experimentally ascertained facts. On the other hand, a scientific theory can never be conclusively shown to be true (verified) — it can never be more than strengthened by testing (corroborated). But the more often a theory or hypothesis has withstood strenuous attempts at falsification, the greater is its degree of corroboration.

Do hypotheses of phylogenetic relationship have this character of scientific hypotheses i.e. do they show a marked asymmetry between the impossibility of being verified by facts, and the possibility of being falsified by them? Are hypotheses of phylogenetic relationship, as scientific statements, compatible with Popper's demarcation criterion and can they be made subject to the hypothetico-deductive method of testing?

In approaching this problem, I shall ignore the fact that the question of the demarcation of scientific activities is itself the object of sharp discussions in the philosophy of science (Lakatos & Musgrave 1970, 1974; Beatty 1982, Scheibe 1983). Leaving these discussions aside, two seemingly irreconcilable positions about the character of hypotheses of phylogenetic relationship are diametrically opposed to each other. By some authors the epistemological views of Popper have been adopted completely in the study of historical relationships. Such authors hold that the hypotheses formulated by phylogenetic systematics are potentially falsifiable (Bonde 1975, 1977; Wiley 1975, 1976, 1981; Engelmann & Wiley 1977; Løvtrup 1977; Platnick & Gaffney 1977, 1978; Gaffney 1979; Cracraft 1979; Eldredge & Cracraft 1980).

Other authors have opposed the application of the hypothetico-deductive method to the practice of discovering relationships between the products of phylogenetic development (Patterson 1978a; Rieppel 1980; Arnold 1981; Cartmill 1981). In essence, these authors regard hypotheses about the phylogenetic relationship between the real unities of living Nature as being, in principle, neither verifiable nor falsifiable.

The crucial point on which these controversies turn is differing usage of the word falsification. If the expression "to falsify" is used in the widespread meaning of "to refute totally" or "to prove definitively false", then hypotheses of phylogenetic relationships are in fact not falsifiable. But if the expression "to falsify" is intentionally taken in a limited sense as meaning "to weaken" or "to reject" (Cracraft 1979, p.34; Eldredge & Cracraft 1980, p.69; Farris 1983, p.9), then the mode of procedure of phylogenetic systematics fits basically into the framework of the hypothetico-deductive method of scientific argument.

I base this position on the following assertions. Because of its coherent logical concepts and its consistent methodology, phylogenetic systematics is in a position:

1) to formulate objective hypotheses about the phylogenetic relationships between real unities of Nature;
2) to subject rival hypotheses of relationship to an empirical testing procedure;
3) to make probability judgements, based on experience, between rival hypotheses of relationship as a result of which some hypotheses are favoured over others;
4) to expose empirically grounded hypotheses of relationship to critical and objective testing, in which they can be strengthened or weakened but never definitively proved false;
5) to place the hypotheses of relationship favoured by testing into a consistently phylogenetic system.

On this basis I shall sketch out the logical path which leads from the formulation of hypotheses of phylogenetic relationship, *via* the testing of rival hypotheses, to the placing of the favoured hypotheses in a phylogenetic system.

The formulation and testing of hypotheses of phylogenetic relationship are carried out by applying the theory of evolution. I repeat here, in a few words, the essential axioms already presented in the introduction: evolutionary novelties show themselves in the lineages of successive populations of species; evolutionary species arise by the splitting of species or the fusion between parts of species; each closed descent community in Nature is the product of the splitting of a stem species common to it alone.

The origin of evolutionary novelties in the lineages of evolutionary species is the basic precondition for arranging the products of phylogenesis in order. The erection of a consistent phylogenetic system, on the other hand, is fundamentally independent, both in logic and methodology, from the problems of how and when such novelties evolve in the lineages of species. In setting up such a system such problems are purely subsidiary.

1) The practice of phylogenetic systematization of organisms is independent of all suppositions about the causal mechanisms of evolution. This is true, whether the suppositions come from the hypothetical edifice of "the synthetic theory of evolution" (Dobzhansky, Ayala, Stebbins & Valentine 1977; Mayr 1982c), from expansions or modifications of this theory, or even from alternatives to it (Riedl 1975, 1977; Løvtrup 1977; Rosen & Buth 1980; Gutmann & Bonik 1981a,b; Wiley & Brooks 1982; Wuketits 1981, 1982, 1983; etc.).
2) Phylogenetic systematics is likewise independent of the question whether evolutionary novelties arise continuously in the lineages of evolutionary species, appear suddenly in the initial phase of species splits (Eldredge & Gould 1972; Gould

& Eldredge 1977), or mainly in a late phase of the sympatric coexistence of reproductively isolated sister species (Bock 1979). The vehement current debates
about alternative models of evolution—gradualism *versus* punctuationalism—are
in principle irrelevant to solving the problems of phylogenetic systematics.

II. The formulation and testing of hypotheses of relationship, assuming the dichotomous splitting of species

The central research objective of phylogenetic systematics is the discovery of the
first degree of relationship (the closest possible) between real unities in Nature. The
corresponding search for two sister species or monophyletic sister groups, which
together go back to a stem species common to them alone, makes the following
assumption: dichotomous splittings of stem species into two descendent species are
real processes in living Nature. And this is entirely independent of whether, in any
particular case, the splits follow the pattern of allopatric, parapatric or sympatric
speciation.

As to the continuing debate on the "principle of dichotomy" (Hull 1979), I base
my position on one simple consideration. Phylogenetic systematics can show that sister species and monophyletic sister groups exist in Nature by proving apomorphous
agreements shared by two particular taxa alone—agreements which can indisputably
be interpreted as synapomorphies. The mere fact that such groupings can convincingly be proved, time and time again, in the most different groups of organisms,
shows conclusively that the assumption that species split dichotomously is more than
a mere heuristic methodological principle. On the contrary, the existence of such
groupings implies that the splitting of a stem species into two descendent species
must be a widespread mode of origin for closed descent communities in Nature. This
argument, of course, does not put in question the possible occurrence of multiple
species splits—I shall treat these at length in the next section.

On the basis of the concept of dichotomy, I shall now formulate a first series
of rival hypotheses of relationship. The three species A, B and C are supposed
all to derive from a stem species v common to them alone and to have arisen by
two dichotomous splits. Under these assumptions, there are three, and only three,
competing hypotheses of relationship (Fig. 9).

— A and B are adelphotaxa, being the sole descendent species of the stem species
  x.
— A and C are adelphotaxa, being the sole descendent species of the stem species
  y.
— B and C are adelphotaxa, being the sole descendent species of the stem species
  w.

The testing of these three hypotheses begins by ascertaining empirically the existence of heritable differences and agreements between the three evolutionary species.
Differences arise by the transformation of pre-existing features or by the acquisition of new features in the lineages of evolutionary species. As I have repeatedly
stressed, however, phylogenetic systematics is less interested in the differences be-

50

tween evolutionary species than in the similarities and agreements between them. This is legitimate in logic and fact, "because both phenomena behave relative to each other like positive and negative of the same picture" (Hennig 1980, p.15).

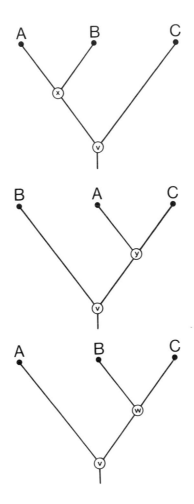

Fig. 9. The three rival hypotheses as to the phylogenetic relationship between the three evolutionary species A, B and C, assuming dichotomous splitting of the stem species.

Possible forms of agreement

I shall now focus on the evolutionary agreements between different species. There are three possibilities which differ in principle. In what follows, I shall explain these and illustrate each using the "three-taxon relation" of species A, B and C (Fig. 10).

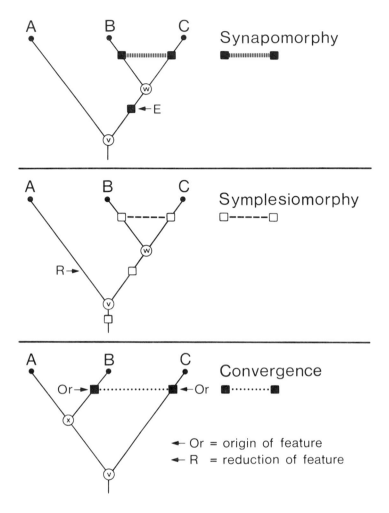

Fig. 10. The three forms of evolutionary agreement between different taxa, illustrated by using the three evolutionary species A, B and C.

1) Synapomorphy

This can be defined as follows:

"An agreement between adelphotaxa (evolutionary species, monophyletic species groups) in a feature which arose in their shared stem lineage as an evolutionary novelty, and which first existed as an autapomorphy in the stem species common to them alone; the feature was taken over from this stem species."

In the simplest conceivable example, the closed descent community with the species B and C has a stem lineage consisting of a single evolutionary species— the stem species w (cf. p.44). A particular agreement between the species B and C is supposed to have arisen as an evolutionary novelty in the lineage of the stem species w. This apomorphy, which was primarily lacking in the next older stem species v, may represent the evolution of a feature that did not previously exist, or obviously it may represent the evolutionary transformation of a feature already present or even its complete suppression. In any case, the apomorphous (advanced) feature of the evolutionary species w is handed on, in the process of species splitting, to become a synapomorphy of the descendants B and C.

2) Symplesiomorphy

This is definable thus:

"An agreement between evolutionary species or monophyletic species groups in a feature which did not arise in their common stem lineage but was taken over from stem species that were even more remote."

A particular agreement between the species B and C was already expressed as a feature in the stem species v of the three species A + B + C. In the splitting of the stem species v, the feature was taken over without change into the lineage of the evolutionary species w and then later, when this species in turn split, was passed, again unchanged, to the descendent species B and C. On the other hand, in the lineage of species A an evolutionary transformation occurred; in the present example the feature is supposed to have been completely lost. If so, the agreement present only between B and C depends on the retention unaltered of a plesiomorphous (primitive) feature. Its common possession, shared by species B and C, is a symplesiomorphy.

3) Convergence

This is definable thus:

"An agreement between evolutionary species or monophyletic species groups in a feature which was absent in their latest common stem species and first evolved, independently, in the lineages of the unities after these lineages had separated from each other."

A particular agreement between B and C did not yet exist as a feature in the latest common stem species v. Rather the feature evolved twice, separately, in the lineages of the species B and C with a similar or even identical form. An agreement in an apomorphous (advanced or derived) feature which arose twice independently is a convergence[1] of the species B and C.

---

[1] Several further terms exist to signify the independent evolution of similar or identical features in different organisms. Such terms include parallel evolution, parallelism, analogy, homoiology and homoplasy. The dispute over whether these numerous words in fact stand for different evolutionary phenomena, or not, is of no importance here. For the aims of phylogenetic systematics, only one question is important: whether apomorphous agreements between different species have arisen only once, or several times. To indicate the former case the term synapomorphy has been created. For the second case, of repeated origin, the traditional word convergence has been generally adopted in the terminology of phylogenetic systematics.

The value of agreements

Given this state of things, the next question is: what significance in the testing of rival hypotheses of relationship is possessed by the three forms of potential agreement?

Symplesiomorphy (Fig. 11)

Factual content: A particular agreement between the species B and C is a symplesiomorphy (feature 1). The common possession of the primitive feature is first inserted in the diagram of the three species A + B + C with unconnected lines and then the symplesiomorphy is placed in the three rival hypotheses of relationship (Fig. 9).

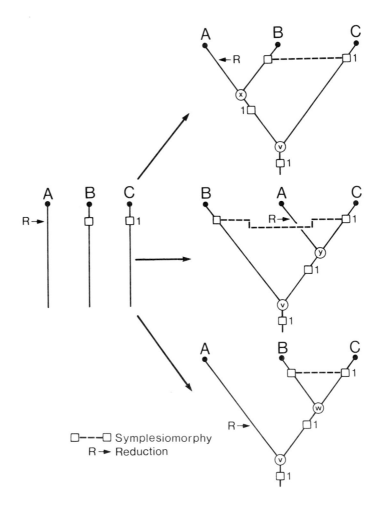

□---□ Symplesiomorphy
R➔ Reduction

Fig. 11. The insertion of a symplesiomorphy between the species B and C into the three rival hypotheses of relationship.

Result: The symplesiomorphous agreement between the species B and C is consistent with any of the three hypotheses of relationship. The evolutionary change of the feature 1 to the apomorphous condition (R = reduction) in the lineage of the species A could occur in any of the cases, irrespective of whether A + B, A + C or B + C are sister species.

Symplesiomorphies are completely worthless for any testing procedure which includes probability judgements between rival hypotheses of relationship.

Convergence (Fig. 12)

Factual content: Another agreement between species B and C is a convergence (Feature 2). The common possession of the apomorphous feature is, again, first placed in the diagram of the three species with unconnected lines and then the convergence is inserted in the three hypotheses of relationship.

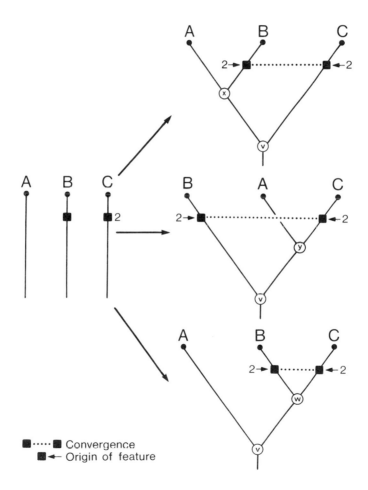

Fig. 12. The insertion of an apomorphous agreement between the species B and C into the three rival hypotheses of relationship, if the agreement is interpreted as a convergence.

Result: The convergent agreement between species B and C in feature 2 is consistent with any of the three hypotheses of relationship. Its separate evolution in B and in C is completely independent of whether these two species are adelphotaxa or not.

Convergences are not merely worthless for phylogenetic systematics. Worse still, they are apt to hinder the discovery of phylogenetic relationships.

Synapomorphy (Fig. 13)

Factual content: A third agreement between species B and C is a synapomorphy (Feature 3). Once again, the common possession of the apomorphous feature is inserted into the diagram with unconnected lines. According to the definition of the word synapomorphy, the agreeing feature is required to have arisen once only as an evolutionary novelty (autapomorphy) in the stem lineage of the species B and C. Consequently, the agreement can only be transposed into one hypothesis of relationship—namely into the hypothesis in which B and C are descendants of the stem species w.

Result: The synapomorphous agreement in feature 3 validates the hypothesis of an adelphotaxon relationship between the species B and C.

Synapomorphies are the decisive agreements for the purposes of phylogenetic systematics. Probability judgements between rival hypotheses of relationship can be made only on the basis of synapomorphies.

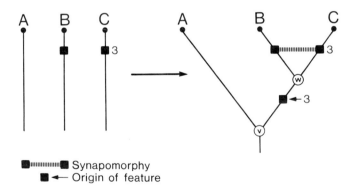

Fig. 13. The transposition of an apomorphous agreement between species B and C into the only possible hypothesis of relationship if the agreement is interpreted as a synapomorphy. The synapomorphy is the basis for favouring this hypothesis of relationship.

Deciding between symplesiomorphy, synapomorphy and convergence

Out of the three forms of evolutionary agreement, therefore, only synapomorphy matters. And this brings us, inescapably, to a new problem, which is that: before testing hypotheses of phylogenetic relationship, it is always necessary to determine the precise character of similarities and agreements between species. If these cannot

56

be shown to be synapomorphies, they must consistently be excluded from analyses of relationships.

Probability judgements between symplesiomorphy, synapomorphy and convergence fall into two logically coupled steps.

1) For every agreement between different species, the question must first be asked whether the **agreeing feature is plesiomorphous or apomorphous**. A complex of arguments is available for making an empirically orientated judgement about the alternatives of plesiomorphy and apomorphy—these arguments will be dealt with in detail in the next chapter.

2) If the probability judgement favours apomorphy then, and only then, do **the alternatives of synapomorphy and convergence** come under discussion. In deciding between them, there is in the methodology of phylogenetic systematics fundamentally no empirical yardstick. Instead, the probability judgement between synapomorphy and convergence is a logical operation made in the framework of **the principle of parsimony**. There are endless discussions about the value of this principle in scientific argumentation, and in systematics the controversy continues till this day (Andersen 1978; Panchen 1982; Brady 1983; Farris 1983; Sober 1983)[1]. But concerning this problem, the following statement is valid in every case. When an agreement interpreted as being apomorphous exists between different species, it is always more parsimonious to assume that it has arisen uniquely as an evolutionary novelty, than to postulate repeated independent development giving an identical phenotype. In accordance with the principle of parsimony, it is first necessary to interpret apomorphous agreements between different species as having evolved only once. It is not logically legitimate to conclude *a priori* that convergence has occurred. The deduction that an apomorphous agreement is convergent can only be forced on us *a posteriori* when, from the distribution of apomorphies in the taxa being compared, insoluble conflicts of interpretation have arisen. I shall illustrate this in the following examples and return to the same problem later (p.142).

The favouring of particular hypotheses of relationship

Using three cases of increasing complexity, I shall now illustrate the possible ways of deciding between rival hypotheses of relationship. Then I shall give reasons for the statement already made that hypotheses of phylogenetic relationship are in principle not falsifiable (i.e. cannot be finally refuted).

---

[1] While this book was in press, I received from A. G. Kluge a paper entitled: "The relevance of parsimony to phylogenetic inference" (1984). In this impressive treatment of the problem, Kluge sharply distinguishes between: the baseless supposition "that evolution is parsimonious" (evolutionary parsimony); and the logically unobjectionable application of the principle as a methodological tool for the study of phylogenetic relationships (methodological parsimony).

My use of the principle of parsimony in making probability judgements between synapomorphy and convergence, and in favouring the simplest hypothesis of relationship among several rival hypotheses, exactly corresponds to the style of methodological parsimony advocated by Kluge in solving the problems of phylogenetic systematics.

First case

To begin with the simplest case, there is a single apomorphous agreement between two of the three species being compared. Turning back to the Fig. 13 already discussed, in the three-taxon relationship of A + B + C there is an agreement between B + C which can be interpreted as an apomorphy but there are no corresponding agreements between the species A + B or A + C. According to the principle of parsimony, the apomorphous agreement between B + C can in this instance be interpreted, without any contradictions, as a synapomorphy. It justifies the hypothesis of relationship that B + C are adelphotaxa and that they have arisen from a stem species w shared by them alone. By one and the same decision, a particular hypothesis of relationship is favoured over two other, ungrounded hypotheses.

Second case

In the next example, three well marked agreements are observed — one between the species B + C, another between A + C and yet another between A + B

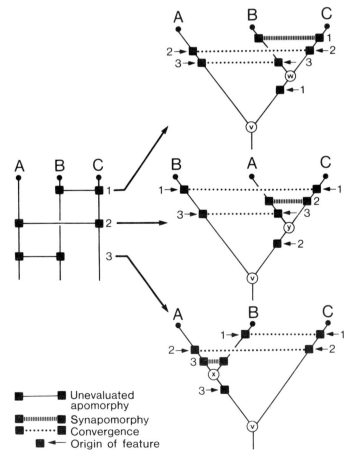

Fig. 14. The neutralization of three rival hypotheses of relationship by the demonstration of an apomorphous agreement respectively between species B + C, A + C and A + B.

58

(Fig. 14). It is supposed that all three features can be interpreted as evolutionary novelties acquired within the monophyletic group A + B + C. In this instance, feature 1 is an apomorphous agreement between B + C, feature 2 an apomorphous agreement between A + C and feature 3 an apomorphous agreement between A + B.

In the face of these facts, the unevaluated apomorphies are first inserted neutrally in the diagram with unconnected lines. If, following the principle of parsimony, we now try to explain the agreements as synapomorphies we necessarily run into an irresolvable conflict. For when the agreements are inserted in the rival hypotheses of relationship, it is evident that only one of them can represent a synapomorphy. Thus, if agreement 1 is taken to be a synapomorphy between B + C, we are forced *a posteriori* to regard agreements 2 and 3 as convergences. If agreement 2 is hypothesized as a synapomorphy between A + C, then features 1 + 3 can only be interpreted as convergences. And finally, if agreement 3 is seen as a synapomorphy between A + B, then apomorphies 1 and 2 must be convergences.

In this schematic example, I ignore the fact that complex agreements, rich in structure, have more significance than simple common features (p.143). Leaving that fact aside, there is numerical equality of apomorphies between the three pairs of species and it is impossible to favour one particular interpretation by using the principle of parsimony. The assigned distribution of apomorphous agreements neutralizes the three rival hypotheses of relationship. This being so, the question of the first degree of phylogenetic relationship within the species group A + B + C cannot be decided.

Third case

Not letting ourselves be discouraged, we subject the three species to a further critical analysis of features. This reveals features 4–6 which are agreements between B + C and can be interpreted as apomorphies (Fig. 15). It is now possible to make a rationally arguable probability judgement by using parsimony. Four of the six apomorphies can, with mutual consistency, be seen as synapomorphies. Logically this higher number is given preference, argues in favour of an adelphotaxon relationship between B and C and favours the corresponding hypothesis of relationship. There is an inevitable consequence—if, with valid logic, the four agreements between B + C are seen as synapomorphies, then we are forced *a posteriori* to regard apomorphy 2, between A + C, and apomorphy 3, between A + B, as convergences (cf. Gaffney 1979a).

Result

And so now I can draw a conclusion from these examples. There is only one empirical source of data in making the probability judgements of phylogenetic systematics between competing hypotheses of relationship, such as:

1) A + B form adelphotaxa;
2) A + C form adelphotaxa;
3) B + C form adelphotaxa.

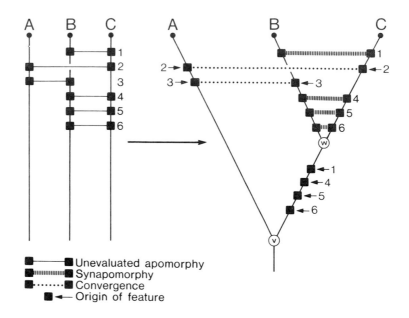

Unevaluated apomorphy
Synapomorphy
Convergence
Origin of feature

Fig. 15. The establishment of four apomorphous agreements between the species B + C and one apomorphy each between A + C and A + B. Parsimony favours the interpretation of the four agreements between B + C as synapomorphies, and at the same time favours the hypothesis of an adelphotaxon relationship between these two species.

This sole source of data is the proof of agreements between two of the three taxa respectively.

If an agreement between a particular species pair (B + C) is evaluated as a synapomorphy favouring hypothesis of relationship 3, then this same process of decision discredits the rival hypotheses 1 and 2. If, as the analysis of features continues, further agreements between B + C are discovered which can be interpreted as synapomorphies, then the favoured hypothesis of relationship is strengthened or even confirmed, and its rivals are further weakened or even rejected. Expressed in other words, the hypothesis of an adelphotaxon relationship between species B + C has been corroborated by the recognition of several agreements which can, with mutual consistency, be seen as synapomorphies. On the other hand, it is in fact and logic impossible to assert that the two rival hypotheses have been disproved by this process. They are, and they remain, nothing else than hypotheses less likely at the present stage of investigation. It may be that tomorrow, on a hitherto neglected level of investigation—such as ultrastructure, biochemistry or behaviour—a whole series of apomorphous agreements between taxa A + B will be discovered. Hypothesized as synapomorphies, they may, by their number and complexity, overcome the hitherto favoured, confirmed, corroborated hypothesis 3. The initial probability decision would have to be thought through again—under the new conditions, relationship hypothesis 1 would be favoured and this would automatically entail the rejection of hypotheses 2 and 3.

As already said, the hypothetico-deductive method favours attempts at rejection over attempts at confirmation, but this cannot be extended to the probability decisions between hypotheses of phylogenetic relationship. The process of favouring and corroborating a particular relationship hypothesis, and the corresponding process of weakening and rejecting alternative hypotheses, are carried out inseparably in one and the same judgement.

Occasionally this decision process is portrayed as if: 1) a particular hypothesis of relationship could be established as the least rejectable (i.e. the most favoured) hypothesis on the basis of a first series of apomorphous agreements; 2) these agreements could then be excluded from consideration; and 3) independently of them, the true process of testing could be continued on the basis of other agreements.

I must state my position on this. Obviously it is possible to favour a hypothesis of relationship by means of a first series of agreements, or even by means of a single apomorphous agreement — and obviously it is also possible that a hypothesis favoured at first may at any time be rejected. There is, however, no testing procedure independent of the first probability decision. All agreements, without exception, remain in the running. And all apomorphous agreements known at the particular state of the investigation are weighed alongside each other in the critical search for probability judgements that can be based empirically.

In seeking to favour particular hypotheses of relationship, it is true to say, with Hennig (1980, p.21; 1984, p.51) that: "The criterion of truth is coherence." Objective knowledge, in the sense that our insights approach the facts which Nature has created in a unique process of historical development, can only be reached through the mutual compatibility of judgements about the three possible forms of agreement between different species. To establish an adelphotaxon relationship between two taxa, one agreement is, in principle, enough if it can be interpreted convincingly and without contradictions as a synapomorphy. The probability that a hypothesis of the first degree of phylogenetic relationship between the two taxa is true increases as more and more apomorphous agreements between them are detected, which are interpretable, compatibly, as synapomorphies.

III.   Multiple species splits

The alternative to dichotomous splitting is a multiple speciation with the **fragmentation of a stem species into more than two descendent species**.

How could a "simultaneous" origin of several descendants from one stem species be empirically demonstrated as likely in the dimensions of geological time? I shall ignore this controversial question, since difficulties in methodologically establishing a particular, theoretically possible process give no indication whether, and to what extent, it has actually occurred in Nature. I assume, as already explained, that dichotomous splitting is widespread. But I shall also accept that the simultaneous splitting of a stem species into several descendants is a further possibility in the origin of closed descent communities. And I shall examine what these multiple splits imply for the study of phylogenetic relationships.

For this discussion, I shall formulate the theoretical preconditions under which the simplest case — **a trichotomy** — can happen (Fig. 16a).

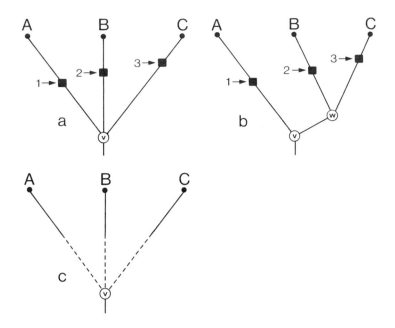

Fig. 16. a) True trichotomy. The simultaneous splitting of a stem species into three descendent species. b) Pseudotrichotomy—the sequence of two dichotomous splits cannot be recognized if the apomorphous features are distributed as shown. c) Correct representation of the unresolved relationship between the three species which together derive from a stem species shared by them alone.

1) It is supposed that the total range of an evolutionary species v is divided by a natural catastrophe into three mutually separated sub-areas so that the three sub-populations thus isolated from each other can no longer pass over the eco-geographical barriers that have just arisen. 2) It is likewise supposed that environmental conditions change in different directions in the sub-areas. 3) This situation is supposed to result in the evolution of contrasting new features in the three sub-populations. In other words, three new evolutionary species, each with its own apomorphies, develop from a single stem species. But logically there can be no synapomorphous agreements between any two of the three species, neither between A + B, nor between A + C nor B + C.

Under these assumed conditions, there is in fact a true trichotomy. There is no sister-species relationship between any two of the three species A, B and C.

Using correct methodology, therefore, an analysis of relationships must come correspondingly to the result that here there is a closed descent community, not further analysable, comprising three species.

Reversing the argument, however, would lead to a fatal logical error—if, that is to say, the lack of any agreements which could be seen as synapomorphies between any two of the three species were held to suggest that some process like the trichotomy sketched above had in fact happened in Nature.

For a three-taxon relation which cannot be analyzed by the methods of phylogenetic systematics may obviously arise in quite other ways. This can be exemplified by what may be called **pseudo-trichotomy** (Fig. 16b). The stem species v gave rise to the two descendent species A and w. Species A evolved an autapomorphy 1— the isolating mechanism which separated its populations reproductively from the species w. The latter had only a short life-span. It became involved in a speciation before ever an evolutionary novelty could develop in its lineage. After the splitting of species w there then arose new features 2 and 3 in the lineages of the descendent species B and C respectively.

Thus, although the species A, B and C are the products of two dichotomous species splits, and there is in reality an adelphotaxon relationship between B and C, there is nonetheless no prospect of recognizing the sequence of two dichotomous speciations, nor any possibility of discovering the true phylogenetic relationships using the methods at present available to phylogenetic systematics. When no agreements exist that can be seen as synapomorphies, then phylogenetic relationships cannot be discerned.

What do these **statements imply for the study of phylogenetic relationships?** It is sometimes asserted, without qualification, that they imply a serious limitation of our work and put basically in doubt the possibility of building a consistently phylogenetic system. A realistic view of the situation gives a different result. It suggests that, if Nature provides no key to unlock the phylogenetic relationships between three or more species, then these species groups should simply be treated as unities of relationship which cannot be further resolved at present. The next step would be to search for the adelphotaxa of these unities. Of course, the struggle should never be abandoned too early. We have a duty to search obstinately for apomorphous agreements within every unresolved three-taxon relationship. By doing so, some synapomorphy or other between two or the three taxa may perhaps be made probable.

It is worthless, and therefore superfluous, to draw a diagram of relationship of an unresolvable three-species unit. If anyone, nevertheless, wishes to do so, then the lines of the three species should not be extended down to the symbol of the stem species, for this might pretend that a trichotomy had occurred (Hull 1979). It is logically correct, and better suits the indeterminate factual situation, if the unresolved relationships between the three taxa are visually expressed by using dashed lines.

IV. The problem of "identity" between stem species and particular daughter species

General survey

In the course of the argument, I find it convenient to follow the discussion of multiple speciation by next considering the problem of the persistence of the genotype and phenotype of stem species in particular daughter species. As I have already explained when presenting the species as the working unit of evolution (p.18), the peripatric, parapatric and sympatric models of speciation may imply that, in dichotomous splitting, one of the two daughter species will at first remain unchanged with respect to the stem species.

I shall discuss this phenomenon using the example of the three species A + B + C and also assuming that, as the result of two dichotomies, a sister-species relationship exists between B and C. (Naturally, the example could be replayed assuming adelphotaxon relationships between A + B or A + C—the results would be congruent in each case.) If all the conceivable possibilities are exhausted, the following models may be developed for the relationships between the three species.

1) **The feature pattern of a stem species persists unchanged through two species splits** (Fig. 17a)

Subsequent to the splitting of stem species C, the daughter species A develops an evolutionary novelty 1, whereas the second daughter species C' remains unaltered. This unchanged daughter species undergoes the next splitting process and so, in turn, becomes a stem species. And now its daughter species B evolves an autapomorphy 2, while the other daughter species C'' retains the feature pattern of the stem species C'.

In this model, the evolution of autapomorphies 1 and 2 provides the minimum of necessary isolation mechanisms for the separate existences of the evolutionary species A, B and C''.

2) **The feature pattern of the older stem species persists unchanged after the first split, while that of the youger stem species persists after the second split** (Fig. 17b)

The older stem species, now symbolized by A, transmits its genotype and phenotype unchanged to the daughter species A'. In the stem lineage of the species B + C', autapomorphy 1 arises, and this, after the second split, becomes a synapomorphy of B and C'. The younger stem species C transmits its phenotype unchanged, after this second split, to the daughter species C' while the other daughter species B develops the evolutionary novelty 2.

These autapomorphies 1 and 2 once again represent the minimum number of necessary isolation mechanisms for the reproductive separateness of the species A', B and C'.

Interpretation

What are the consequences of these models for the analysis of phylogenetic relationships? I shall take the two examples in turn.

The first model represents a third case—in addition to true trichotomy and the already discussed "pseudo-trichotomy"—of a unit of relationship, comprising three species, which is not further resolvable. For if a stem species transmits its feature pattern unchanged through two successive splitting processes, then no synapomorphous agreements can exist between the daughter species that have just arisen. The sister-group relationship which in fact exists, in Nature, between two of the three species (B and C') can, under these conditions, not be empirically detected using the methods of phylogenetic systematics.

The second model presents a totally different situation. Although the two stem species A and C have transmitted their feature pattern unchanged to two different daughter species A' and C', this phenomenon does not disturb the analysis of phylogenetic relationships in any way. Why not? Because, with the acquisition of evolutionary novelty 1 in the stem lineage of B + C', there exists after the second split a synapomorphy between these two species. Methodologically it is therefore possible to recognize the adelphotaxon relationship between B and C' and to establish it empirically. This result is independent of whether the daughter species C' retains the feature pattern of the stem species C completely unchanged, or not.

The probability judgements of phylogenetic systematics are therefore disturbed only in the case of the first model. And even with this model, we could probably assume that the transmission unchanged of the feature pattern of a particular stem species through two successive speciation processes is the exception rather than the rule. All this justifies the view that the possible identity in genotype and in phenotype, as between stem species and particular daughter species, will have, at most, only a slight deleterious effect on the analysis of phylogenetic relationships. This agrees with Schlee's view (1981) concerning "the survival of stem species", though he used other arguments.

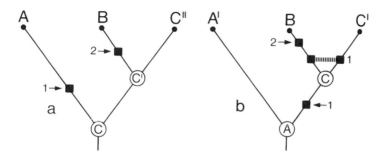

Fig. 17. Models of the possible identical transmission of the feature pattern of stem species to particular daughter species. a) Transmission of the phenotype of a stem species through two successive speciations. b) Transmission of the feature pattern of the older stem species past the first speciation, and that of the younger stem species past the second speciation.

As in Figs. 2 and 3, the agreement in genotype and phenotype between stem species and particular daughter species is emphasized by the use of similar symbols. Compared with the stem species A and C (or C'), the corresponding new but unchanged daughter species are indicated merely by the symbols A' and C' (or C'').

Example: *Duplominona* Karling (Proseriata, Plathelminthes)

These theoretical considerations can be illustrated by the concrete example of the plathelminth taxon *Duplominona* Karling (Fig. 18). In the sandy beaches of the Galapagos Islands, four very similar species are found living: *D. galapogoensis, D. karlingi, D. krameri* and *D. sieversi* (Ax & Ax 1977; Ax 1977). Their status as reproductively isolated evolutionary species is established by their sympatric coexistence in the littoral of a single island (Santa Cruz).

The "*galapagoensis* species group" is characterized by two important apomorphous features: a) the evolution of a tube in the spine-covered cirrus of the

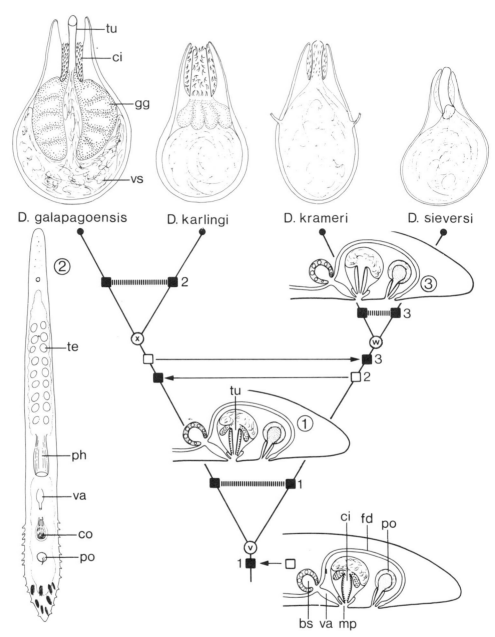

Fig. 18. Diagram of the phylogenetic relationships between the four species of the plathelminth taxon *Duplominona* Karling which are found living in the Galapagos Archipelago. Above: copulatory organs of the four species based on squash preparations. Left: habitus and structure of *D. karlingi*. Right: reconstructions in sagittal section of the genital organs; upper right—ground pattern of the species pair *D. krameri* and *D. sieversi*; middle right—ground pattern of the four Galapagos species which have a tube in the cirrus and the vagina united with the male pore; bottom right—*D. kaneohi*, from Hawaii, with no tube in the cirrus.

Abbreviations: bs = bursa, ci = cirrus, co = copulatory organ, fd = female duct, gg = granule glands, mp = male pore, ph = pharynx, po = prostatoid organ, te = testis, tu = tube of cirrus, va = vagina, vs = vesicula seminalis.

copulatory organ; and b) the union of the vagina with the male genital pore. The former feature is shared by the *galapagoensis* group with the species *D. septentrionalis* Martens from the North Sea and with *D. stilifera* Sopott-Ehlers & Ax from the North American Pacific coast. The second feature is shared with *D. kaneohei* Karling, Mack-Fira & Dörjes from Hawaii and *D. canariensis* Ehlers & Ehlers from the Canary Islands and Bermuda (Karling, Mack-Fira & Dörjes 1972; Ehlers & Ehlers 1980b; Martens 1983; Sopott-Ehlers & Ax 1985; Ax & Sopott-Ehlers 1985).

This distribution of apomorphies leads to a contradiction. Only one of the two apomorphous features could be an apomorphy of the "*galapagoensis* species group". For if a) the existence of the tube in the cirrus is seen as an autapomorphy of the *galapagoensis* group (feature 1 in Fig. 18), then the comparable structure in *D. septentrionalis* and *D. stilifera* must have arisen convergently. And b) if, on the other hand, the union of the vagina and the male pore is viewed as an autapomorphy of the *galapagoensis* group, then the similar feature in *D. kaneohei* and *D. canariensis* must be a convergence. In any case, however, one or other of the two features can be interpreted as an autapomorphy of the *galapagoensis* group which justifies the following hypothesis: namely, that the four species *D. galapagoensis*, *D. karlingi*, *D. krameri* and *D. sieversi* are closer related with each other than any of them is with any species from outside Galapagos waters.

On this basis, the next step is to discover the relationships within the monophyletic "*galapagoensis* group of the taxon *Duplominona*". In this connection, two stem lineages can be distinguished.

The species *D. galapagoensis* and *D. karlingi* agree in having the testicular follicles arranged in two longitudinal rows (feature 2), whereas the testes in *D. krameri* and *D. sieversi* are grouped together in a median strand as in other species of the taxon *Duplominona*. This is only a simple difference in position, but the arrangement of testicular follicles in double rows can be seen as the apomorphous alternative, and the agreement between *D. galapagoensis* and *D. karlingi* can be accordingly be regarded as a synapomorphy.

D. krameri and *D. sieversi*, correspondingly, lack the complex of granule glands which is normally interpolated between the seminal vesicle and the cirrus. The existence of such glands in *D. galapagoensis*, *D. karlingi* and all other *Duplominona* species is undoubtedly a plesiomorphy and their absence in *D. krameri* and *D. sieversi* is the apomorphous alternative of feature 3. Using the parsimony principle, I postulate the loss of the granule glands as a unique evolutionary occurrence in the lineage of the stem species w that was common to them alone. In other words, I interpret the lack of granule glands as a synapomorphy of *D. krameri* and *D. sieversi*.

The result of this empirical analysis of features can be summed up in the following hypotheses: The four *Duplominona* species that live in the Galapagos Archipelago derive from a stem species v shared by them alone. Within this monophyletic group of species, there are two pairs of sister species i.e. *D. galapagoensis* and *D. karlingi* are adelphotaxa with the stem species x shared by them alone; and *D. krameri* and *D. sieversi* are adelphotaxa with the stem species w shared by them alone.

This result suggests the following thoughts on the possible transmission of the feature pattern of stem species to particular daughter species:

1) The common stem species v of the monophylum "*galapagoensis* group of the taxon *Duplomonina*" could not have transmitted its feature pattern unchanged to any of the four species. This possibility can be excluded because both of the stem lineages which begin from stem species v display an evolutionary novelty which the stem species could not have possessed.
2) On the other hand, the genotype and phenotype of the stem species x may very well persist in *D. galapagoensis* or *D. karlingi.*
3) In the same way, the feature pattern of stem species w may persist unchanged in one of the two daughter species *D. krameri* or *D. sieversi.*

There is no way of deciding between these alternatives, but the unsolved problems do not affect the phylogenetic analysis in any way. The agreements in the features 2 and 3 have both been interpreted as synapomorphies by which the species pair *D. galapagoensis* + *D. karlingi* can be hypothesized as adelphotaxa, as well as the species *O. krameri* + *O. sieversi.* In setting up and justifying these hypotheses of relationship, it does not matter in the slightest whether or not the feature pattern of the stem species x and w has been transmitted unchanged, after they split, to one of their two daughter species.

V. Exposing taxa to the testing procedure

Imagine a group of three evolutionary species or three monophyletic taxa. We wish to make empirically based probability decisions about their mutual adelphotaxon relationships by testing rival phylogenetic hypotheses. What conditions must the group fulfil to make this possible?

Leaving aside the possible case of trichotomy, I return to a statement that I have made already: in a comparison of any three organisms, two species are, in principle, always closer related to each other than either of them is to the third. Having established the importance of apomorphous agreements for phylogenetic systematics, I can now justify the hypothesis that *Lumbricus terrestris* and *Melolontha melolontha* are closer related to each other than either of them is to *Homo sapiens* (p.34). Among these three forms, only the earthworm and the cockchafer show marked agreements in such features as the ventral "rope-ladder" nervous system, or the teloblastic production of metamery, which in Man are lacking. These agreements can be derived, without any contradictions, from a common stem species. In other words, I postulate the former existence of a stem species for *Lumbricus* and *Melolontha* which had a rope-ladder nervous system and teloblastic segment formation — a stem species which was not, at the same time, a stem species of *Homo.*

This hypothesis is logically valid and can be empirically justified. It would, however, be a grave mistake to translate it into a diagram of phylogenetic relationship as shown in Fig. 19. For this diagram automatically makes two completely false assertions about the phylogenetic relationships of the three species. Thus, firstly, it is self-evident that *Lumbricus terrestris* and *Melolontha melolontha* are not the only two descendent species of a stem species w shared by them alone. And it is likewise self-evident that *Homo sapiens* is not the adelphotaxon of the species pair *L. terrestris* + *M. melolontha.* Any attempt to establish these three species together as

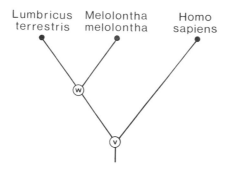

Lumbricus   Melolontha   Homo
terrestris   melolontha   sapiens

Fig. 19. An impermissible diagram showing the phylogenetic relationship between arbitrarily chosen species which lack stem species common to them alone.

a monophyletic taxon with a stem species v shared by them alone would fly in the face of all experience. Here we simply lack what is basically needed in formulating hypotheses about the sister-species relationships between two of the three taxa.

This argument may seem trivial. Unfortunately, however, I cannot forgo it because such impermissible representations of the phylogenetic relationships between any taxa whatever have become deplorably fashionable.

Løvtrup (1977) refers to a statement of the phylogenetic relationship between three taxa A, B and C as a "basic classification" and gives the result as a dichotomous dendrogram (Fig. 20a). So far, so good! In searching for the "twin taxon" (adelphotaxon) of the vertebrates he then examines a series of "three-taxon relationships" between arbitrarily chosen taxa and displays the results as corresponding diagrams or basic classifications. An example is the hypothesis which he favours of the relationships between the Echinodermata, Mollusca and Vertebrata as shown in Fig. 20b. So as not to interrupt the argument, I shall not discuss Løvtrup's

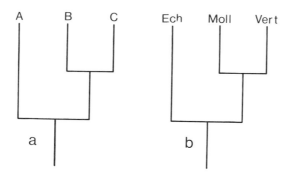

A   B   C   Ech   Moll   Vert

a   b

Fig. 20. a) A basic classification according to Lovtrup, presented as a dichotomous diagram (after Lovtrup 1977, Fig. 2.3). In so far as A, B and C are evolutionary species or monophyletic taxa, which together derive from a stem species shared by them alone, this diagram corresponds to the diagrams of phylogenetic relationship in the present book.

b) Impermissible diagram to show the phylogenetic relationships between the three monophyletic taxa Echinodermata, Mollusca and Vertebrata (after Lovtrup 1977, Fig. 3.19). Any attempt to derive the three taxa from a stem species common to them alone is baseless.

69

astounding assertion that the Vertebrata are closer related to the Mollusca than to the Echinodermata. For that hypothesis, as such, is logically just as legitimate as the hypothesis already discussed of the mutual phylogenetic relationship between earthworm, cockchafer and Man. The fatal error begins when the hypothesis is translated into a diagram of phylogenetic relationship. If this representation has any sense, each of the two furcations must correspond to a speciation process, whether the stem species in question is marked with a special symbol or not. In any case, and whatever the relationship of the three taxa to each other is finally decided to be, the whole diagram makes no sense as a portrayal of a phylogenetic systematization, so long as no empirical arguments are given that the Echinodermata + Mollusca + Vertebrata are the sole descendants of a stem species common to them alone—as the diagram necessarily implies for the lower furcation. Such arguments are lacking here, as they also are for all Løvtrup's other three-taxon relationships in which the vertebrates are compared with other arbitrarily chosen taxa.

As a result of this discussion, the **problem of how to expose taxa to the testing procedure of phylogenetic systematics** is theoretically within grasp. The **preconditions** already discussed can be expressed in the following three points:

1) It is the first degree of relationship between different organisms which is crucial for phylogenetic systematization. In searching for it, it is totally useless to expose arbitrarily chosen taxa to the testing process.
2) In analyzing the relationships between three or more evolutionary species, these species must derive together from a stem species common to them alone. In other words, they must form a monophyletic taxon.
3) This requirement is equally valid in examining the phylogenetic relationships between three or more supraspecific taxa—they must together be the descendants of a single stem species common to them alone. In this case, however, a second condition also arises—each single supraspecific taxon which is to be exposed to the testing of rival phylogenetic hypotheses must itself be established as a monophylum.

No doubt these formulations will soon come to be contradicted and these rationally indisputable requirements will supply welcome ammunition to the opponents of phylogenetic systematics. For do they not amount to saying that the empirical search for a consistent phylogenetic system is bankrupt? Do they not assume as a presupposition what we cannot at all know *a priori*? For at the outset nobody can know the monophyly of the species groups within which adelphotaxon relationships are to be established by testing rival hypotheses of phylogenetic relationship.

What can be done in this situation? We have no other choice but to put up with various **provisional orderings**, got from various sources.

On the one hand, we are confronted after more than a hundred years of "phylogenetic research" with a glut of the most diverse classifications. They cannot simply be ignored. Of course, the various logical and methodological operations by which they have come into existence are a matter of indifference. But if we suspect that a particular traditional classification may be a provisional arrangement meeting our requirements, then we accept it in the first place as an *ad hoc* hypothesis and expose it to the testing procedure of phylogenetic systematics.

On the other hand, no-one is prevented from using his own experience and intentionally choosing species or species groups for test if they present agreements which may be apomorphies shared by them alone.

In any case, it is legitimate to begin work with such **suspect provisional orderings**, whether with the smallest possible groups of three evolutionary species or with three supraspecific taxa of any size whatever. It must then be shown whether, in the course of test, the provisional arrangements can be fitted, upwards and downwards, into a hierarchical system of mutually compatible hypotheses of relationship. If this does not succeed for an *ad hoc* hypothesis chosen as a provisional arrangement, then this arrangement must simply be thrown out. And then the search begins for a new empirical approach, using another combination of taxa which together are suspected to form a closed descent community.

I have now provided a basis on which I can illustrate the logical path and methodological procedure of phylogenetic systematics using concrete examples. I shall do this at the levels both of evolutionary species and of supraspecific taxa. The stages are: suspected provisional arrangement→inserting the taxa→testing rival hypotheses of relationship→favouring a particular hypothesis of relationship→transposing the favoured hypothesis into a consistent phylogenetic system→continuing the systematization at other levels of the hierarchy.

VI.   Pongo—Pan—Gorilla—Homo

Ever since T. H. Huxley's "Evidences as to Man's Place in Nature" (1863) no subject of phylogenetic research has excited broader interest than the question of our own position in the system of organisms.

Which other taxon among recent organisms is the adelphotaxon of *Homo sapiens*? Is it a single species (sister species) or a supraspecific taxon (sister group).

Obviously I cannot discuss this problem exhaustively since I have to treat it in a few pages and within the scope of this book. Its literature easily fills entire libraries and its complexity has led Kluge (1983) to the pessimistic conclusion that a phylogenetic systematization of Man and his relatives is now farther off than in Darwin's day. In spite of all "conflicts of evidence" I intentionally choose this example so as to illustrate the logical operations of phylogenetic systematics by comparing species which every biologist knows. For this purpose, I focus on the hypothesis of relationship favoured by numerous authors in the work recently edited by Ciochon & Corrucini (1983) entitled "New interpretations of ape and human ancestry". Ciochon has himself summarized the arguments for this hypothesis (1983).

No unprejudiced observer can escape the impression that, of all organisms, the great man-like apes of the Old World share the largest number of similarities with Man. Consequently in this instance I shall exceed the required minimum of three species and choose five evolutionary species for exposure to testing in a suspect provisional arrangement. They are: the orang outang (*Pongo pygmaeus*), the two chimpanzee species (*Pan troglodytes* and *Pan pygmaeus*), the gorilla (*Gorilla gorilla*) and *Homo sapiens* himself. In what follows, I shall use only the four "generic" names, since the two chimpanzee species can be treated as a unity. This has the benefit of simplifying the comparison to four taxa only.

The suspect provisional arrangement has weight. In a phylogenetic analysis based on the methods of phylogenetic systematics, Delson, Eldredge & Tattersall (1977) specify those features of the dentition and post-cranial skeleton which, by comparison with the Hylobatidae (gibbons), can be postulated as evolutionary novelties of a stem species common only to *Pongo, Pan, Gorilla* and *Homo*. Furthermore I refer the reader to the apomorphous features listed by Ciochon (1983) which are hypothesized to have evolved in the common stem lineage of these same four taxa. One such apomorphy is the chromosome number 2n = 48, although *Homo sapiens* has only 46 chromosomes—I shall return later to these two alternative character states (p.79).

To begin with, therefore, the hypothesis of a species group comprising the great man-like apes and Man has been established empirically. This fulfils the necessary precondition for exposing the four taxa to the testing process of phylogenetic systematics. And now, pursuing the thoughts set out in the introduction, I shall stress the impressive number of agreements between orang, chimpanzee and gorilla. I list this series of well marked common features below, alongside the alternative character-states found in Man.

Pongo + Pan + Gorilla	**Homo**
Stance pronograde.	Stance orthograde.
Vertebral column in a simple curve.	Vertebral column with s-shaped curve.
Locomotion quadrupedal.	Locomotion bipedal.
Pelvis relatively flat.	Pelvis basin-shaped.
Facial skeleton elongate, without a chin.	Facial skeleton vertically beneath cranium, with chin.
Foramen magnum far posterior in cranium.	Foramen magnum far anterior, in middle of skull base.
Jaws long, with U-shaped dental arcades.	Jaws short, with parabolic dental arcades.
Diastemata present. In the upper jaw between the second incisor and the canine; in the lower jaw between the first premolar and the canine.	No diastemata. Closed tooth rows in upper and lower jaws.
Average brain weight 400–500 gm (range 300–720 gm).	Average brain weight 1450 gm (range 900–2200 gm).
Chromosome number 2n = 48.	Chromosome number 2n = 46.

Without doubt, these agreements between the man-like apes have to be seen as a solid basis for a new, extremely suspect, provisional ordering. And in fact this ordering gives rise to a hypothesis of relationship which nowadays appears as a corresponding classification in almost every handbook and textbook. Thus *Pongo, Pan* and *Gorilla* are united into a family Pongidae while *Homo sapiens* is placed opposite them, bearing the same rank, as the sole recent representative of the family Hominidae.

If this proposed subdivision were the result of a consistent phylogenetic systematization, it would have the following logical implications. Orang outang, chimpanzee and gorilla would be seen as the sole descendants of a stem species u shared by them alone. Farther back in time, a stem species t would have split into two lineages, being respectively the stem lineages of the great man-like apes and that of Man (Fig. 21).

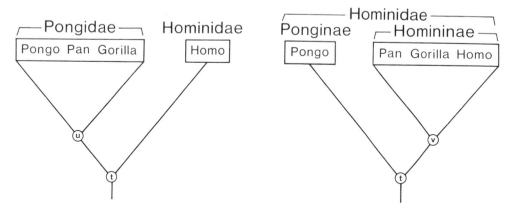

Figs. 21 and 22. Two rival hypotheses as to the phylogenetic relationships between orang outang, chimpanzee, gorilla and Man. Fig. 21: *Pongo, Pan* and *Gorilla* form a monophyletic taxon. Fig. 22: *Pan, Gorilla* and *Homo* form a monophylum. In both diagrams the three taxa are combined into a block which is not further analyzed.

But however well grounded a provisional ordering may be suspected to be, in order to become the basis for testable hypotheses of relationship it needs proof of shared apomorphies for the taxa in question. On the basis of the anatomical agreements just listed, such proof cannot be brought. For throughout the list, *Pongo, Pan* and *Gorilla* show the plesiomorphous condition while *Homo* is always characterized by the alternative apomorphous condition. The only purposeful attempt to establish the great man-like apes as a monophylum by the methods of phylogenetic systematics is that of Kluge (1983). After an analysis of older researches (Schultz 1936, Strauss 1939; etc.), Kluge hypothesizes various agreements of the urethro-vaginal region, as well as the chromosome number of 48, as synapomorphies of the taxa *Pongo* and *Pan + Gorilla*, though of course he also admits the possibility of other interpretations (symplesiomorphy, convergence).

Are there well based alternatives to the hypothesis that the taxon Pongidae, comprising the man-like apes orang, chimpanzee and gorilla, forms a monophylum?

In the classical work of Huxley, published as early as 1863, there is a famous sentence: "Whatever systems of organs be studied, the comparison of their modification in the ape series leads to one and the same result—that the structural differences which separate Man from the Gorilla and the Chimpanzee are not so great as those which separate the Gorilla from the lower apes (i.e. monkeys)."

In the language of phylogenetic systematics, this sentence suggests the following hypothesis of relationship: *Pan, Gorilla* and *Homo* together form a monophyletic species group with a stem species v common to them alone (Fig. 22). Since Huxley's day, this view has been enthusiastically advocated by Weinert (1932, 1944) as the "summoprimate" hypothesis. He did this on the basis, among other things, of the unique interrelationships of the frontal sinuses in *Pan, Gorilla* and *Homo* as well as the fusion, common to these taxa alone, of two carpal bones—the os centrale and the scaphoid—in early development (cf. p.123). Further morphological agreements which can be interpreted as synapomorphies of the taxa *Pan* + *Gorilla* and *Homo* have been collated by Ciochon (1983).

The hypothesis of a monophyletic species group comprising *Pan, Gorilla* and *Homo* has been impressively confirmed at the biomolecular level. Serological studies of the specificity of antigen-antibody reactions, the sequencing of the amino-acid residues of proteins (haemoglobin, fibrinopeptides, cytochrome c, serum albumin), DNA hybridization etc. point basically in one and the same direction (Doolittle, Wooding, Lin & Riley 1971; Dayhoff 1972; Sarich & Cronin 1976; Goodman 1975, 1977; Bruce & Ayala 1979; Goodman, Baba & Darga 1983; Cronin 1983; Sarich in Cronin 1983; Ciochon 1983). The biomolecular agreements between *Homo* and the African man-like apes *Pan* + *Gorilla* are always greater than the agreements between the latter two taxa and the Asiatic orang outang—even though their interpretation as synapomorphies is not always obvious or has even been contested in some points (Kluge 1983). In addition, there are identical karyological features. From the survey by Seuánez (1979), I emphasize, for example, that strongly fluorescent chromatin is observed only in the chromosomes of *Pan, Gorilla* and *Homo*, but not in *Pongo* nor in any other primates.

If we accept this hypothesis (Ciochon 1983), which has been established from many different points of view and is now widely favoured, we can now expose the taxa *Pan, Gorilla* and *Homo* to the testing procedure of phylogenetic systematics. Three rival hypotheses of relationship confront each other (Fig. 23), and there are empirical data in support of each of them.

1) *Pan* and *Homo* are adelphotaxa

At a macromolecular level, chimpanzee and Man seem to be almost identical (King & Wilson 1975) as confirmed by DNA hybridization (cf. Ax 1985a).

2) *Gorilla* and *Pan* are adelphotaxa

The remarkable knuckle-walking gait, with correlated agreements in the anatomy of the hand, is interpreted as a synapomorphy of *Pan,* and *Gorilla* (Tuttle 1969, 1975; Delson, Eldredge & Tattersall 1977; de Bonis 1983; Ciochon 1983).

3) *Gorilla* and *Homo* are adelphotaxa

The basis of this hypothesis is a series of specific chromosomal agreements (Miller 1977; Seuánez 1979).

I shall now focus on the chromosomes as a complex of features. According to Seuánez (1979), apomorphous agreements exist between only two of the three pairs of species.

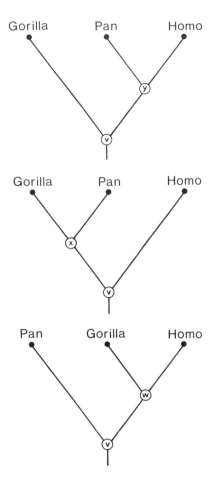

Fig. 23. The three rival hypotheses of relationship concerning the phylogenetic connections between gorilla, chimpanzee and Man.

a) *Pan—Homo:*   no apomorphies.
b) *Gorilla—Homo:* four apomorphies i.e. the longer arm of the y chromosome has strongly fluorescent chromatin; a pair of autosomes have strongly fluorescent chromatin in the centromere region; secondary constrictions exist in another pair of autosomes; identical sections with constitutive heterochromatin (repetitive DNA sequences) occur in several homologous chromosomes.
c) *Pan—Gorilla:*   two apomorphies i.e. terminal Q and C banding; two pericentric inversions in two homologous chromosomes.

According to the parsimony principle, therefore, the quantitatively dominant agreements between *Gorilla* and *Homo* should be hypothesized as synapomorphies while the apomorphies shared by *Pan* and *Gorilla* should be seen as convergences. In the meanwhile, Mai (1983) has stated that such conclusions are incompatible with an extensive analysis and evaluation of the structural rearrangements in the chromosomes of *Pongo, Pan, Gorilla* and *Homo.* The hypothesis that "Man's closest relative may be the gorilla" (Miller 1977, Seuánez 1979) is rejected by Mai as karyologically baseless.

The supposition that there is an adelphotaxon relationship between *Gorilla gorilla* and *Homo sapiens* furthermore runs counter to a series of advanced morphological agreements between the two African man-like apes (Ciochon 1983). In this connection the unique knuckle-walking gait should be especially emphasized. The view that the bipedalism of Man could be derived from a knuckle-walking stem species shared by *Pan, Gorilla* and *Homo* (Washburn 1968, Franzen 1972) finds no support in the anatomy of *Homo sapiens,* nor do the fossil representatives of the taxon Hominini (*Australopithecus, Homo habilis, Homo erectus* (p.210)) show, in the bones of the hand and wrist, any of the features associated with knuckle walking in the African man-like apes. Consequently the complex of structural and functional features correlated with knuckle-walking is nowadays interpreted as a convincing synapomorphy of *Pan* and *Gorilla* (Ciochon 1983). The postulate that apomorphous knuckle walking evolved uniquely in the stem lineage common only to the African man-like apes favours the view that there is an adelphotaxon relationship between *Pan* and *Gorilla* as against the two rival hypotheses of relationship.

I shall now transpose the result of the testing procedure into a consistently phylogenetic system and place alongside each other the two possible equivalent ways of displaying the supposed connections i.e. as a diagram of phylogenetic relationship and as a hierarchical tabulation (Fig. 24, bottom).

Pan and *Gorilla,* as adelphotaxa with a stem species x shared by them alone, are united to form the monophylum Panini. Only *Homo sapiens,* among present-day species, confronts the Panini as an adelphotaxon of equal rank with its own name of the Hominini. The Panini and the Hominini arose from the stem species v common to them alone. Correspondingly the monophylum Homininae is created for them at the next level of the systemic hierarchy. The adelphotaxon of the Homininae is the Ponginae, with the single recent species *Pongo pygmaeus* (Fig. 24, middle). Logically enough, therefore, the Ponginae and the Homininae are united together at the next level of superordination to form the monophyletic taxon Hominidae. These steps complete the systematization of the four taxa exposed to test i.e. *Pongo, Pan, Gorilla* and *Homo.*

In choosing the "Family" name Hominidae for these four taxa, and the names Ponginae (for *Pongo*) and Homininae (for *Pan* + *Gorilla* + *Homo*) with the categorical rank of "Sub-family", I follow Goodman's proposals (1975). The application of the "Tribe" names Panini (for *Pan* + *Gorilla*) and Hominini (for *Homo*) follows Schwartz, Tattersall & Eldredge (1978). In this connection I must stress that the taxon Ponginae, monotypic at the present day with the sole species *Pongo pygmaeus,* necessarily takes the same rank as its supraspecific adelphotaxon Homininae. Likewise, the taxon Hominini, now monotypic with the sole recent species *Homo sapiens,* takes the same rank as its supraspecific adelphotaxon Panini. This is a matter of purely formal classificatory constraints. These correspond exactly to the situation which I have already explained in assigning a taxon Ornithorhynchidae for the platypus *Ornithorhynchus anatinus* and a monotypic taxon Tubulidentata for the aardvark

Orycteropus afer (p.24–41). The expansion of the taxon Hominini to a supraspecific taxon by including fossil species will be treated below (p.210).

For the rest, I emphasis here again that the choice of the names of supraspecific taxa is in principle totally meaningless for phylogenetic systematization. So also is the assignation of categories of the Linnaean hierarchy—such as Family Hominidae, Subfamily Homininae, Tribe Hominini (p.234). The only things that count are to display the adelphotaxon relationships in an unmistakeable manner and clearly to indicate the identical ranks of the adelphotaxa.

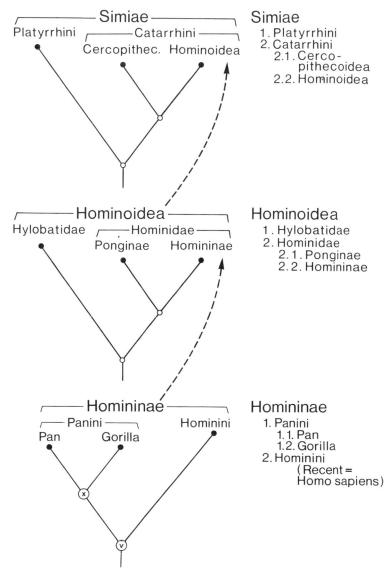

Simiae
1. Platyrrhini
2. Catarrhini
   2.1. Cerco-
       pithecoidea
   2.2. Hominoidea

Hominoidea
1. Hylobatidae
2. Hominidae
   2.1. Ponginae
   2.2. Homininae

Homininae
1. Panini
   1.1. Pan
   1.2. Gorilla
2. Hominini
   (Recent =
   Homo sapiens)

Fig. 24. Some parts of the phylogenetic system of the Simiae, displayed in two equivalent forms—the diagram of phylogenetic relationships (left column) and the hierarchical tabulation (right column).

I now wish to outline the phylogenetic system of the taxon Simiae (Anthropoidea) beyond the level of the Hominidae (Fig. 24). In doing so, I shall neither formulate nor test rival hypotheses of relationship but briefly justify the respective sister-group relationships that have been postulated.

The sister group of the Hominidae is formed by the Asiatic gibbons as the taxon Hylobatidae (*Hylobates, Symphalangus*), for which the great elongation of the arms represents an autapomorphy. Well marked apomorphous agreements between the two taxa are the "*Dryopithecus* pattern" of the lower molars, with five cusps and a Y-shaped pattern of furrows, the secondary lack of a tail and the existence of a vermiform appendix. These can be interpreted, with no inconsistencies, as synapomorphies and provide a basis for uniting the Hominidae and the Hylobatidae to form the next higher monophylum Hominoidea.

At a still higher level, the Hominoidea and the Cercopithecoidea (African monkeys) are grouped together under the name Catarrhini. The dental formula gives a convincing basis for this adelphotaxon relationship. The Catarrhini are characterized by the apomorphous dental formula $\frac{2123}{2123}$ as against the plesiomorphous condition $\frac{2133}{2133}$ shown by the South American Platyrrhini. This catarrhine condition can be seen as an evolutionary novelty (autapomorphy) for the latest common stem species of the recent Hominoidea and Cercopithecoidea. And on this basis the monophyly of the taxon Catarrhini can be postulated.

Finally, the New World Platyrrhini stand in a sister-group relationship to the Catarrhini of the Old World. There are essential agreements in the brain (central sulcus, parieto-occipital fissure) and in the eyes (frontal position, macula lutea). Interpreted as synapomorphies, these agreements are the basis for constituting the monophylum Simiae.

I have thus formulated a portion of the phylogenetic system. By consistently showing the hypothesized adelphotaxon relationships, this formulation is a store of precise information for particular stages of phylogenesis, beginning with the common stem species of all Simiae and going down to the latest stem species shared between the African man-like apes and *Homo sapiens*. Summarizing the diagrams in Fig.24 gives the following hierarchical tabulation.

Simiae
    Platyrrhini
    Catarrhini
        Cercopithecoidea
        Hominoidea
            Hylobatidae
            Hominidae
                Ponginae
                    (*Pongo pygmaeus* in the recent fauna)
                Homininae
                    Panini
                        Pan
                        Gorilla
                    Hominini
                      (*Homo sapiens* in the recent fauna)

I shall later discuss in detail the methodology of deciding between the alternatives of plesiomorphy and apomorpy. I now anticipate this discussion by taking a difference mentioned already—that between **2n = 48 chromosomes** (*Pongo, Pan, Gorilla*) and **2n = 46 chromosomes** (*Homo*). Which chromosome number is original for the four taxa and which represents the derived alternative?

Out-group comparison is a central methodological tool for phylogenetic systematics (p.115), but in this instance it does not allow any properly grounded decision. Why not? Because in the adelphotaxon Hylobatidae (gibbons), which is the immediate outgroup, neither 46 nor 48 chromosomes occur. All *Hylobates* species possess, with one exception, 44 chromosomes, whereas *Hylobates concolor* has 52 and *Symphalangus syndactylus* has 50 chromosomes (Starck 1974, Kluge 1983). By comparison with the Hylobatidae, therefore, both chromosome numbers—46 and 48—would have to be an advanced condition. Outgroup comparison, however, allows no statement as to whether the latest common stem species of *Pongo + Pan + Gorilla + Homo* had 46 or 48 chromosomes as an autapomorphy.

In this situation, we could attempt to use a plausible explanation of the direction of change in the feature, in constructional or functional terms, as a help in making a decision (p.125). A widespread explanation supposes that the number 46 in *Homo sapiens* arose by the fusion of two acrocentric chromosomes to form chromosome 2 (Dutrillaux 1975; Seuánez 1975; Mai 1983; Ciochon 1983). According to this interpretation, 48 chromosomes could be postulated for the latest common stem species of *Pongo + Pan + Gorilla + Homo*. If inserted in the hypothesis of relationship favoured above, the shared possession of 48 chromosomes would become a synapomorphy of the adelphotaxa Ponginae and Homininae (Fig. 25, top). The three man-like apes would have retained the chromosome number 48, whereas a reduction by fusion happened in the lineage of *Homo*. The chromosome number 46 would thus be an autapomorphy of the species *Homo sapiens.*

However, there is another possible explanation, no less plausible. Kluge (1983) suggested that a structural change could equally well have happened in the opposite direction, from 46 to 48 chromosomes—by a splitting of the chromosome which is given the number 2 in Man. This explanation would postulate 46 chromosomes for the latest common stem species of *Pongo + Pan + Gorilla + Homo*. In the lineage of *Homo*, the number 46 would thus have remained unchanged, whereas in the stem lineage common to the three man-like apes a single change to 48 chromosomes would have occurred (Fig. 25, bottom). In other words, the agreement of having 48 chromosomes would be a synapomorphy between the taxon *Pongo* and a taxon comprising *Pan + Gorilla*. The feature could thus be taken as a basis for establishing a monophylum Pongidae.

There is no definite reason for preferring one of these explanations of the change in chromosome number over the other (Kluge 1983). Consequently, in an analysis of the phylogenetic relationships between the taxa *Pongo, Pan, Gorilla* and *Homo*, the alternative of 48 or 46 chromosomes must for the moment be left on one side.

It is worth adding that the identical number of 48 chromosomes becomes a worthless symplesiomorphy under both possible explanations when *Pongo, Pan,* and *Gorilla* are being compared with each other. I have indicated this in Fig. 25 at the appropriate level of comparison. This obligatory change in significance from

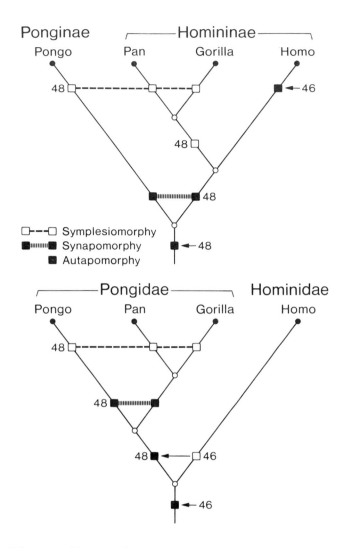

Fig. 25. The possible ways of interpreting the alternatives of 48 chromosomes (*Pongo, Pan, Gorilla*) and 46 chromosomes (*Homo*). Explanation in text.

synapomorphy to symplesiomorphy in passing from one level of the systemic hierarchy down to the next level of subordination is discussed in detail later (p.144).

VII. Monotremata — Marsupialia — Placentalia[1]

I have already used the taxon Mammalia above, to exemplify the hierarchic structure of the phylogenetic system (p.39). I now wish to justify in detail the phylogenetic

---

[1] Monotremata (= Prototheria, Non-Theria). Marsupialia (= Metatheria), Placentalia (= Eutheria). Instead of the rather stereotyped synonyms in brackets, I use the older, more expressive group names. The insertion of the names Prototheria and Monotremata for two different levels of rank is redundant, as I explain later (p.223).

hypothesis assumed above, i.e. that there is an adelphotaxon relationship between the Marsupialia and the Placentalia.

I recall, first, the two preconditions for exposing supraspecific taxa to the testing procedure: the three groups under comparison—the Monotremata, Marsupialia and Placentalia—and also the taxon Mammalia which includes them, must all be validated as monophyla.

I shall describe the necessary provisional ordering very briefly and stress only a few well marked agreements. The totality of the mammals can be suspected to be a monophylum because, for example, they have hair and feed their young with milk. To justify the monophyly of the Monotremata, I mention a remarkable spur which is developed identically on the hind leg in the male of the platypus and the echidnas. For the Marsupialia, I shall for the moment be content with the brood pouch that gives them their name (Latin *marsupium* = pouch). For the Placentalia I correspondingly select the chorio-allantoid placenta by which the embryo is tightly connected to the mother. A summary of the important autapomorphies of all these taxa is given below (p.87), but the superficial provisional arrangement here presented will, for the moment, suffice in formulating the conceivable hypotheses of phylogenetic relationship between monotremes, marsupials and placentals and in exposing these hypotheses to the procedures of empirical test. The hypotheses are:

1) Monotremata and Placentalia are adelphotaxa;
2) Marsupialia and Placentalia are adelphotaxa;
3) Monotremata and Marsupialia are adelphotaxa.

First hypothesis of relationship

There is not one single agreement between the monotremes and the placentals which can be interpreted as an apomorphy common to them alone. So far as I know, the hypothesis that the Monotremata and the Placentalia might have a sister-group relationship to each other has never been advocated. Nevertheless, it remains a logically legitimate hypothesis of relationship. But since, at the present stage of investigation, it has no empirical basis, I shall not consider it further.

Second hypothesis of relationship

By contrast, the marsupials and the placentals have a whole series of features which are common to these two groups alone. By comparison with the Monotremata, these features can be interpreted with all desirable certainty as apomorphies. I would mention, to start with, a new crest of bone (the scapular spine) on the shoulder blade, the separation of the openings of the intestine and the urino-genital system, the fact that the milk glands discharge into nipples, the development of poorly yolked eggs with holoblastic cleavage and the birth of live young (vivipary).

In accordance with the principle of parsimony, these five apomorphies can be seen, with no mutual inconsistencies, as synapomorphies of the Marsupialia and the Placentalia. This interpretation can be inserted in the corresponding diagram of phylogenetic relationship (Fig. 26) and tends to support, provisionally, the hypothesis of an adelphotaxon relationship between the Marsupialia and the Placentalia. In consequence of this hypothesis, these two groups are united into a monophylum with the name Theria.

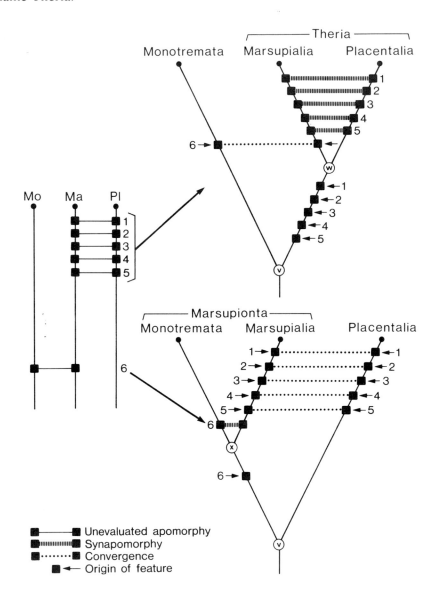

Fig. 26. Two rival hypotheses of the phylogenetic relationships between the three mammalian taxa of Monotremata, Marsupialia and Placentalia. Several apomorphous agreements between the Marsupialia + Placentalia (features 1–5) are interpreted as synapomorphies (upper diagram) or as convergences (lower diagram). One apomorphous agreement between the Monotremata + Marsupialia (feature 6) is interpreted as convergence (upper diagram) or as synapomorphy (lower diagram). Further explanation in text.

Third hypothesis of relationship

The hypothesis of relationship favoured above will, of course, only endure if it can still be favoured when confronted with the third hypothesis—that of an adelphotaxon relationship between the Monotremata and the Marsupialia. And here, in fact, there is a conflict. For there are three remarkable agreements between the Monotremata and the Marsupialia which have been interpreted as synapomorphies (Kühne 1973, 1975). On this basis the two groups have been postulated to form a monophylum with the name Marsupionta (Gregory 1947). These three features must therefore be discussed in more detail.

The brood pouch (marsupium)

Among the Monotremata, only the echidnas (*Tachyglossus, Zaglossus*) form a temporary brood pouch to protect the eggs and shelter the newly hatched young. The platypus (*Ornithorhynchus*) lays its eggs in a burrow—a brood pouch is lacking.

Similarly, a brood pouch is not always present in the Marsupialia. It is completely lacking in the Caenolestidae as well as in several representatives of the Didelphidae and the Dasyuridae.

Also there are differences in development. The brood pouch of *Tachyglossus* develops as an unpaired fold of skin in the umbilical region whereas that of *Didelphis* arises from paired anlagen near the groin (Starck 1978a,b).

Discontinuous distribution in the two taxa and differing ontogenies do not allow us to insist that the brood pouch is an autapomorphy of a stem species common to the Marsupialia and the Monotremata alone. At the present stage of investigation it seems more likely that brood pouches evolved convergently and separately within the Monotremata and within the Marsupialia. It has even been suggested that the marsupium evolved independently several times within the Marsupialia (Marshall 1979). As a presumably convergent apomorphy between parts of the Monotremata and parts of the Marsupialia this morphological agreement must be firmly excluded from the analysis of relationship.

The marsupial bones (ossa marsupii)

A pair of marsupial (epipubic) bones in the Monotremata and the Marsupialia provide attachment points for muscles in the pelvic region. Despite the name, there seems to be no functional interdependence between the marsupial bones and the marsupium (brood pouch) (Starck 1978, 1979; Marshall 1979). For in monotremes and marsupials, marsupial bones, when they exist, are always characteristic of both sexes. In phylogenetic analysis, therefore, it is legitimate to consider the feature "marsupial bones" separately from the feature "marsupium".

All Monotremata and nearly all Marsupialia possess marsupial bones. Among the latter, they are absent in *Notoryctes, Thylacinus* and the South American Borhyaenidae, but this absence must certainly be interpreted as secondary (see below). In the first instance, therefore, the presence of a marsupial bone can be postulated for the latest common stem species of recent Monotremata as well as for the latest common stem species of the recent Marsupialia.

This, however, does not say very much. The crucial argument in interpreting the alternatives "marsupial bones present" (Monotremata + Marsupialia) and "marsupial bones absent" (Placentalia) is supplied by fossils. For marsupial bones have been shown to exist in the †"Tritylodontia", which are representatives of the stem lineage of the Mammalia, in the †"Multituberculata" and the †"Eupantotheria", which are representatives of the stem lineage of the Theria (p.217), and even in †*Asioryctes*, which is a fossil representative of the stem lineage of the Placentalia (survey by Marshall 1979). There is only one rational conclusion from these empirically proven facts — marsupial bones were already present in the latest common stem species of all recent Mammalia. From this, they passed into the stem lineages of the Monotremata and the Theria. Subsequently, after the splitting of the latest common stem species of the recent Theria, they passed into the stem lineage not only of the Marsupialia, but also into that of the Placentalia. Within this latter stem lineage, they were, of course, lost. Since marsupial bones are lacking in all recent Placentalia, the loss in this case can be interpreted as a single evolutionary event. Finally, the marsupial bones were convergently lost in a few Marsupialia.

This analysis shows, therefore, that the shared possession of marsupial bones is a worthless symplesiomorphy of the Monotremata and the Marsupialia. It therefore fails as a basis for the asserted sister-group relationship between these two taxa.

Replacement of a single tooth in ontogeny

The final agreement between the Monotremata and the Marsupialia demands an assessment of the alternative character states "diphyodonty" and "monophyodonty" which occur respectively in the Placentalia and Marsupialia.

In the ground pattern of the diphyodont dentition of the Placentalia, the incisors, canines and premolars were first formed as a temporary milk dentition and later replaced by the permanent teeth. The molars were always formed once only. By contrast, in the Marsupialia only a single element of the dentition is produced at first as a milk tooth and then later as a permanent tooth, this being the last premolar. All the rest of the dentition arises only once in the form of permanent teeth — a phenomenon which may be subsumed under the rather liberal term of monophyodonty.

Which of these alternative features is the apomorphy? Considering that, among other vertebrates, a repeated, polyphyodont tooth replacement is widespread, it would be difficult to derive the diphyodonty of the Placentalia from the condition in Marsupialia. On the basis of what is observed in fossils, such a derivation can be rejected with reasonable certainty. Among the †"Eupantotheria" as representatives of the stem lineage of the Theria, a milk-tooth dentition has been shown to exist (Butler, Green & Krebs 1973). The conclusion from this situation is that the latest common stem species of the recent Theria must have had a diphyodont dentition. This was taken over unchanged into the stem lineage of the Placentalia, whereas in the stem lineage of the Marsupialia it was transformed to an almost monophyodont state by extensive suppression of the milk dentition. The monophyodonty of the marsupials is the apomorphous character state among the pair of alternatives.

84

This brings me to the situation in the Monotremata which, as adults, are well known to be secondarily toothless. In the echidnas (*Tachyglossus, Zaglossus*) there is no trace of tooth anlagen even in ontogeny. *Ornithorhynchus anatinus*, however, produces a whole series of teeth in the upper and lower jaws which are then shed in the course of post-embryonal life. The earliest identification of the vestigial cheek teeth (Green 1973) was rejected by Kühne (1973) who replaced it with the following re-interpretation of the dental formula (Fig. 27):

incisors $\frac{0}{5}$, canines $\frac{1}{1}$, premolars $\frac{1}{1}$, molars $\frac{4}{4}$.

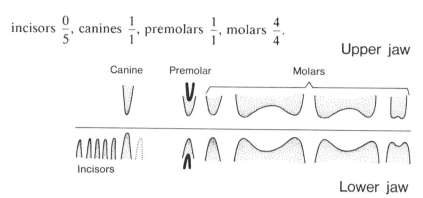

Fig. 27. Interpretation of the embryonic dentition of the duck-billed platypus *Ornithorhynchus anatinus* (Monotremata). Redrawn from Kühne (1973).

And now comes the crucial point. The single premolar in the upper and the lower jaw of *Ornithorhynchus* is always laid down twice—as a milk tooth and as a permanent tooth, according to Kühne. As a logical consequence, this tooth is seen as equivalent to the last premolar of the Marsupialia. This would suggest a decisive agreement between the dentition of the platypus and of the marsupials—the solitary replacement of a single tooth. It would be fully legitimate to attempt to interpret such an apomorphous agreement as a synapomorphy between *Ornithorhynchus anatinus* and the Marsupialia. What would be the consequences of this attempt? An almost monophyodont dentition would be required to exist in the stem species of the Monotremata as well as in that of the Marsupialia, and both taxa would have to be derived from a stem species x common to them alone. Fig. 26 takes account of this consideration by placing the feature as a synapomorphy (feature 6) in the lower diagram.

However, this diagram of phylogenetic relationship at once shows the weakness of such an attempt. For the five apomorphous agreements listed above in support of hypothesis 1 must now be counted as convergent apomorphies. But that is not all. According to Marshall (1979) there are, in total, about 20 apomorphies shared between the Marsupialia and the Placentalia (cf. p.88). It would indeed be hopeless to try to establish each of these, point by point, as resulting from convergent evolution.

As regards the shared feature of the replacement of a single premolar, there is an inevitable consequence. If Kühne's anatomical interpretation is correct, which has been questioned (M. Griffiths 1978), then we are forced to see this apomorphous agreement between *Ornithorhynchus* and the Marsupialia *a posteriori* as a convergence.

Result

The outcome of this empirical testing of the three rival hypotheses of relationship can be expressed as follows: the "Theria" hypothesis of an adelphotaxon relationship between the Marsupialia and the Placentalia can be excellently validated; it is based on about twenty apomorphous agreements which, in accordance with the principle of parsimony, can be interpreted without mutual contradiction as synapomorphies of the two taxa Marsupialia and Placentalia. Far behind comes the "Marsupionta" hypothesis; as support it can call only on a single apomorphous agreement in tooth replacement which could in principle be interpreted as a synapomorphy of the Monotremata and the Marsupialia. This hypothesis, however, comes inescapably into conflict with the Theria hypothesis and must therefore be rejected. Favouring the Theria hypothesis requires that the replacement of a single cheek tooth be interpreted as a convergent apomorphy between the Monotremata and the Marsupialia. The hypothesis of a sister-group relationship between the Monotremata and the Placentalia is logically a third possibility but, at the present stage of investigation, must be rejected as baseless.

Thus the supposition of an adelphotaxon relationship between the Marsupialia and the Placentalia emerges from the testing procedure of phylogenetic systematics as a clear favourite.

Transposition of the Theria hypothesis into the system and the complete validation of the Mammalia and its highest sub-groups as monophyla

For any given part of the consistently phylogenetic system, a complete validation of the monophyly of all supraspecific taxa is required. Correspondingly, before a favoured hypothesis of relationship can be transposed into a particular segment of the phylogenetic system, all previous suspect provisional arrangements must be replaced by empirically founded hypotheses of monophyly.

The transposition of the favoured Theria hypothesis into the system is largely a repetition of what has been dealt with already (p.81) and can therefore be fairly brief. The segment of the system in which the Monotremata and the Theria form the highest ranking adelphotaxa within the Mammalia is shown here again with its two equivalent representations, i.e. as relationship diagram and hierarchical tabulation alongside each other (Fig. 28, below). But I now slightly modify the diagram of phylogenetic relationship, in that the positions of the three supraspecific taxa Monotremata, Marsupialia and Placentalia are represented by their stem species x, y and z.

On this basis I indicate the five taxa as being monophyla by means of five blocks of autapomorphies (■). From the surveys of Marshall (1979) and Hennig (1983), I abstract a series of striking features which must have arisen as evolutionary novelties in the stem lineages of the individual descent communities and would thus have been present in the latest common stem species of their recent representatives. The plesiomorphous alternative features of the sister groups are shown in brackets or their primary absence is indicated by a minus sign.

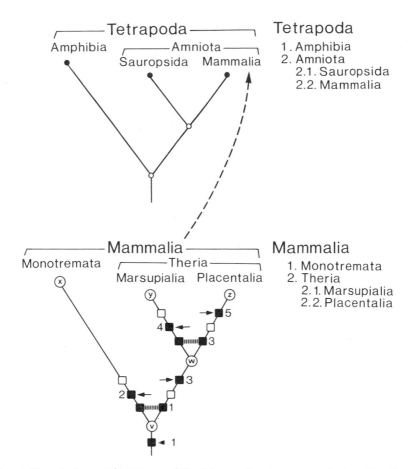

Fig. 28. (Below.) The phylogenetic system of the Mammalia shown as relationship diagram and as hierarchical tabulation. This part of the system is completely validated by five groups of autapomorphies. Explanations in text. (Above.) Continuation of the phylogenetic system at the levels of the Amniota and Tatrapoda.

1) ■ **Autapomorphies of the Mammalia,** present in the latest common stem species v of all recent species of mammals

= **synapomorphies of the adelphotaxa Monotremata and Theria.**

(Symplesiomorphies of the sister group Sauropsida are given in brackets.)
— Synapsid skull with a single lateral temporal foramen. (Primarily anapsid skull without temporal foramen.)
— Skull connected to the first cervical vertebra by means of two occipital condyles. (Skull with a single occipital condyle.)
— Secondary palate consisting of maxillaries and palatines. (—.)
— Secondary jaw articulation between the squamosal bone and the dental. (Primary jaw articulation between the articular and the quadrate.)
— Middle ear with three auditory ossicles, i.e. stapes (columella), malleus (articular) and incus (quadrate). (Middle ear with a single auditory ossicle, the columella.)

87

— Angular incorporated in the middle-ear region as the tympanic bone. (Angular a bone in the lower jaw.)
— Cochlear of the inner ear slightly curved. (Cochlea straight.)
— Heterodont dentition with regional differentiation into incisors, canines, premolars and molars. (Homodont dentition.)
— Diphyodont tooth replacement with a milk dentition for the incisors, canines and premolars. (Polyphyodont dentition.)
— A basic number of seven cervical vertebrae. (Variable number of cervical vertebrae.)
— Basic number of phalangeal joints 23333. (Higher number of phalangeal joints 23453.)
— Marsupial (i.e. epipubic) bones present. (−.)
— Hair present. (−.)
— Homoiothermy. (Poikilothermy.)
— Mature erythrocytes have no nucleus. (Erythrocytes with nucleus.)
— Muscular diaphragm present. (−.)
— Sweat glands and sebaceous glands present. (−.)
— Facial musculature and muscular lips present. (−.)
— Male copulatory organ with a glans and a prepuce. (−.)
— Milk glands for feeding the young. (−.)

2) ■ **Autapomorphies of the Monotremata,** present in the latest common stem species w of all recent monotremes

= **synapomorphies of the adelphotaxa Tachyglossidae and Ornithorhynchidae.** (Plesiomorphies of the sister group Theria are given in brackets.)

— Thigh gland and spur on the hind limb of the male. (−.)
— Secondary absence of teeth. (Teeth present.) The value of this feature is uncertain; convergent loss in echidnas and platypus seems to be possible.

3) ■ **Autapomorphies of the Theria,** present in the latest common stem species w of all recent therians

= **synapomorphies of the adelphotaxa Marsupialia and Placentalia.** (Plesiomorphies of the sister group Monotremata in brackets.)

— Coracoid in the shoulder girdle reduced. (Procoracoid and metacoracoid present.)
— Scapular spine as a new lamella of bone on the scapula. (−.)
— Teeth with prismatic enamel. (Teeth with non-prismatic enamel.)
— Tribosphenic molars. (−.)
— Cochlea of the inner ear with at least one spiral whorl. (Cochlea only curved, without a whorl.)
— Whiskers (vibrissae) on the face. (−.)
— Separation of urogenital sinus from anal opening. (Common opening into the cloaca.)

88

- Openings of the milk glands united into nipples. (Mammary fields without nipples.)
- Milk extracted from the nipples by pumping and sucking movements of the tongue and the intra-oesophageal epiglottis. (Simple licking of milk from the mammary fields.)
- Yolk-poor eggs and holoblastic cleavage. (Yolk-rich eggs and meroblastic discoidal cleavage.)
- Vivipary. (Ovipary.)
- Placenta as a yolk-sac placenta. (−.)

4) ■ **Autapomorphies of the Marsupialia**, present in the latest common stem species y of all recent marsupials[1]

= **synapomorphies of the highest ranking adelphotaxa Marsupialia** which are not yet satisfactorily identified (Kirsch 1977a,b; Keast 1977; Patterson 1981b). (Plesiomorphies of the Placentalia in brackets.)

- Monophyodonty; only the last premolar developed as milk tooth and permanent tooth. (Diphyodonty.)
- Prismatic enamel of the teeth with canaliculi. (Prismatic enamel without canaliculi.)
- Pseudovagina; medial unpaired canal between the primary paired vaginae. (Pseudovagina not present.)

5) ■ **Autapomorphies of the Placentalia**, present in the latest common stem species z of all recent placentals

− **synapomorphies of the adelphotaxa Edentata (Xenarthra) and Epitheria** (according to McKenna, 1975).
(Plesiomorphies of the sister group Marsupialia in brackets.)

- Connection of the dorsal region of the neocortex in the hemispheres of the brain by way of the corpus callosum. (Connection by way of the dorsal parts of the anterior commissure.)
- Retinal rods of the eye are simple, without oil droplets. (Retinal rods double, with oil droplets.)
- Ground pattern of the dental formula $\frac{3143}{3143}$. (Ground pattern of the incisors $\frac{5}{4}$; the homologies of the cheek teeth between Marsupialia and Placentalia are disputed; cf. Westergaard 1983.)
- Marsupial bone lost. (Marsupial bone present.)
- Penis simple, with a single termination. (Penis with bifid end.)
- Monodelphy; unification of the oviducts into an unpaired vagina. (Didelphy; separation of the paired oviducts.)
- Chorio-allantoic placenta. (Yolk-sac placenta.)

---

[1] The brood pouch, which I selected in the provisional arrangement, cannot be considered as certain in this connection. As argued above, a marsupium was possibly not present in this stem species but was first evolved within the Marsupialia.

— Blastocyst with external trophoblast and internal embryoblast. (Formative cells corresponding to the embryoblasts situated superficially in the wall of the blastocyst (Starck 1978a,b)).

Closing remarks

Having brought the autapomorphies together into series of features, the following statements are justified. The five groups of autapomorphies arose in the stem lineages of the five closed descent communities of the Mammalia, Monotremata, Theria, Marsupialia and Placentalia, respectively between the oldest stem species and the latest common stem species of the recent representatives. Take, for example, the more than 60 autapomorphies of the Mammalia (Hennig 1983). There were about 150 million years available for the evolution of these features, beginning with the early differentiation of the synapsid skull and finishing with the shift of the jaw articulation and the insertion of the articular and the quadrate into the middle ear. During this span of time, the stem lineage of the Mammalia existed in Nature. The lineage began with the splitting of the latest common stem species of the Amniota in the Carboniferous and finished with the latest common stem species of recent mammals. The splitting of this stem species near the Jurassic-Cretaceous boundary was the start of the adelphotaxa Monotremata and Theria (p.217).

Continuing the systematization at the next higher level of the hierarchy

I shall finish this section by searching for the sister group of the Mammalia. For this purpose, it is necessary to produce a new suspected provisional arrangement in which, as well as the Mammalia, at least two other presumably monophyletic taxa of the vertebrates must be inserted. In addition to this, it must be suspected that these three supraspecific taxa together derive from a stem species shared by them alone.

Under these conditions, the Amphibia and the Sauropsida must be considered, in the first place, in building up a new "three-taxon relation". I postpone the justification of their monophyly to the next section because, in the case of the Sauropsida, it is problematic and represents an instructive conflict in phylogenetic systematization. The monophyly of all three taxa together, on the other hand, can be established simply and convincingly on the basis of a single apomorphous agreement. As is well known, the amphibians, sauropsids and mammals have paired walking limbs with detailed common features in the skeleton beginning at the insertion in the pectoral or pelvic girdles and extending out to the distal phalanges. I postulate the existence of pentadactyl anterior and posterior limbs for the latest common stem species of all recent quadrupeds[1] and on these I base a monophylum Tetrapoda (Fig. 28, above).

In comparing the Amphibia, Sauropsida and the Mammalia, there are again three rival hypotheses of relationship:

---

[1] Jarvik's hypothesis that the tetrapod limb has evolved twice independently (i.e. separately in the Urodela and in the Anura) has several times been refuted with convincing arguments. Surveys are given by Szarski (1977), Gaffney (1979b), Rosen *et al.* (1981). Furthermore, De Saint Aubain (1980) has recently demonstrated subtle agreements in the ontogenies of the limbs of urodeles and anurans.

1) Amphibia and Sauropsida are adelphotaxa;
2) Amphibia and Mammalia are adelphotaxa;
3) Sauropsida and Mammalia are adelphotaxa.

The formulation of these hypotheses is logically legitimate. In the present case, however, competition between them exists only on paper. Neither between the Amphibia and the Sauropsida, nor between the Amphibia and the Mammalia does any single agreement exist that can be interpreted as a synapomorphy. On the other hand, the third hypothesis of relationship can be extremely well validated by abundant apomorphous identical features present in the Sauropsida and the Mammalia. The development of the embryo in a closed amniotic cavity is an obvious synapomorphy between the Sauropsida and the Mammalia. Moreover, it gives the name for the two groups as united together into the monophylum Amniota.

VIII.   The position of the Chelonia in the system of the Amniota—a case of conflict in phylogenetic systematization

The systematization of the recent species of Hominidae and of the highest ranking supra-specific taxa of Mammalia were chosen deliberately. For in both examples the logical path and methodological stages of formulating rival hypotheses of relationship, of favouring a particular hypothesis and of inserting it into a consistent phylogenetic system can be made generally understandable.

The effectiveness of the rational argumentation of phylogenetic systematics, however, can also be demonstrated in unsolved problems. I shall now discuss the systematic position of the chelonians (tortoises and turtles) within the Amniota as an example of such an unresolved conflict. Here I do not seek to favour a particular hypothesis come what may. Rather I shall narrow the problem down, so that instead of being infinitely open-ended it is reduced to a few alternatives which can be formulated precisely. It is between these alternatives that a probability decision can and must finally be made.

I shall first deal briefly with the highest ranking sister-group relationships within the Tetrapoda, taking up the thread from the end of the preceding section. In this connection, the Amniota are not the only well validated monophyletic taxon. For, contrary to widespread opinion, the monophyly of the Amphibia is also satisfactorily grounded on autapomorphies which they alone possess.

Amphibia

— Fore limbs with only four fingers in the Urodela and the Anura. By contrast with the basic pattern of the pentadactyl tetrapod limb, metacarpal 5 and also the fifth finger are secondarily absent (in the Gymnophyona the limbs are lost completely).
— The occipital bones with a double condyle, unlike the primarily single condyle of the Amniota.
— The subdivision of the teeth into a distal crown and a basal pedicel by means of a ring-shaped furrow in the dentine (Parsons & Williams 1962, 1963; Schultze 1970).

— The papilla amphibiorum as a particular sensory termination in the inner ear. It has no equivalent in other vertebrates.

Amniota

— The thick horny skin. The formation of horny scales and claws.
— The embryonic membranes of the amnion and the serosa. The allantois.
— Hard-shelled, yolk-rich eggs; meroblastic cleavage.
— Direct development without metamorphosis; no aquatic larva.
— The vertebral centra arise from the pleurocentra and the intervertebral discs from the hypocentra.
— Differentiation of the first two cervical vertebrae to form the atlas-axis complex.

Since the Chelonia show all these features, they obviously form a sub-taxon of the Amniota. Before we can begin to search for their sister group, I shall characterize the Chelonia themselves by way of autapomorphies and likewise all those amniote taxa which can be established as monophyla independent of the possible position of the Chelonia. (The Mammalia have been dealt with already.) I shall therefore present a few obvious autapomorphies for the Chelonia, Lepidosauria, Rhynchocephalia, Squamata, Archosauria, Crocodilia and Aves (G. Peters 1978; Hennig 1983).

Chelonia

— Rigid trunk armour consisting of bony and horny plates with a dorsal shield (carapace) and a ventral shield (plastron).
— Shoulder girdle transposed to beneath the ribs.
— Toothless jaws, covered with a horny sheath.

Lepidosauria (Rhynchocephalia + Squamata)

— Teeth firmly fused to the edges of the jaws.
— Pre-formed points of fracture in the centra of several vertebrae in the base of the tail.
— The stapes ear-ossicle is thin and elongate.
— Cloacal fissure transverse.

Rhynchocephalia

— Upper jaw with a beak-shaped process.
— Quadrate and pterygoid have grown together to form a plate of bone perpendicular to the long axis of the skull.
— Lower jaw lacks the splenial.
— Ribs distally expanded antero-posteriorly.

Squamata

— Cathapsid skull. Only the upper temporal opening is surrounded by bone. The

92

lower temporal opening gapes downwards as a result of the complete loss of the quadrato-jugal and the partial reduction of the squamosal.
— Quadrate freely moveable.
— Paired male copulatory organs.

Archosauria (Crocodilia + Aves)

— Thecodont dentition, i.e. teeth inserted in deep pits (among Aves in some fossil forms only e.g. †*Archaeopteryx*, †*Hesperornis*). Derived from the sub-thecodont ground pattern of the Amniota convergently with the Mammalia.
— Loss of the palatal teeth.
— Loss of the pineal foramen.
— Pre-orbital opening developed, in addition to the two post-orbital temporal openings.
— Mandibular foramen (dental foramen) in the lower jaw between the dental, angular and supra-angular (lost in recent crocodiles).
— Triradiate pelvis, i.e. pubis and ischium not forming a plate but developed as long rods directed posteriorly. Together with the ilium, this produces a three-rayed structure.
— Fifth toe of the hind limb lost.
— Union of the Eustachian tubes in the median plane.
— Jacobson's organ lost (possible synapomorphy with the Chelonia).
— Completely separated heart chambers.
— Pulmonary diaphragm. The primarily single pleuro-peritoneal cavity is divided by a septum into the unpaired peritoneal cavity and a pair of pleural cavities.
— A mobile ear valve made of skin.
— Long lagena.
— Absence of a post-embryonic urinary bladder.

Crocodilia

— Nostrils united to form an unpaired opening at the tip of the snout.
— Pre-orbital foramen lacking (present only in fossil crocodiles).
— Reduction of the epipterygoid.
— Secondary palate in connection with a semi-aquatic mode of life. The pre-maxillaries, maxillaries, palatines and pterygoids form a roof of bone beneath the choanae.
— Clavicles reduced.
— Long, laterally compressed tail as the main swimming organ.
— Pupils vertically elongate.
— Cloaca an elongate fissure in the median plane.

Aves

Most of the autapomorphies of birds are connected with the acquisition of flying ability.

— Homoiothermy.
— Development of feathers out of horny scales.
— Formation of the bill by elongation of the pre-maxillaries.
— Loss of the teeth. They are replaced functionally by the horny sheath of the bill and the muscular gizzard.
— Fusion of the two post-orbital temporal openings with each other and with the orbit.
— Quadrate freely mobile.
— Reduction of the ectopterygoid.
— High mobility of the neck as a result of the development of saddle-shaped articulations between the cervical vertebrae.
— Rigid trunk without regional subdivisions.
— Synsacrum. Several pre- and post-sacral vertebrae fused with the two primary sacral vertebra.
— Pygostyle. The last 4—6 caudal vertebrae are fused to form a single "ploughshare bone".
— Scapula long and slender.
— Clavicles united to form a furcula.
— Fusions in the pelvis. The ilium, pubis and ischium are united to form a single bone (the innominate bone).
— Reduction of the ventral ribs. Flattening of the thoracic ribs and evolution of the uncinate process.
— Fore limb transformed to carry the wing of the hand and arm. Reduction of fingers 4 and 5. Reduction of the number of phalangeal joints in the remaining fingers and fusions in the metacarpal region.
— Hind limb shows a more advanced development of the intertarsal joint of the Sauropsida (see below). Articulation between two complex bony elements (the tibio-tarsus and the tarso-metatarsus). There is a functional connection with the bipedalism of birds.
— Hallux of the hind limb opposable.
— Greatly enlarged eyes.
— Development of the syrinx as a sound-producing organ.
— Lungs with air-passages (parabronchi, mesobronchi) and air sacs. Pneumatization of the bones in connection with these.

The search for the sister group of the Chelonia

In attempting to identify the adelphotaxon of the Chelonia, three possibilities must be discussed. I have expressed these in three diagrams as rival hypotheses of relationship (Figs. 29–31). In these diagrams, the autapomorphies of the taxa characterized above have intentionally been left out. This is done so as to leave the diagrams clear for features important in considering the Chelonia.

The position of the Chelonia in a phylogenetic system of the Amniota has recently been discussed independently by G. Peters (1978) and Gaffney (1980). A weakness in the argumentation of both these authors, however, is that they have not considered hypothesis 3 as given below.

Hypothesis of relationship 1 (Fig. 29)

a) The Chelonia are the adelphotaxon of all other amniotes.
b) These "Amniota ceteri", for which there is no accepted name, can be separated into the Diapsida and the Mammalia.

G. Peters (1978; in Kaestner 1980, Fig. 4) defends hypothesis 1, above all by considering the various expressions of the **atlas-axis complex** (feature 1 in the diagrams).

As is well known, the two most anterior neck vertebrae are differentiated in a striking form in all amniotes to subserve the mobility of the head. Except in chelonians, the following situation exists in all: the atlas consists of a ring-shaped structure comprising the neural arch and intercentrum of the first vertebra. The pleurocentrum morphologically associated with this vertebra is liberated from the anlage of the atlas and becomes fused with the pleurocentrum of the axis. The second cervical vertebra thus includes pleurocentra 1 and 2. Pleurocentrum 1 extends forward, as the odontoid process, into the ventral part of the ring of the atlas.

In Chelonia, on the other hand, pleurocentrum 1 is not fused with the axis. There is no odontoid process and the pleurocentra of both these cervical vertebrae articulate with each other. Apart from this, pleurocentrum 1 may articulate freely with the atlas or may be fused with it to form a single piece of bone (Guibé 1970).

Peter's interpretation is that the atlas-axis complex existed as a structural and functional unity, and was an evolutionary novelty, in the latest common stem species of all recent Amniota. In that stem species, however, pleurocentrum 1 was not yet fused with the axis. This condition was retained as a plesiomorphy only among the Chelonia. In a second stem lineage, common to all other amniotes, pleurocentrum 1 became fused to the axis, so forming the odontoid process. I have inserted this interpretation into the first diagram of relationship. Under this hypothesis, the feature "axis with odontoid process" is seen as an autapomorphy of all non-chelonian amniotes.

In support of his position, G. Peters alludes to another pair of alternative features, namely that the **chelonian skull is anapsid** (i.e. without temporal openings) whereas the skulls of all other amniotes have temporal openings. Nowadays the view that the anapsid skull is a plesiomorphy within the amniotes is accepted almost everywhere. On the other hand it is by no means settled whether any particular expression of the temporal openings—as, for example, one pair only—can be assigned to the common stem species of all other amniotes. Peters himself casts doubt on the value of these alternative features by accepting the widespread view that the diapsid condition of the Lepidosauria and of the Archosauria, with two pairs of temporal openings, and the synapsid skull of the Mammalia (= Synapsida) with its single pair of temporal openings, have arisen independent of each other from the primarily anapsid skull of the Amniota.

Gaffney likewise favours the first hypothesis of relationship, but in seeking to justify a monophyletic taxon Diapsida + Mammalia (= Synapsida), he has to admit, with reference to the temporal openings, that: "Unfortunately ... there is no specific fenestrated condition that can be stated as being typical of this group" (1980, p.605). There is, however, the possibility that the existence of the lower temporal opening can be seen as a synapomorphy of the Diapsida and the Mammalia. If so it could

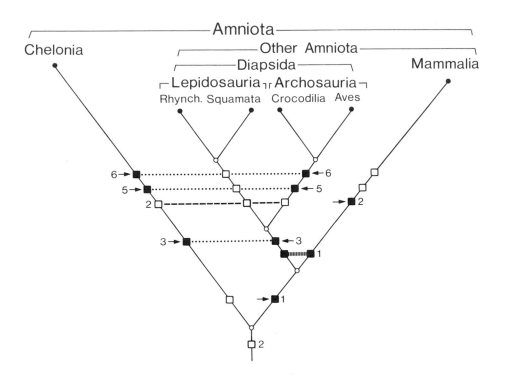

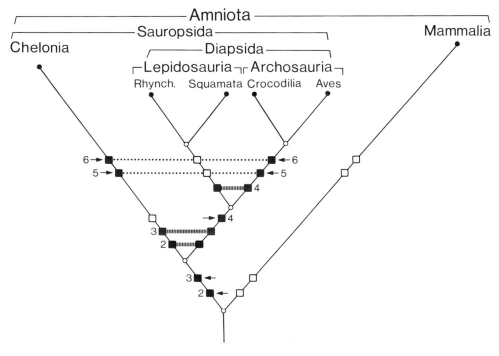

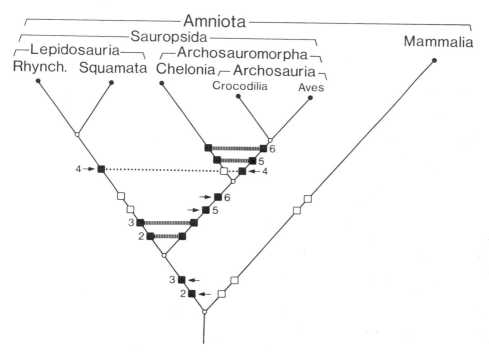

Figs. 29–31. Possible positions of the Chelonia in the phylogenetic system of the Amniota. Fig. 29. the Chelonia as the adelphotaxon of all other Amniota. Fig. 30. the Chelonia and a group Diapsida as the highest ranking adelphotaxa within the Sauropsida. Fig. 31. the Chelonia and the Archosauria as subordinate adelphotaxa within the Sauropsida.

Six features have been inserted in the diagrams. with the following interpretations of their alternative expressions:
1) Atlas-axis complex—□ = no odontoid process. ■ = axis with odontoid process.
2) Subdivision of the ventral aorta—in Figs. 30 and 31. □ = division into aorta and pulmonary artery. ■ = division into left aorta, right aorta and pulmonary artery—in Fig. 29. □ = division into three vessels, ■ = division into two vessels.
3) Hind limb—□ = fifth metatarsal normal, mesotarsal articulation lacking, ■ = fifth metatarsal hook-shaped. mesotarsal articulation present.
4) Skull structure—□ = skull anapsid. ■ = skull diapsid with two pairs of post-orbital openings.
5) Origins of the subclavian arteries—□ = from the aortas. ■ = from the carotid arches.
6) Fate of the pleurocentrum of the pro-atlas—□ = fusion with the odontoid process of the axis. ■ = fusion with the occipital condyle.

be argued that the upper pair of temporal openings arose later, as an evolutionary novelty in the stem lineage of the Diapsida.

Because of the uncertainty in interpreting the temporal openings, Gaffney throws more stress on another feature—**Jacobson's organ**. Following Parsons, he interprets the complete lack of such an organ in the Chelonia as a plesiomorphy. And he considers that the organ, which is widespread in other amniotes, evolved only once, in a stem lineage unique to the Diapsida + Mammalia.

Three complexes of features, therefore, favour the first hypothesis of relationship. For the Chelonia are thought to possess in each case the plesiomorphous condition, while the corresponding alternatives are seen as autapomorphies of a stem species common to the Diapsida + Mammalia alone. These features are as follows:

Chelonia	Other Amniota
Pleurocentrum of the atlas not developed as an odontoid process.	Pleurocentrum of the atlas fused with the axis.
Primarily anapsid skull.	Skull with temporal openings (?one pair of ventral openings).
Primarily without Jacobson's organ.	Jacobson's organ present in the ground pattern of the Diapsida + Mammalia.

These interpretations conflict with those of other features which are the basis for hypotheses 2 and 3.

Hypothesis of relationship 2 (Fig. 30)

This hypothesis includes two sub-hypotheses, as follows:
a) The Amniota can be divided into the adelphotaxa Sauropsida and Mammalia (Goodrich 1916, 1930).
b) The Chelonia are the sister group of all other Sauropsida—these being united in the taxon Diapsida.

In attempting to justify sub-hypothesis a , I first stress a well known pair of alternatives thrown up by the controversies on the evolutionary history of the heart and circulation in Amniota. These are: 1) **two aortic arches**, arising from the left and right heart chambers (Chelonia, Lepidosauria, Crocodilia); and 2) **a single aortic arch** arising from the left heart chamber (Aves, Mammalia). These are shown as feature 2 in the diagrams 29–31.

In the embryological development of the Amniota, the aortas and the pulmonary artery arise from a single anlage vessel—the ventral aorta (Holmes 1975). In Chelonia, Lepidosauria and Crocodilia this anlage comes to be divided into three vessels, being split first by the aortico-pulmonary septum and later by an additional aortic septum. These splits produce the right and left aorta and the pulmonary artery (Fig. 32 above). In the Aves and Mammalia, on the other hand, the differentiation of the aortic septum fails to happen. In both groups the ventral aorta becomes divided into two vessels only—an aorta and the pulmonary artery (Fig. 32 below).

For the birds, this difference is not difficult to interpret. Since the sister-group relationship of the Crocodilia and the Aves is well established, the existence of a single aorta can only be the apomorphous condition. Its origin can be explained by the suppression of the aortic septum in the embryonic subdivision of the ventral aorta (Holmes 1975).

In interpreting the comparable situation in the Mammalia, on the other hand, there are two possibilities: a) The division of the ventral aorta into an aorta and a pulmonary artery represents the original condition of all amniotes and was retained as a plesiomorphy in the stem lineage of the Mammalia. The subdivision into three vessels would first have happened within the Amniota in a stem lineage shared by

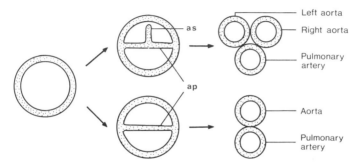

Fig. 32. Transverse sections of the ventral aorta with a schematic representation of its sub-division in the course of embryonic development.
Upper series: Sauropsida (except Aves). Subdivision into left aorta, right aorta and pulmonary artery.
Lower series: Aves and Mammalia. Subdivision into an aorta and the pulmonary artery.
ap = aortico-pulmonary septum, as = aortic septum (after Holmes 1975, slightly altered).

the Chelonia, Lepidosauria and Archosauria. Under this view, the existence of a right and a left aorta is an autapomorphy which justifies the union of the three taxa into a group Sauropsida (Figs. 30 and 31).

Or b) the division into three vessels already existed in the latest common stem species of the Amniota. The development of two vessels alone is an advanced condition of the Mammalia (Fig. 29) and originated, independent of the Aves, by a comparable suppression, within the stem lineage of the Mammalia, of the aortic septum of. the ventral aorta . If so, in the comparison of the Chelonia with the Lepidosauria and the Archosauria, the subdivision into two aortas and a pulmonary artery would be a worthless symplesiomorphy.

Another striking agreement between the Chelonia, Lepidosauria and Archosauria is the differentiation of a **mesotarsal articulation** in the hind limb, with the principal axis of flexure between the proximal and the distal tarsal bones (feature 3 in the diagrams). In the Crocodilia the line of articulation has inserted itself between the astragalus and the calcaneum (Krebs 1974). In the Aves, the mesotarsal articulation has evolved into the extremely apomorphous intertarsal articulation, situated between the two compound bones known as the tibiotarsus and tarso-metatarsus. In conjunction with the mesotarsal joint, the **fifth metatarsal** occurs in a characteristic form as a very short bone with a strong proximal curvature.

The mammals lack a comparable differentiation in the hind limb (see p.104, however) and consequently the presence of the complex of features "mesotarsal joint with hook-shaped fifth metatarsal" can be seen, on good grounds, as an apomorphy that first arose within the Amniota. In accordance with the principle of parsimony, it is also legitimate to postulate that this complex of features evolved only once, in a stem lineage common to the Chelonia, Lepidosauria and Archosauria alone. Interpreted as an autapomorphy, this complex could be used to support the Sauropsida hypothesis. This interpretation, however, has been called in question, for Starck (1979, p.589) has suggested that the mesotarsal joint has evolved convergently in the Lepidosauria and the Archosauria.

Another complex of specific agreements between the Chelonia, Lepidosauria and Archosauria concerns the structural elements in the ear region of the skull (Watson

1954). A tall quadrate and a slender stapes, compared with the converse situation in extinct representatives of the stem lineage of the Mammalia (low quadrate, massive stapes), were seen by Gaffney (1980) as apomorphies. At the same time, however, he discounted them by regarding them as convergences of the Chelonia and the Diapsida.

The proposed monophyly of the Sauropsida is therefore based mainly on a particular interpretation of different conditions in two features, as follows:

Mammalia (plesiomorphies)	Sauropsida (autapomorphies)
Subdivision of the ventral aorta into an aorta and the pulmonary artery.	Subdivision of the ventral aorta into two aortas as well as the pulmonary artery.
Primary lack of a mesotarsal joint.	Evolution, once only, of the mesotarsal joint and a hook-shaped fifth metatarsal.

Both these interpretations have been disputed, and therefore the hypothesis of the monophyly of a taxon Sauropsida cannot be seen as convincingly grounded. It must, however, remain under discussion, since it alone, without inconsistencies, is compatible with the third hypothesis of relationship.

This brings me to sub-hypothesis b. The Lepidosauria and the Archosauria are characterized by a **diapsid skull**. In both taxa, the bony arches between the temporal openings are formed in the same manner by the post-orbital and part of the squamosal.

As against this, the ground pattern of the Chelonia includes a primarily **anapsid skull**. According to hypothesis 2, this condition was carried as a plesiomorphy into the stem lineage of the taxon Sauropsida. If so, the identification, discussed under hypothesis 1, of the temporal opening of Mammalia (= Synapsida) with the ventral temporal opening in Lepidosauria and Archosauria must be rejected. Rather, the second hypothesis requires the following conclusion: the apomorphous condition of the diapsid skull evolved only after the separation of the stem lineages of the Mammalia and of the Sauropsida, within the Sauropsida — synapsid and diapsid conditions must have originated, independent of each other, from an anapsid condition.

Under the assumptions of hypothesis 2, the existence of two temporal openings can be considered, in accordance with the principle of parsimony, as a synapomorphy of the Lepidosauria and the Archosauria (feature 4 in Fig. 30). This unites them into a monophylum Diapsida which forms the adelphotaxon of the Chelonia.

Hypothesis of relationship 3

This, again, comprises two sub-hypotheses, as follows:
a) The Amniota are divided into the adelphotaxa Sauropsida and Mammalia.
b) Within the Sauropsida, the Chelonia form the sister group of the Archosauria (De Beer 1937, v. Hofsten 1941, Remane 1967, Løvtrup 1977).

Here also, as in hypothesis 2, a monophylum Sauropsida is postulated to comprise the Chelonia, Lepidosauria and Archosauria. But hypothesis 3 rejects the supposition that there is a sister-group relationship between the Chelonia and a taxon Diapsida. It points out some subtle agreements between the Chelonia and the Archosauria which can be seen as synapomorphies.

A first remarkable agreement concerns the **course of the subclavian arteries** which supply the fore limbs. In the Chelonia, Crocodilia and Aves they originate, in exactly the same fashion, from the carotid arches. Unlike an origin from the aortas, the rostral shift of the subclavian arteries onto the carotids occurs only in these three taxa and can be seen, with reasonable certainty, as an apomorphy. At the same time, it seems extremely unlikely that this shift could have happened twice convergently. According to the principle of parsimony, it must be interpreted as a unique evolutionary novelty within the Amniota which was present in a stem species shared by the Chelonia and Archosauria alone (feature 5 in Fig. 31).

Another remarkable agreement concerns the **fate of the pro-atlas** — a vestigial vertebra laid down in the embryo in front of the atlas in all amniotes. In Lepidosauria and Mammalia, the pleurocentrum of the pro-atlas fuses in ontogeny with the odontoid process of the axis. In the Chelonia and the Archosauria, on the other hand, the pleurocentrum enters into the occipital part of the head and becomes part of the occipital condyle in exactly identical manner.

Assuming that hypotheses 1 or 2 are correct, then in both cases it necessarily follows that the identical alternative fates of the pleurocentrum of the pro-atlas must have arisen convergently either in the Lepidosauria + Mammalia or in the Chelonia + Archosauria. I have inserted the latter conclusion, as feature 6, in the relationship diagrams of Figs. 29 and 30.

Under hypothesis 3 there is a much simpler explanation of the facts. The identical behaviour of the pleurocentrum of the pro-atlas in the Lepidosauria and Mammalia is the plesiomorphous condition, and the fusion of the pleurocentrum to the occiput is the apomorphous alternative. It evolved only once, within the Sauropsida, and existed as an evolutionary novelty (autapomorphy) in a stem species shared by the Chelonia and Archosauria alone (interpretation of feature 6 in Fig. 31).

Further agreements between the Chelonia and the Archosauria include, among other things, the ontogenetic connection of the stapes with Meckel's cartilage by way of the pars interhyalis or the form of the nictitans tendon and its connection with the pyramidal muscle.

In the light of these agreements, the absence of **Jacobson's organ** in the Chelonia can readily be seen as a secondary loss (by contrast with hypothesis 1) and can thus be made consistent with an adelphotaxon relationship between the Chelonia and the Archosauria.

In the crocodiles and the birds, Jacobson's organ is always absent in the adult though laid down in both taxa in the embryo as a ventro-median evagination of the nasal cavity. In the Chelonia, on the other hand, there is not the slightest trace of an embryonic anlage (Parsons 1970, Gaffney 1980). Under hypothesis 3, the most parsimonious explanation of this situation would be as follows: Jacobson's organ was lost within the Sauropsida once only — it was no longer present in a stem species common to the Chelonia and Archosauria alone. The absence of Jacobson's organ is, in other words, a synapomorphy of the Chelonia and Archosauria. In the latter, there is still an embryonic anlage whereas the organ has been completely eliminated from the genetic information of the Chelonia. Contrary to Gaffney's supposition

(1980), the absence of an anlage in ontogeny does not prove conclusively that the organ was primarily lacking among the turtles and tortoises.

The hypothesis of a sister-group relationship between the Chelonia and Archosauria, and the consequent union of both taxa in a monophylum Archosauromorpha, can thus be based on a whole series of agreements which, compatibly with each other, can be seen as synapomorphies.

The consequences of favouring a particular hypothesis of relationship

All three hypotheses of relationship claim, as their empirical foundation, several agreements which exist, respectively, only between particular taxa of Amniota. In logic, however, only one of the three hypotheses can correspond to the truth—in so far as no further hypotheses are proposed and advocated. It is therefore necessary to consider critically the consequences of favouring each hypothesis of relationship. For this purpose, I intentionally reverse the sequence.

Chelonia and Archosauria as adelphotaxa

The only serious obstacle to accepting this hypothesis is the anapsid skull of the chelonians. If we favour this hypothesis, we must accept one of the two following assertions:
— The anapsid condition of the Chelonia is an apomorphy. Two pairs of post-orbital temporal openings would be part of the ground pattern of the Sauropsida. It was lost in the stem lineage of the Chelonia, after the stem lineages of the Chelonia and Archosauria had separated. There are no empirical grounds for this supposition. The skull of the mesozoic †*Triassochelys* is completely anapsid. Moreover, reduction of the skull roof in the taxon Chelonia always proceeds from the edge, never by the formation of temporal openings (Starck 1979).
— Two pairs of post-orbital temporal openings have arisen twice, convergently, within the Sauropsida i.e. once within the stem lineage of the Lepidosauria and once within the stem lineage of the Archosauria after this had separated from the lineage of the Chelonia (interpretation of feature 4 in Fig. 31).

Chelonia and Diapsida as adelphotaxa

Unlike the last assertion of the preceding hypothesis, we can now begin from the more parsimonious supposition that two pairs of post-orbital temporal openings evolved once only within the Sauropsida and were present as an autapomorphy in the common stem species of the Lepidosauria and Archosauria (which was not at the same time the stem species of the Chelonia). This interpretation is shown as feature 4 in the relationship diagram of Fig. 30.

However, the already mentioned series of apomorphous agreements between the Chelonia and the Archosauria make an almost insurmountable obstacle to favouring this hypothesis. Hypothesis 2 demands convergent evolution of such remarkable apomorphies as the rostral shift of the subclavian arteries onto the carotids and the fusion of the pleurocentrum of the pro-atlas into the skull (interpretation of features 5 and 6 in Fig. 30). There are no empirical indications that would make this seem reasonable. And the interpretation of the diapsid skull as a possible synapomorphy of Lepidosauria and the Archosauria is not convincing enough, as a single feature, to require such suppositions *a posteriori.*

Chelonia and all other Amniota as adelphotaxa

In addition to the convergent evolution of the apomorphous agreements between the Chelonia and the Archosauria, this hypothesis postulates convergent evolution of the mesotarsal joint and the hook-shaped fifth metatarsal. Also, the division of the ventral aorta into three vessels is assigned by this hypothesis to the ground pattern of the Amniota. It therefore sees the mammalian condition, with the vessel divided only into an aorta and a pulmonary artery, as an apomorphy (interpretation of the alternative features 2, 3, 5 and 6 in Fig. 29).

The view that the anapsid skull of Chelonia is a plesiomorphy can also be made compatible with hypotheses 2 and 3 and the absence of Jacobson's organ in Chelonia can be seen as an apomorphy. That being so, the only remaining support for hypothesis 1 is the interpretation presented above of the atlas-axis complex. But there are other interpretations even for this complex of features.

Thus, firstly, it could be that the pleurocentrum of the atlas became fused to the axis more than once in the phylogeny of the Amniota. As concerns the Mammalia, this supposition can be supported empirically by fossils. Thus in the †"Pelycosauria" (*Dimetrodon*) and the †"Therapsida" (Gorgonopsida), which are representatives of the stem lineage of the mammals, forms are known in which the pleurocentrum of the atlas lies free in front of the axis (Kemp 1969; cf. Starck 1979, Fig. 43).

Secondly it is possible that the odontoid process was secondarily liberated from the axis in the stem lineage of the Chelonia, as was already argued by Ihle et al. (1927, p.98). When the occipital process became a ball joint, new types of movement became possible in the chelonians which made movement between the atlas and the axis superfluous.

Result

In my judgement, the probability that one of the three hypotheses correctly reflects the phylogenetic relationship of the Chelonia decreases in passing from hypothesis 3 to hypothesis 2 to hypothesis 1.

My aim in this discussion, however, as I have already stressed, was not so much to favour a particular hypothesis of relationship. Rather it was to present the rational arguments of phylogenetic systematics as they concern a particular problematic case. The clear presentation of unsolved problems is the best starting point for new, well aimed researches.

Addendum

In the preceding discussion, I have not considered the recent work of Gardiner (1982) on tetrapod classification because of its fallacious arguments and partly false data. The central point of his classification is the hypothesis of a sister-group relationship between the Aves and the Mammalia. The apomorphous agreements between the Crocodilia and the Aves, which argue against this interpretation, remain mostly unmentioned by Gardiner—thus Hennig (1983) should be compared with Gardiner's assertion that: "These groups do share one or two unique features". The remarkable apomorphies shared between the Chelonia and the Archosauria, which are the basis for hypothesis 3 above, are likewise damaging to Gardiner's hypothesis and are either not considered (as with the fate of the pleurocentrum of the pro-atlas) or mistakenly interpreted as autapomorphies of the Chelonia + Crocodilia + Aves + Mammalia (stapes unites with Meckel's cartilage; tendon of nictitans to pyramidalis muscle; subclavian arteries displaced cephalad).

A certain lack of care in dealing with hypotheses of synapomorphy is also clear in other examples. Thus the occurrence of a hooked metatarsal in a few marsupial and placental species is taken by Gardiner as sufficient grounds to see this structure as an autapomorphy of the stem species of all Amniota (and not of the Sauropsida alone, as in hypotheses 2 and 3 above). But one seeks in vain for any explanation of why, in that case, the hook-shaped metatarsal has transformed back into a "normal" metatarsal in the Monotremata, within the Marsupialia and within the Placentalia.

Among the autapomorphies of the Amniota it is likewise disquieting to find the feature "pentadactyl fore limb". Completely counter to the facts, Gardiner plainly interprets the four-rayed hand of the amphibians as the plesiomorphous condition of the tetrapod fore limb. In point of fact, however, it is one of the most convincing synapomorphies between the Urodela and the Anura (cf. p.91).

And finally I must devote a few words to the supposition, central to Gardiner's argument, that the corresponding homoiothermy of the birds and mammals and various phenomena correlated with it, are synapomorphies of the two groups. The evolution of a constant body temperature requires the existence of a thermally insulating body cover as well as effective mechanisms for obtaining and digesting food so as to maintain the necessary high metabolic rate. In birds, of course, these are the feathers, the crop and the gizzard. While in mammals they are hair, the secondary palate and the heterodont teeth with a diphyodont tooth replacement. If the phenomenon of homoiothermy is hypothesized as an autapomorphy of a stem species shared by birds and mammals alone, then it is necessary to suppose, for this stem species, particular arrangements for thermal insulation and treatment of food from which the peculiarities, in principle so different, of birds and mammals may equally be derived. I know of no arguments which take account of these considerations.

G Features as Sources of Information for Phylogenetic Research

Every evolutionary species, and every closed community of descent, possesses a characteristic species-specific or group-specific pattern of features. This pattern always consists of a variegated **mosaic of heterochronously evolved characteristics**. The combination, in every real unity in Nature, of features of early origin, which are thus relatively primitive, with more recently evolved, comparatively advanced features, forms the material basis for constructing a phylogenetic system. Why? Because the phylogenetic relationships created by Nature cannot be discovered by the use of arbitrarily chosen agreements between the unities of Nature. Such relationships can be revealed solely and exclusively on the basis of the shared possession of apomorphous features.

Since organisms consist of a mosaic of plesiomorphous and apomorphous features, but only apomorphies are of value for present purposes, then one conclusion infallibly results. Namely that, for each agreement between two or more taxa, before any element in the patterns of features of real unities in Nature can be used to show phylogenetic relationships, one question must always be put—is the shared possession of a feature primitive or derived? This probability decision between the alternatives of plesiomorphy and apomorphy is a basic procedure in the methodology of phylogenetic systematics. I shall therefore discuss in detail the possible ways in which the decision can be justified empirically.

1. The feature and its bearer

"Ein Merk-Mal ist ein Mal oder Zeichen, das man bemerkt, sich merkt und auf das man andere aufmerksam macht. Eine Merkmal ist mit anderen Worten eine Einheit, die man beobachtet, festlegt und mitteilt."

"A 'Merk-Mal' [feature] is a 'Mal' or sign, that one remarks, which marks itself out and which one remarks to others. A feature is, in other words, a unity which one observes, establishes and communicates."

This **definition of the word "feature"** in scientific biology (Werner 1970, p.30) corresponds to the basic requirement of phylogenetic systematics that features be separable (Hennig 1984). Only "a separable, comprehensible, delimitable peculiarity or characteristic" (Werner) can validly be a feature of an organism which can be distinguished from other corresponding units, or perhaps better, elements, of one and the same organism.

The word **"feature"**, and likewise the word "criterion" in its meaning of "indication", refers to those **objects in an organism** which an observer notices, understands, recognizes, or establishes as separable elements. In the organism itself these are referred to in speech as **attributes, special traits, characters or characteristics**. For the purposes of phylogenetic systematics, it does not matter whether the conceptual fixation of the distinguishable elements of an organism begins from the standpoint of their bearer or from that of the observer. In searching for a uniform terminology,

I consistently prefer the term feature, which is widespread in systematics, together with its derivatives, explained below, of "feature state" or "feature expression". Correspondingly I refer to the totality of all the mutually delimitable elements of an organism as its "**pattern of features**".

This brings me to the **bearer of features**, or "**semaphoront**" as Hennig called it (1950, 1966). Organisms, as the subject of our investigations, always take the form of single individuals. There can never be any doubt that the individual is the bearer of the features that provide the data for the study of phylogenetic relationships .

But this naked statement, of course, needs to be made more precise. In any individual, particular features only appear at particular phases of the life cycle—as, for example, only during the cleavage of the embryo, only in the larval phase or in the adult (in the case of indirect development), or, perhaps, only in one sex in the adult. In the pattern of features that is proper to a whole individual, the single components are expressed in temporal sequence. Hennig therefore requires that the term "semaphoront" should always refer to a narrowly delimited stage in the individual's life cycle. Accordingly the semaphoront is the individual at a particular time, theoretically infinitesimally short, of its existence.

Thus phylogenetic systematics operates methodologically with individuals from particular portions of their life cycles, during which they respectively display constant dimensions. Phylogenetic systematics is less interested, however, in the genealogical connections between individuals than in the phylogenetic connections between evolutionary species and between the descent communities made up of evolutionary species. Phylogenetic systematics must therefore advance by way of the totality of features that are separable in single individuals, to the **species-specific feature patterns of evolutionary species**, and onwards to the **group-specific feature patterns of closed descent communities**. This course is completely valid, since we have already recognized species and closed descent communities as real unities in Nature, with the essential characteristics, ontologically, of individuals.

Comparison of a representative number of individual organisms is the basis for working out the **feature pattern of an evolutionary species**. These should be taken from the significant phases of the life cycle and from different populations in the species-specific area of distribution. In this comparison, all features that constantly agree, and which can be constantly delimited from other features, enter the feature pattern of the species. On the other hand, all features or marks which are recognized as individual variants, or as special to particular local populations, must be excluded.

Working out the species-specific pattern of features does not merely create the preconditions for analysing the phylogenetic relationships between evolutionary species. For it also prepares the way to determining the **feature patterns of supraspecific taxa**. In the totality of those features which the different species have in common, there exist the **constitutive features** by means of which monophyletic taxa can be erected as valid reflections of the closed descent communities in Nature (p.153). From the totality of plesiomorphous and apomorphous agreements between different species the feature patterns of the stem species of monophyletic taxa are reconstructed. And these will be shown to be the basic patterns of features, or **ground patterns**, of closed descent communities in Nature (p.147).

This brings me back to the requirement in systematics that **features be separable in an organism**. What counts as a delimitable element in the feature pattern of an individual or, furthermore, in the feature pattern of an evolutionary species cannot be decided by sitting at a desk. It must be determined by practical work on the real unities in Nature.

As an example, I shall consider three phenomena from the pattern of features in the Australian echidna or spiny anteater, the evolutionary species *Tachyglossus aculeatus*. 1) Probably nobody will deny that the complex acustico-static organ is, in its totality, a feature which can be clearly delimited from other elements of the spiny anteater. Without doubt, however, the component parts of the organ can also be considered as separate features—as, for example, the labyrinth, with three semi-circular canals, the middle ear with three auditory ossicles, or the inner ear with its slightly curved cochlea. 2) In the hind limb, in addition to the claws of the toes, there is a prominent spur connected with a gland on the thigh (p.88). The spur and thigh gland are laid down as anlagen in both sexes but suppressed in the female. Although, in the adult, they are thus only present in the male, they definitely belong, being well delimited elements, to the pattern of features of *Tachyglossus aculeatus*. 3) Turning finally to the surface of the body, it is very evident that each single spine is a concrete structural part of the species which is very effective in defence. Nevertheless, a single spine cannot be recognized as a separable element in the pattern of features of the spiny anteater. That can only be done for the coat of spines in its totality.

Completely different is the question whether, and when, **a particular element of the pattern of features** of an evolutionary species is of **interest for research** into **phylogenetic relationships**. This question can be given a general answer and I repeat the following basic precondition. A single feature is always worthy of attention when it occurs, in similar or identical expression, in two of at least three taxa which together are suspected to form a monophyletic group of species. In these circumstances, the question arises whether the shared possession is based on a plesiomorphous or an apomorphous feature. If the shared possession can be shown to be probably an apomorphy, then the next question is to decide the alternatives of synapomorphy or convergence.

Having said this, I shall now discuss once more the already mentioned features of the spiny anteater:

— The feature "**spiny coat**" is interesting at the lowest possible level of the systematic hierarchy, i.e. in comparing the three recent species of the Monotremata. *Tachyglossus aculeatus* and *Zaglossus bruijni* agree in having a spiny coat, whereas the duck-billed platypus *Ornithorhynchus anatinus* has a coat of soft fur. In comparing these three species, is the feature "spiny coat" a plesiomorphy or an apomorphy? If it is probably an apomorphy, then the next question is: Can the apomorphous agreement in the feature "spiny coat" be seen, without inconsistencies, as a synapomorphy of *Tachyglossus aculeatus* and *Zaglossus bruijni* and thus be used to establish a monophylum comprising these two species?

— A comparison of the component parts of the acustico-static organ is of interest at a much higher systematic level. The three auditory ossicles of the Mammalia are a well marked alternative to the single sound-conducting bone of all other tetrapods. One auditory ossicle is the original condition, while the existence of three auditory ossicles is the advanced expression of the feature "**sound-conducting apparatus in the middle ear**" (p.87). In accordance with the principle of parsimony, a middle ear with three auditory ossicles can be seen as an autapomorphy, and counts as one of the constitutive features in setting up a monophylum Mammalia.

— The significance of the feature "**labyrinth with three semi-circular canals**" is much more remote. All Gnathostomata possess semi-circular canals in all three planes of space. In the Cyclostomata, on the other hand, two canals, at most, are developed. Which condition is the apomorphous alternative? I shall discuss the methodologically possible answer below, in detail.

— The feature "**acustico-static organ**" in its totality, on the other hand, demands attention when the Vertebrata are compared with the Acrania and the Tunicata within the taxon Chordata. I interpret the shared absence of this organ in the Tunicata and the Acraniata as a symplesiomorphy, and the existence of an acustico-static organ as an autapomorphy, or ground-pattern feature, of the monophylum Vertebrata.

I shall use this example to explain more fully the application of the terms "**feature state**" and "**feature expression**" in relation to the term "feature".

So long as we are concerned with a single evolutionary species such as *Tachyglossus aculeatus*, all its distinguishable elements are simply features. From the terminological point of view it does not matter in the least whether they occur only in this species or in others also.

But, in comparing the patterns of features of different species, a distinction must, under certain conditions, be made — as follows:

1) If a delimitable element occurs in some of the various species compared, but not in the rest, then in both cases the word "feature" can be used unchanged. Thus spur and thigh gland become group-specific features of the Monotremata because corresponding features are absent in the Marsupialia and the Placentalia.

2) If a delimitable element occurs in all the species under comparison, but is not everywhere identical, then it is appropriate to speak of different states or expressions of a particular feature. Thus the feature "cochlea" occurs as a component of the auditory organ of all mammals; in the Monotremata it is present as a curved pouch, while in the Marsupialia and the Placentalia, on the other hand, it is in the state of a spiral.

However, in setting up monophyletic groups in the system, it does not matter if, in comparing different species, a particular element is regarded sometimes as a feature and sometimes as a feature state. Thus the features of the spur and thigh gland can together be used as a complex constitutive feature in establishing the monophyly of the taxon Monotremata. But equally, the feature state of a spiral cochlea can be used as a constitutive element in setting up a monophylum Theria to comprise the Marsupialia and the Placentalia.

108

II. The evolutionary origin of the mosaic of features

Given that a mosaic of plesiomorphous and apomorphous features is a basic pre-condition for discovering phylogenetic relationships, then we now need to understand the evolution of the corresponding species-specific and group-specific pattern of features.

"**Mosaic evolution**" (De Beer 1954b) and the resulting "**heterobathmy of features**" (Takhtajan 1959) are inevitable epiphenomena of the successive splittings of species in time.

Species

In justifying this statement I shall begin with the phylogenesis of a closed descent community comprising three **evolutionary species** (Fig. 33). The species A, B and C are supposed to form a monophylum α. Between their stem species v and the stem species u of the adelphotaxon β at least one difference in feature must have arisen. It would indeed have been this difference which, in splitting the next older stem species t, was responsible for producing a reproductive barrier between them. It does not matter at all what sort of a feature this was—whether a divergence in structure, a change in behaviour, the evolution of a biochemical difference,

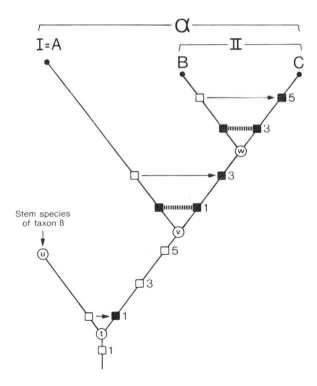

Fig. 33. Phylogenetic relationship diagram for the three recent species A, B and C. Evolutionary novelties 1, 3 and 5 arose in the lineages of v, w and C.

or whatever. Suppose that feature 1 were changed in the lineage of species v, while the sister species u retained the plesiomorphous feature state of the common stem species t. The species v would thus show an evolutionary novelty (autapomorphy) which, in the course of further phylogenesis, would become a synapomorphy of its daughter species A and w.

Again, in the splitting of species v into the lineages of the species A and w at least one feature must have changed. The alteration is supposed now to have affected feature 3 in species w where it forms an autapomorphy which, consequently, later became a synapomorphy of daughter species B and C. Finally, between these two latter species there must, again, have arisen at least one difference of feature. This is inserted in the diagram as the evolution of a new feature 5 in species C.

Thus, the origin of a closed descent community comprising three reproductively isolated species requires, in theory, no more than the evolutionary alteration of three features or else the evolution of three new features. Methodologically speaking, autapomorphy 1 establishes the monophyly of the supraspecific taxon α (I + II), while autapomorphy 3 does likewise for the monophyly of the supraspecific taxon II (B + C). Alternative feature 3 also differentiates species A from taxon II, while the difference in feature 5 distinguishes between the sister species B and C.

Note that, in this model, the species A and B possess not a single unique feature. Nevertheless, it may seem likely that these species also, in the course of millions of years of separate existence, would themselves have acquired specific autapomorphies. In theory, however, it must always be reckoned that evolutionary species may potentially have no features of their own. For the differences which exist between sister species (B and C) may theoretically have originated in the lineage of only one of them, while the other species retains unchanged the feature pattern proper to the common stem species (w). I have already discussed this problem of the persistence of the genotype and phenotype of stem species in particular daughter species (p.63). Whether or not both sister species have changed relative to their stem species could only be discovered if the pattern of features for the stem species were completely known, at least for all observable differences. Even with the best fossil record, this requirement seems difficult to fulfil.

Closed descent communities

In turning to **species groups**, the situation is in principle different. If it is possible to replace certain evolutionary species with supra-specific taxa, which can each be hypothesized as monophyla on the basis of constitutive features, the diagram, shown in Fig. 34, must then represent a real stage in the phylogenesis of three closed descent communities arising from species A, B and C. But, unlike the first model, the species A and B now both have their own proper feature (their own autapomorphy), having respectively acquired the evolutionary novelties 2 and 4.

Under these conditions, the species A, B and C can now be replaced by the latest common stem species x, y and z of the recent Monotremata, Marsupialia and Placentalia (Fig. 35). In selecting the required minimum of five autapomorphies I intentionally make an arbitrary selection from the five series of constitutive features of the mammals and their highest ranking sub-taxa (p.87–89). I do so to avoid the

110

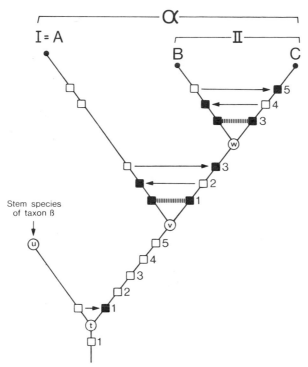

Fig. 34. Diagram of phylogenetic relationship for the three recent species A, B and C. Completion of the pattern of features by the origin of evolutionary novelties 2 and 4 in the lineages of the species A and B.

possible misunderstanding that I am concerned here with those evolutionary novelties which, in the phylogenesis of the Mammalia, produced the reproductive barriers between the respective stem species. These features are unknown, and there is no prospect that they ever will be known.

I begin with the auditory ossicles of the middle ear as my choice from the features of series 1. After the splitting of the latest common stem species t of the Sauropsida and the Mammalia, two bones of the jaw articulation were profoundly transformed in the stem lineage of the mammals and taken into the sound-conducting apparatus of the middle ear as additional auditory ossicles. Autapomorphy 1 ("middle ear with malleus, incus and stapes") subsequently became a synapomorphy of the sub-taxa Monotremata and Theria. On the other hand, the stem species u of the Sauropsida took over unchanged, from the stem species t of the Amniota, the plesiomorphous feature state with a single auditory ossicle in the middle ear.

After the splitting of the stem species v, the evolutionary novelty "spur and thigh gland" arose in the stem lineage of the Monotremata. This complex of features was present in the latest common stem species x of the recent Monotremata and became a synapomorphy of the taxa Tachyglossidae and Ornithorhynchidae. On the other hand, in the stem lineage of the Theria, the evolutionary novelty "vivipary" evolved, with the birth of live young, while the Monotremata retained the plesiomorphous alternative "ovipary" (feature 3). With the splitting of stem species w, the autapomorphy "vivipary" became a synapomorphy of the Marsupialia and the Placentalia.

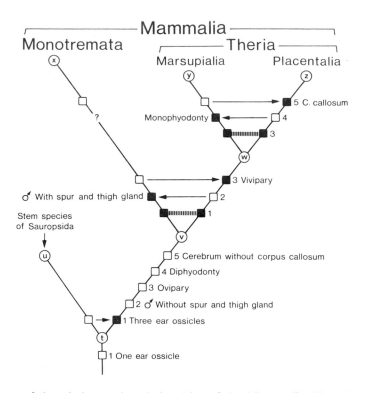

Fig. 35. Diagram of the phylogenetic relationships of the Mammalia. Note the successive evolution of five autapomorphies in the stem lineages of the Mammalia (1), Monotremata (2), Theria (3), Marsupialia (4) and Placentalia (5).

The phenomenon "monophyodonty" evolved in the stem lineage of the Marsupialia while the Placentalia retained the original diphyodont tooth replacement (feature 4). In the stem lineage of the Placentalia there arose the "corpus callosum" as a new connection between the cerebral hemispheres (feature 5). With the splitting of stem species y, the feature state "monophyodonty" became a synapomorphy of the highest ranking sub-groups of the marsupials. And with the splitting of stem species z, the feature "corpus callosum" became a synapomorphy of the highest ranking adelphotaxa among the placentals.

Summing-up

I shall now sum up this discussion of the heterobathmy of features, as it occurs in the species-specific and group-specific feature patterns of real unities in Nature. For purposes of illustration, I have discussed the phylogenesis of three closed descent communities derived from a stem species common to them alone, with the least number (five) of novelties which our methodology requires. Even in this simple case, the differing ways of combining one constant feature with four alternative features will produce a mosaic, varying from taxon to taxon, of primitive and derived features (Fig. 36).

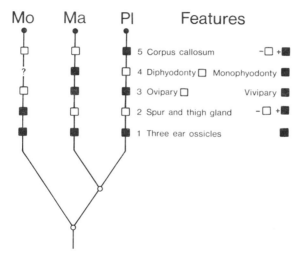

Fig. 36. Differing combinations of plesiomorphies and apomorphies, with one constant feature (1) and four alternatives (2–5), in the feature patterns of the Monotremata (Mo), the Marsupialia (Ma) and the Placentalia (Pl).

As a rule, however, evolution has created far more than one novelty in the stem lineage of a closed descent community. Nature offers a richly varied pattern of features for evaluation by phylogenetic systematics, although the list of autapomorphies on which monophyletic taxa can be based is not always so luxuriant as in the Mammalia or the Theria.

There is another side to the penny. As stated, evolutionary species may exist that have no features proper to them alone. It may also be, therefore, that closed descent communities exist in Nature whose representatives possess not a single apomorphy common to them alone. This would be true when a particular evolutionary species, which later became the stem species of a closed descent community, acquired not one evolutionary novelty during its own period of existence (cf. p.63, Fig. 16b). But there is no way of knowing whether, or to what extent, such cases exist in Nature. For with our methodology it is impossible to discover a closed descent community if it occurs in Nature with no autapomorphy.

III.   The alternatives of plesiomorphy and apomorphy

1.   The basis of evaluation

I now wish to give precision to the two possible kinds (1 and 2 below) of pairs of alternative features—both of these possibilities are used in comparing evolutionary species and closed descent communities. I shall explain these possibilities on the basis of a few groups-of-three which are suspected to be monophyla and in which two taxa share one alternative while the third taxon has the other.

1) Positive feature v. negative feature. A feature is either present or completely lacking. For example:

113

— Embryonic membranes; Sauropsida + Mammalia have amnion and serosa while the Amphibia have no corresponding embryonic membranes.
— Corpus callosum; the Placentalia have a commissure between the two cerebral hemispheres while the Marsupialia and the Monotremata have no such connection.
— 2nd. antenna; the Crustacea have a head appendage in the premandibular segment while Myriapoda + Insecta have no corresponding head appendage.
— Marsupial bones; Monotremata + Marsupialia have a marsupial (epipubic) bone while the Placentalia have no such bone.

2) Alternative feature states. A feature is present throughout but occurs in two distinguishable states or expressions. For example:
— Auditory ossicles; Sauropsida with one auditory ossicle but Monotremata + Theria with three auditory ossicles.
— Cochlea of the inner ear; Monotremata with a curved cochlea but Marsupialia + Placentalia with a spiral cochlea.
— 1st. antenna; Myriapoda + Entognatha (Insecta) with segmented antennae but Ectognatha (Insecta) with annulated antennae.
— Mouth parts; Myriapoda + Ectognatha with ectognathous mandibles and maxillae but Entognatha with entognathous mouth parts.

For the aims of phylogenetic research, it is a question, under both 1) and 2), of **alternative features forming pairs with the same significance** in every case. In other words, absence can also be made use of. It does not matter whether the absence is primary, or is the secondary result of reductive evolution. However this may be, the negative feature takes on, with respect to the positive feature, the same significance as a particular feature state with respect to the alternative feature state. Thus amnion and serosa can only be used as an apomorphy in setting up a taxon Amniota within the vertebrates because the absence of comparable membranes in amphibia and fish contrasts with it as a plesiomorphy. Conversely, the lack of a second antenna can be claimed as a constitutive feature in uniting the Myriapoda and Insecta into a monophylum Tracheata because in this case the negative feature can be seen as the apomorphy and the existence of the appendage in the Crustacea as the plesiomorphous alternative.

This brings me to a further consideration. When a **particular feature**, within a supposedly monophyletic group, occurs as **alternatives**, then one alternative is always an evolutionary novelty (apomorphy). That must needs be so, because the stem species of the group cannot have possessed in its pattern of features both alternatives together. It is usually tacitly assumed that the other alternative always represents the plesiomorphy, derived unchanged from the common stem species, but does this necessarily follow?

1) When the alternatives are a positive feature *versus* a negative feature, the answer is definitely "Yes". It does not matter whether the existence or the absence of a particular feature is the apomorphy—in either case the alternative

114

must be the plesiomorphy, persisting in the form which once existed in the stem species.

2) Given alternative states of a feature, the answer is more complex. Here also, the second state will as a rule represent the plesiomorphous condition of the common stem species. Why? Because it will not normally happen that some particular feature of a stem species, after that species has split into the lineages of the two daughter species, will have changed in both lineages. Nevertheless, such a possibility must always be reckoned with. And in fact there are examples where both alternative feature states must be seen as apomorphies compared with the plesiomorphous state in the stem species of the taxon (p.170).

In making probability judgements concerning two states of a feature, therefore, it must always be considered whether the **alternatives represent plesiomorphy and apomorphy or two alternative apomorphies**.

2.   Arguments towards a probability judgement

In evaluating alternative features for research into phylogenetic relationships, there are no laws or criteria by which the course of evolution in the pairs of features or pairs of feature states can be read off or proven directly. A number of "directional arguments for evolutionary change" (de Jong 1980) are available, however, by means of which logical deductions on the direction of evolutionary change can be made.

In analyzing critically the essential arguments used in making the probability judgement between the alternatives of plesiomorphy and apomorphy, the following question is centrally important: Does the individual argument allow empirically testable hypotheses to be formulated, or does it allow nothing more than an explanation, perhaps plausible but not objectively testable, of the course of evolutionary changes? In my view, it is more important to answer this question for arguments that have often been used, than to recount all the arguments that have been put. The applicability of some of the arguments is sometimes very limited, as has been documented in recent surveys (de Jong 1980; Stevens 1980; Arnold 1981).

Arguments based on the distribution of features

a) Out-group comparison

This can be expresssed as follows:
"If a feature occurs as alternatives in a supposedly monophyletic group of species, then that feature state which also occurs outside the group is probably the plesiomorphy."

In evaluating the alternative features in a supposedly monophyletic group of species, consisting of at least three taxa, then we search in an out-group comparison for the "closest phylogenetic relatives". For the purposes of comparing such a

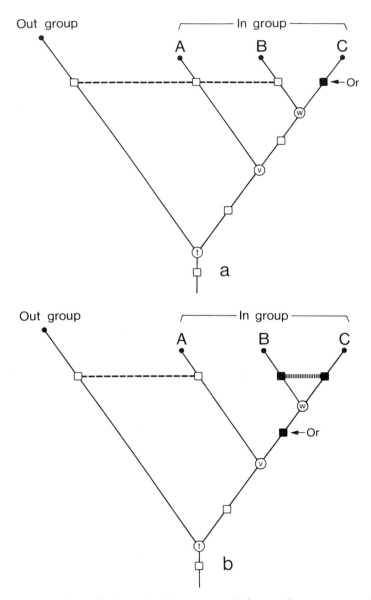

Fig. 37. An evaluation of alternative features as plesiomorphy or apomorphy, by means of the out-group comparison. a) One feature state occurs in the out-group and in taxa A + B, while in taxon C the other state is found. On the basis of this distribution, the first state is postulated for the stem species t + v + w, and its existence in taxa A + B is seen as a plesiomorphy. The alternative state thus arose as an apomorphy in the lineage of C.

b) Only taxon A has the feature state which, by comparison with the out-group, is seen as the plesiomorphy. Taxa B and C share the apomorphy. Its shared possession indicates either that it evolved once, in the lineage of the stem species w (as shown in the figure), or twice, convergently, in the lineages of B and C.

feature, this out-group may be seen as a very coarse preliminary arrangement and may remain so. It need not necessarily represent the adelphotaxon of the "in-group" nor itself be established as a monophylum. It is enough, if the out-group consists of organisms that possess one of the alternative features of the in-group.

To take a schematic example, an out-group shares the stem species t with an in-group comprising species A + B + C (Fig. 37). If a feature, or feature state, occurs in the out-group, and likewise in part of the in-group (in one or two of the three species), then it would very probably have been present in the stem species t and have passed from it into the stem lineages of the in- and out-group. Logically, this feature must represent a plesiomorphy within these stem lineages. In Fig. 37a, the species A and B, together with the out-group, have retained the original feature. The apomorphous alternative developed as an evolutionary novelty only in the lineage of species C. In Fig. 37b, the plesiomorphy is retained only in A while the apomorphy had already arisen in the lineage of the species w. It became a synapomorphy of species B and C.

As a universal method (Watrous & Wheeler 1981), out-group comparison can be applied to every form of feature, from total shape to ultrastructural detail, from behaviour to molecular process. I shall now consider it in relation to the pairs of features cited above.

The first group of alternatives = positive feature v. negative feature

— The embryonic membranes of vertebrates

A simple out-group comparison shows that the amnion and serosa are absent, not only in the Amphibia, but also, in the primarily aquatic vertebrates. We can therefore postulate that the stem species of the Tetrapoda had no embryonic membranes. The negative feature "absence of amnion and serosa" in Amphibia is the plesiomorphy. while the existence of the embryonic membranes in the Sauropsida and Mammalia is the apomorphous alternative.

— Corpus callosum in the brain of Mammalia

The commissure between the two cerebral hemispheres which exists in the Placentalia is absent in the Monotremata and Marsupialia, and in all other vertebrates as the "out-group". Within the taxon Mammalia, the lack of a corpus callosum must be seen as a plesiomorphy, while the positive feature in the Placentalia would be an apomorphy.

— Second antenna in the arthropod head

Within the taxon Arthropoda, the chelicerae of the Chelicerata are equivalent, judging by their position and innervation, to the second antennae of Crustacea. Chelicerata can be taken as the out-group for the Mandibulata (Crustacea and Myriapoda + Insecta). The common stem species would have had a pair of premandibular appendages which passed into the stem lineages of the Chelicerata and the Mandibulata. If so, the existence of the appendage (antenna 2) in the Crustacea is the plesiomorphy, and its absence in Myriapoda and Insecta is the apomorphous alternative.

117

—Marsupial bones of Mammalia

Monotremata and Marsupialia share the possession of a marsupial (epipubic) bone in the pelvis, while such a bone is absent in Placentalia. In this instance, all other vertebrates could serve as the out-group, but among them it is vain to seek for a marsupial bone. To be consistent, the negative feature of the placentals should be seen as the plesiomorphy, while the existence of the marsupial bone in monotremes and marsupials should be taken as the apomorphous alternative (but see p.83,119).

The second group of alternatives = two states of a feature

—Auditory ossicles of the vertebrates

Concerning the alternatives of one or three auditory ossicles, the comparison can be extended beyond the Sauropsida to the rest of the land vertebrates. In so doing, the state with one auditory ossicle recurs in the Amphibia. This distribution suggests that the condition of the middle ear in the Sauropsida represents the plesiomorphy, while the uniform development of three auditory ossicles in the Monotremata and the Theria is the apomorphous alternative.

—Cochlea of the inner ear in the Mammalia

In the out-group of the Sauropsida the cochlea is found, as in the Monotremata, in the form of a slightly curved pouch. Consequently, the state in the monotremes can be taken as the plesiomorphy and the spiral cochlea of the Marsupialia and the Placentalia would be the apomorphous state.

— First antenna in the arthropod head

In interpreting the pair of features "segmented antenna v. annulated antenna" (Fig. 51) within the Tracheata (Myriapoda and Ectognatha + Entognatha), an out-group comparison reveals a segmented antenna in the Crustacea. This suggests that the segmented antenna of the Myriapoda and Entognatha is the plesiomorphous state, while the annulated antenna of the Ectognatha is the apomorphous alternative.

— Mouth parts in the arthropod head

The pair of features here is: "mouth parts freely inserted in the head v. mouth parts inserted inside a pouch in the head capsule" (Fig. 51). Here again, the out-group Crustacea suggests a well based probability judgement since they show the first-mentioned alternative. Logically, therefore, the freely inserted ectognath mouth parts of the Myriapoda and Ectognatha can be seen as the plesiomorphy while the entognath mouth parts of the Entognatha, sunk into the head capsule, would be the apomorphy.

General results

From these examples, two important results of a general nature emerge:

— The out-group comparison leads to empirically based, and therefore testable, hypotheses concerning the direction of evolutionary transformations. The obtained resulting probability judgements represent objective hypotheses which can legitimately be exposed to the other methodological procedures of phylogenetic systematics.

— As to the question whether a pair of feature states corresponds to the simple alternatives of plesiomorphy or apomorphy, or represents two alternative apomorphies (p.114), outgroup comparison allows the following answer: If one of the feature states occurs identically in the outgroup, then the alternatives represent plesiomorphy and apomorphy — all the above examples refer to situations of this sort. But if, in the outgroup, a third feature state occurs, different from either, then alternative apomorphies may be suspected. I shall deal with examples of this sort later (p.170).

On the other hand, the **limitations of the method** emerge, in the following two points:

— Out-group comparison is no defence against mistakenly interpreting the negative feature as a plesiomorphy. Despite impeccable outgroup arguments, the absence of a feature in part of the ingroup may always be the result of secondary evolutionary loss. In the example of the marsupial bones, an extension of the out-group comparison to fossils of the stem lineage of the mammals leads to a correction (see below). The negative feature "absence of a marsupial bone" is not the plesiomorphous, but the apomorphous alternative.

— Out-group comparison does not function when a feature occurs as a positive feature with two states, within a supposedly monophyletic group of species, but is not found outside the group. I recall here the two alternatives in the acustico-static organ of vertebrates, i.e. at most two semi-circular canals (in Cyclostomata) v. three semi-circular canals (in Gnathostomata). It does not matter, in this connection, whether the Acrania or the Tunicata form the adelphotaxon of the Vertebrata, for both these "out-groups" lack a labyrinth. Even the indication of two semi-circular canals in †"Ostracodermata" as representatives of the stem lineage of the Cyclostomata does not help. By mere use of out-group comparison there is no way of deciding whether the labyrinth in the latest common stem species of the Cyclostomata and the Gnathostomata had two canals or three. And there is, therefore, no way of deciding which of the two conditions has been retained as a plesiomorphy (compare p.121).

b)   Comparison with the stem lineage of the taxon

This can be expressed as follows:
"If a feature, in a supposedly monophyletic group of species, occurs as alternatives, then the alternative found in the stem lineage of the taxon is probably the plesiomorphy."

I shall explain this situation by using a schematic example (Fig. 38). In taxon α,

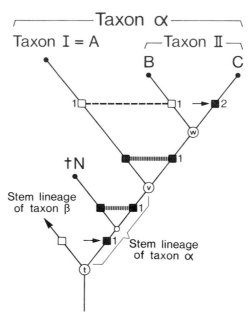

Fig. 38. Recognizing alternative features as plesiomorphy and apomorphy by comparison with fossil representatives of the stem lineage of the taxon. The feature state that occurs in fossil taxon N must be interpreted as a symplesiomorphy of taxa A and B, while the alternative state in taxon C is an apomorphy.

a particular feature is supposed to occur as alternatives 1 and 2, but in the adelpho-taxon β, which is likewise the out-group, it is absent. However, alternative 1 now turns up in a fossil taxon N which can be systematized as a representative of the stem lineage of the taxon α, on the basis of arguments explained later (p.201). In that case, alternative 1 must have evolved in the stem lineage of taxon α as an evo-lutionary novelty and in consequence belonged to the feature pattern of species v, which was the latest common stem species of sub-taxa I and II. In comparing these two sub-taxa, autapomorphy 1 thus becomes synapomorphous agreement 1. And in passing to the next lower level of subordination, there is, necessarily, a change in value which I shall discuss in detail later (p.144). For in comparing A and B, the common possession of alternative 1 becomes a symplesiomorphy, while, at the same time, alternative 2 in C is indicated as the apomorphous state.

I shall apply this train of argument to two concrete examples which have already been mentioned.

To illustrate the alternatives of a **positive** *versus* **a negative feature**, I refer again to the marsupial bones of the mammals. In the fossil record, proof of the marsu-pial bone in the †"Tritylodontia" is of first importance for this problem, since these fossils can be systematized as representatives of the stem lineage of the Mammalia. If A, B and C in Fig. 38 are replaced respectively by the Monotremata, Marsupialia and Placentalia, then the †"Tritylodontia" can take the position of the fossil taxon N. That indicates, however, that the marsupial bones must have arisen as an evolutionary novelty in the stem lineage of the Mammalia and have been present in the latest common stem species v of the Monotremata (A) and of the Theria (B + C). At the same time, the marsupial bones, representing alternative 1, constitutes a synapomorphy of the taxa Monotremata and Theria. And on the next lower level of subordination, in comparing the Monotremata *Tachyglossus, Zaglossus* and *Ornithorhynchus* with one another and with the Marsupialia, it necessarily becomes a symplesiomorphy. The negative feature, as alternative 2, can thus be interpreted as an apomorphy, only explicable by the loss of the marsupial bones in the stem lineage of the taxon Placentalia.

The stem lineage can also supply crucial data for **evaluating two states of some feature which is not found outside the taxon**. The question, which I have left open, as to the original number of semi-circular canals in the labyrinth could be decided if fossils of the stem lineage of the vertebrates were known, i.e. which did not belong to the stem lineage of the Cyclostomata or of the Gnathostomata. The state found in representatives of this stem lineage must be the plesiomorphous condition of the labyrinth.

Thus in comparing fossil and recent representatives of supposedly monophyletic taxa, analysis of how the features are distributed shows the **central significance** of the **time dimension** in working out the sequence of evolutionary changes in the features. However, the stratigraphical (and temporal) sequence of the fossils that bear the features need not automatically correspond to change in the features from plesiomorphy to apomorphy (Schaeffer, Hecht & Eldredge 1972; Cracraft 1974; Hecht & Edwards 1977; Bonde 1977; Eldredge & Cracraft 1980; Wiley 1981). It is enough, in this connection, to recall the heterobathmy of features. Every recent species is made up of a pattern of less or more evolved features, and the same must obviously be supposed for every fossil taxon. That means, however, that older fossils may well carry the apomorphous state of a particular feature, while younger ones retained the plesiomorphous alternative.

c) In-group comparison with an evaluation of the quantitative distribution of various feature states

The allegedly useful technique of in-group comparison can be expressed in the following statement:

"That feature state which has the greatest distribution among the species of a supposedly monophyletic taxon is probably the plesiomorphy."

A feature state which all species of a monophyletic group possess, or which occurs predominantly among the highest ranking sub-taxa of the group, is very probably a plesiomorphy within the group. Seeing that all recent mammals have a middle ear with three auditory ossicles, the only justifiable hypothesis is that this feature state was present in the latest common stem species of recent Mammalia. In other words, the existence of malleus, incus and stapes is a plesiomorphy for every single species of mammal. Seeing that nearly all mammals of the highest ranking subtaxa Monotremata and Theria have hair, and that the alternative negative feature "hairlessness" only turns up here and there in subordinate groups of the Placentalia (Cetacea, Sirenia), then the only deducible hypothesis is to postulate a coat of fur for the latest common stem species of recent Mammalia. And in that case the feature state "hair", seen in the overwhelming majority of mammal species, is a plesiomorphy, and the absence of hair in whales and sea cows is an apomorphy.

These, however, are all self-evident truisms. They can scarcely have been used in formulating the argument which was later dignified as the "Principle of commonality" (Kluge & Farris 1969; Schaeffer, Hecht & Eldredge 1972; Hecht 1976). It is more important to examine the following question: Given a supposedly monophyletic group with sub-taxa which respectively show different states of a feature, can counting the number of species of these sub-taxa help in deciding the plesiomorphy or apomorphy of particular feature states?

I shall later discuss the proboscis organ of the Kalyptorhynchia (Plathelminthes), in which there is an interesting relationship between the four basic forms of the proboscis and the number of species in which these forms respectively occur (120 : 60 : 19 : 6). The decrease in the number of species is here exactly correlated with stages of increasing apomorphy in the probable evolutionary development of the proboscis (p.136).

However, there is no difficulty in citing any number of cases where the situation is totally otherwise. The three monotreme species *Tachyglossus aculeatus*, *Zaglossus bruijni* and *Ornithorhynchus anatinus* all lay eggs. But the Marsupialia, with about 250 species, and the extremely species-rich Placentalia are, without exception, live-bearing. Must vivipary, on the basis of the number of species where it occurs and its distribution throughout two extensive sub-taxa, therefore be seen as the plesiomorphous condition of the Mammalia? Certainly not! On the contrary, out-group comparison with the Sauropsida and all other vertebrates establishes the ovipary of just three species of recent mammals as the plesiomorphous state of the Mammalia in this feature.

In the Insecta, all the "Apterygota"[1] together (i.e. Diplura, Protura, Collembola, Archaeognatha, Zygentoma) number 3,200 species. This is a vanishingly small fraction of the total, compared with more than 750,000 winged Pterygota. Nevertheless, out-group comparison gives good reason to see the absence of wings in the above-named taxa as the plesiomorphous alternative, and the existence of wings in the Pterygota as the apomorphy.

Probability judgements as to the direction of evolution in alternative features could legitimately be based on number of species only if speciation proceeded at comparable rates within different taxa. But that is not the case. In judging between the alternatives of plesiomorphy and apomorphy, this argument is useless.

d) Comparison of the expression of features in ontogeny and in the adult

Here the proposal can be put as follows:

---

[1] See p.185.

"If, within a supposedly monophyletic group, a feature state occurs only in ontogeny in certain species, but also in the adult in the others, then it is probably the plesiomorphous state of the feature."

This phenomenon of ontogenetic character precedence (Hennig 1966) is often dubbed "recapitulation". I shall discuss it with two examples.

First pair of alternatives: positive feature v. negative feature

In the highest ranking sister groups within the Cetacea (whales) I shall discuss the alternatives: teeth present (Odontoceti) v. teeth absent (Mysticeti). The baleen whales are those whale species in which the positive feature "teeth" occurs only in ontogeny. In them, teeth appear as anlagen in the upper and lower jaws but in the course of individual development are completely resorbed again.

This suggests the hypothesis that the existence of teeth, in Odontoceti, is the plesiomorphous alternative while their absence, in Mysticeti, is the apomorphy.

Second pair of alternatives: two states of a feature

In the primate taxon Hominidae, the orang outang has nine carpal bones in the hand, while the carpus of chimpanzee, gorilla and Man contains, in the adult, only eight bones. However, in the ontogeny of these three taxa, nine carpal bones appear as anlagen, but of these the central bone and the scaphoid bone later fuse together. For the alternatives of "eight v. nine carpal bones", *Pan*, *Gorilla* and *Homo* are thus the species in which the nine-bone state occurs only in ontogeny.

This suggests the hypothesis that the common stem species of the four species of Hominidae had nine carpal bones in its hand. This state was retained unchanged in *Pongo* as the plesiomorphy. The feature state of the eight-bone carpus in *Pan*, *Gorilla* and *Homo* is the apomorphous alternative. In their development, these three taxa recapitulate the plesiomorphous feature state (cf. p.74).

Discussion

In the framework of what I have already said, the argument from ontogenetic feature precedence must be brought more sharply into focus with the following question: The above examples are meant to be representative — can it be said of them that the probability judgement between the alternatives of plesiomorphy and apomorphy is indeed based on the ontogenetic recapitulation of one of the two feature states? The answer is a definite "No". Without any ontogenetic data whatever, mere out-group comparison, and the consideration of fossil representatives of the stem lineage, would lead to objective hypotheses on the direction of evolution of the features.

In the case of the Cetacea, a) toothed species are recorded from the stem lineage of the descent community; and b) the dentition of these fossils is heterodont and derivable from the dental formula in the ground pattern of the Placentalia, i.e. $\frac{3143}{3143}$.

In the case of the Hominidae, all other members of the Simiae can be used as the out-group. All of these have nine carpal bones in the hand. Using this comparison, the count of eight bones in the carpus can be hypothesized as the apomorphous alternative without any contradictions. And this would remain true even if, in the ontogeny of *Pan*, *Gorilla* and *Homo*, only eight carpal bones ever appeared.

I know of no case in which a logically correct probability judgement concerning the alternatives of plesiomorphy and apomorphy was reached solely by comparing the ontogeny with the adult. The interpretation of particular states of features in development as being ontogenetic recapitulations of plesiomorphies is reached *a posteriori*. It is either based on previous hypotheses as to the direction of evolution of features—hypotheses set up by using other arguments. Or else it is based on preconceived hypotheses concerning the phylogenesis of the organisms that bear the features.

How can these assertions be justified? The ontogenetic facts are basically ambiguous in their meaning. And, for that simple reason, they cannot be claimed as an independent measuring rod when used in evaluations. Evolutionary changes in features may make themselves manifest at any stage of the life cycle. In this connection, the phenomenon of neoteny is relevant, i.e. the persistence of ontogenetic features into the phase of sexual reproduction or, in other words, the occurrence of sexual maturity at a developmental stage that has juvenile feature states.

I shall exemplify such a situation by using the gills of Urodela. In the great majority of salamanders and newts, as is well known, gills are limited to the larval period; they disappear at metamorphosis. Taxa like *Ambystoma*, *Proteus*, *Necturus*, *Typhlomolge* or *Siren*, on the other hand, are provided with gills even as sexually mature animals. Now if the argument from ontogenetic feature precedence were applied in the taxon Urodela to the alternative of "gills present v. gills absent", then the perennibranchiate condition would come out as the plesiomorphy, while the lack of gills in the adult would be the apomorphy. However, even a simple outgroup comparison with the Anura (frogs and toads) shows that the precise opposite must be true. And in fact we know, from the famous axolotl and other perennibranchiate urodeles, that disturbances in the interactions of the hypophysis, thyroid and tissues will hinder normal metamorphosis. The persistence of gills after sexual maturity is here the apomorphy, and their absence is the plesiomorphous alternative.

Thus a particular state in ontogenesis may represent the plesiomorphy among a pair of alternative features found in sexually ripe organisms, as in the tooth anlagen of baleen whales or the carpal bones of Hominidae. But on the other hand it may, as in the case of the gills of urodeles, represent the apomorphy. Consequently, the argument from ontogenetic feature precedence must *a priori* fail as a method of evaluating the alternatives of plesiomorphy or apomorphy. Obviously, an ontogenetic observation may, *a posteriori*, confirm hypotheses concerning the direction of evolution of features—as when, for example, a state regarded as a plesiomorphy by out-group comparison is passed through during ontogeny by an organism that bears the apomorphy in the adult.

In this argument, therefore, I share the cautious view of Rieppel (1979) and de Jong (1980), contrary to Nelson (1978), Wiley (1981) or Osche (1982). Of course, there is nothing new in this position which was already advocated, with great logical acuity, by Tschulok (1922).

124

Explanation of the constructive and functional alteration of features

Objective hypotheses on the direction of evolution of features can thus be produced by using the distribution of alternative features among the real unities in Nature. However, they give not the least indication of the historical path of the postulated transformations. The naked result of out-group comparison may be enough for discovering phylogenetic relationships, but every biologist would welcome a satisfying explanation of the constructional and functional changes between one alternative feature and the other. Here, however, we are less interested in what seems desirable for a total understanding of evolution. Rather we must ask whether constructional and functional explanations of changes in features can provide a separate method for evaluating the alternatives of plesiomorphy and apomorphy.

In examining this question, I shall again take the example of the antennae of insects with the alternatives of the segmented antenna v. the annulated antenna, and I shall look for a plausible explanation for a transformation in one of the two possible directions. In the segmented antenna of the Entognatha all segments, except the distal one, have a musculature. In the annulated antenna of the Ectognatha, on the other hand, only the basal shaft (scapus) has muscles. The more distal pedicel portion, with Johnston's sense organ as well as the flagellum, is without muscles (Fig. 51).

In the transformation of a segmented antenna into an annulated antenna, two competing possibilities emerge:

— In the stem lineage of the Ectognatha, before the latest common stem species of the recent taxa that have annulated antennae, there was a species whose segmented antennae consisted of only two or three segments. The muscle-free flagellum of the annulated antenna arose by a secondary subdivision of the muscle-free distal segment of this segmented antenna. Even the muscle-free pedicel could be derived from this process. Perhaps, however, it originated from a second, muscle-carrying segment of the segmented antenna and lost its musculature in connection with the evolution of Johnston's organ (Hennig 1969). In that case, an antenna with three segments would be needed as starting point for the transformation, since the scapus obviously corresponds to a separate muscle-bearing segment.
— In the stem lineage of the Ectognatha, before the latest common stem species of the recent taxa that have an annulated antenna, there existed a species whose segmented antennae contained numerous segments. In the transformation to an annulated antenna the musculature became confined to a basal joint (scapus). In this explanation not only the scapus and pedicel respectively correspond to one segment of the segmented antenna, but also the individual portions of the flagellum.

But an evolutionary transformation in the converse direction, from the annulated antenna to the segmented antenna, could also be completely explained. Suppose that Johnston's organ, for whatever reason, became reduced. If so, the musculature of the scapus could extend secondarily into the pedicel and, after the individual portions of

125

the flagellum were expanded to become thicker segments, the muscles extended still farther distalwards, omitting only the most distal segment. Under this explanation, also, a section of the flagellum would correspond to a segment of the segmented antenna.

In choosing this pair of features of "segmented antenna v. annulated antenna" I have intentionally taken an example which allows several plausible explanations for both alternative directions. On the other hand, there are obviously numerous cases where explanations of the constructional and functional changes in a feature are satisfying only for one direction. For example, this holds in the mammals for the passage from ovipary (Monotremata) to vivipary (Theria). It also holds for the transformation of the sound-conducting apparatus with one auditory ossicle into the mammalian middle ear with three auditory ossicles. Probably nobody has attempted to explain the converse process, perhaps picturing how the malleus and incus became disarticulated from the middle ear and inserted into a jaw articulation formed of the dental and squamosal.

Whether or not, however, plausible explanations can be offered in the concrete individual case for transformations in both directions of the pair of alternatives, or only in one direction, such interpretations do not, as a general rule, have the character of objective hypotheses. This is so because—except for particular lucky cases of complete fossilization of the transformational stages—there is no way of testing them empirically. In other words, the particular explanation can be neither confirmed nor refuted. Today, perhaps, only one version of a particular constructional and functional change seems convincing. But tomorrow some new author may appear, asserting that his new explanation of the featural change in question is superior to all previous accounts. Since neither explanation is empirically testable, both of them remain arbitrary speculations.

Despite this epistemological situation, **constructional or functional explanations of changes in features** may, in particular situations, be admitted to have a certain **heuristic value** in deciding between the alternatives of plesiomorphy and apomorphy. Often enough, in practical work, we are plagued with pairs of features which, at the present stage of investigation, do not permit any objective hypotheses on the direction of evolutionary transformation. If their interpretation happens to be crucially important for the discovery of phylogenetic relationships, then it seems legitimate, as a first step, to operate with a plausible explanation of the transformation in a particular direction. The provisional character of such a subjective judgement must, however, be stressed. It can be nothing more than a motor for well aimed efforts to formulate testable hypotheses on the direction of evolution of features. I shall deal below with such a situation, using an example of constructional and functional change in the proboscis of the Kalyptorhynchia (Plathelminthes).

Judging the adaptive condition of different feature states

Arranging the products of phylogenesis, and recording their relationships in a phylogenetic system, do not depend, rationally or methodologically, on knowing the

126

mechanisms which govern evolutionary change in the populations of species or the process of phylogenesis. Objective hypotheses on the phylogenetic relationship between species and closed descent communities can be set up without considering the causality of evolution (p.49).

Despite this, phylogenetic systematics will not throw away any argument which, by way of the distribution of alternatives of features, leads to testable hypotheses on the direction of evolutionary change in features. I therefore turn to the arguments which seek to use the **phenomenon of adaptation** in evaluating the alternatives of plesiomorphy and apomorphy. Such arguments, however, now carry a heavy handicap. For the **synthetic theory of evolution**, in which selection is the mechanism for producing adaptations, is itself increasingly beset by problems and seen as in principle untestable (Brady 1979, Rosen & Buth 1981, Maze & Bradfield 1982; cf. Mayr 1982b).

I shall first summarize the basic connection which this theory postulates between the genetic constitution of a population, the mechanism of selection, and the phenomenon of adaptation. The genetic variability of a population arises by random mutations of the genetic information of single individuals, by the recombination of genes and by gene flow between individuals. Genetic variation shows itself by phenotypic variability in the population. It is the point where selection acts—selection being a result of the interactions of individuals with the factors in the environment. In successive generations, the resulting selection pressure continually increases those variants which, under the given environmental conditions, are the most successful. Selection leads, in this way, to evolutionary changes in particular features in the direction of better adaptation of the population to its environment.

According to these views, in any evolutionary change which is caused by the mechanism of mutation and selection, the apomorphy must always be the better adapted feature state, when compared with the plesiomorphous alternatives. To recognize the direction of evolutionary transformations, therefore, nothing more is needed than an objective measure of the degree of adaptation for alternative feature states.

Adaptations to the demands of the environment can, according to the synthetic theory of evolution, only arise under the control of particular selectional forces. That in no way means, however, that all evolutionary changes need be the result of selectively controlled adaptations. Allometric growth processes, pleiotropic effects of genes, mutational changes in genetic information, none of which are affected by any selectional force, and other processes, may potentially result in non-adaptive evolution (Lewontin 1978). Among such processes, according to Bock (1979), may be the bodily isolating mechanisms which, in the process of species splitting, erect reproductive barriers between the newly evolving unities in Nature.

Therefore, we cannot know *a priori*, in any particular case, whether the apomorphous state of a pair of alternatives arose under the control of selection or not.

For the sake of argument, I shall ignore the ill effects of this additional handicap, and start again from the assumption that a particular set of alternative features is the result of selectively controlled evolution. Logically the next question is as follows: Is there an operational method by which the relative degree of adaptation of the different states of a feature to the demands of the environment can be measured?

I shall begin the answer with Stern's analysis (1970) which recognizes three widespread **applications of the word "adaptation"**:

— Adaptation as something which may be possessed by an individual or population.
— Adaptation as the state of being of an individual or a population.
— Adaptation as an evolutionary process.

The connections between these different meanings of the word "adaptation" are summarized by Stern in two **theoretical definitions** (1970, p.40, 41) as follows:

— "An adaptation will be any characteristic of an organism or population, that causes its possessors to be, on the average, at a higher level of adaptation than would occur in its absence."
— "The process of adaptation will be that which leads to higher levels of the state of being."

Obviously these definitions contain no criterion for recognizing the meanings that they denote. I shall therefore pass on to the following **operational definitions** which Stern formulated separately for the single individual and the population (1970, p.44, 57):

— As to the individual: "An adaptation is any transmissible characteristic of an organism that by its presence permits an interaction with the environment that causes its possessors to produce, on the average, more offspring (i.e., zygotes) than would be produced in its absence."

Referring back to the first theoretical definition, the degree or level of adaptation becomes a relative concept. In the comparison of at least two individuals, that organism has reached a higher degree of adaptation to its environment which produces more offspring. The measure for the height of the degree of adaptation is the number of zygotes produced by an individual.

— As to the population: "An adaptation of a population is any characteristic of that population which causes, on the average, a higher rate of increase in size than would occur in its absence."

At the level of the population, therefore, the degree of adaptation is again a relative matter. In comparing at least two populations, the one with the higher growth rate is defined as the population better adapted to its environment. As in the case of the individual, the methodological **measuring rod is the level of reproduction**. This level is determined either by the relative level of reproductive output as a total of individuals and/or by the relative number of generations in unit time in different populations of a species.

Stern's definitions thus supply operational measures for recognizing adaptation and methodologically determining its degree. In particular cases it may also be pos-

128

sible to establish the phenotypic features of an organism or a population on which the degree of adaptation, determined by way of reproduction, depends.

But what is the use, in the **problems of phylogenetic systematics**, of these operational definitions and methodological statements? Very little! There is a fundamental difference between comparisons at the level of individuals and populations, for which the definitions were intended, and comparisons at the level of species and supraspecific taxa, at which the methodological operations of phylogenetic systematics come into action.

Stern (1970) himself stated that it is difficult to take methodological considerations from the **intraspecific level** and then use them to determine the **adaptive nature of features** fixated in populations of **different species**. "A rough estimate of the adaptive value of these structures can be obtained by comparing members of one species to those of closely related species in which a more primitive condition of the trait is still maintained" (1970, p.45). I agree! If I know what state in the comparison of "closely related" species is the plesiomorphy, then I am able, *a posteriori* to make coarse estimates of the possible adaptive value of the apomorphous feature state. We are looking, however, for the exact converse—for an objective method of determining the alternatives of plesiomorphy and apomorphy *a priori*, by way of the adaptive significance of alternative feature states. The obstacles to doing this were clearly formulated by Stern (p.46): "Yet, unless a feature which is now fixed in the population can be proven to have arisen by a series of changes, each associated with increasing reproductive superiority over the previous, the adaptive nature of its origin and its present status remains in the realm of probability." If evolutionary species, as mutually closed reproductive communities, are being compared, then the methodological testing of a corresponding series of postulated transformations is in principle impossible, just because these evolutionary steps have already been passed through. It may be held possible to determine the degree of adaptation in sister species at the time when they begin to diverge in the speciation process (de Jong, 1980). This, however, would be a very modest field for research into phylogenetic relationships, for which, furthermore, it would be necessary to presuppose some certain way of recognizing the speciation phase of "beginning" sister species. In the case of all evolutionary species whose reproductive isolation from "closely related" species has been established by crossing experiments or by proving sympatric co-existence (p.20), these considerations permit no objective hypotheses on the adaptive condition of the particular systematic unit, nor on the adaptive value of any one of its feature states.

These statements are obviously also valid for feature analysis when comparing closed descent communities. If, in two supraspecific taxa, which together form a monophylum, particular features are expressed in alternative states, then the differences must already have been evolved in the respective stem species of these two taxa. The degree of adaptation or the adaptive value of particular feature states would have to be determined by comparing evolutionary stem species when they were still beginning. These stem species are unknown, and they would have to be compared with respect to past environmental conditions, which are likewise unknown.

This brings me to the following **result**: When comparing evolutionary species and closed descent communities, reproductive output cannot be used to discover the degree of adaptation of features or feature states.

Bock & von Wahlert (1965) have produced a different argument. They assert that the **amount of energy** necessary for successful survival in a particular environment is an **adequate measure** of the degree of adaptation of the organism. With reference to the extensive discussions of the concept in Bock (1977, 1979, 1980, 1981), I shall quote the definition of the "degree of adaptation". "The degree of adaptation, the state of being is defined as the amount of energy required by the organism to maintain successfully the synerg of the stated adaptation with a lower energy requirement indicating a better degree of adaptation" (Bock 1979, p.45; 1980, p.221).

A synerg, in this connection, is the link between an organism and its environment which is created by the selection pressure of a particular environmental factor and the biological role of a feature. The central point of the cited definition is an inverse connection between the degree of adaptation and the energy required to maintain the synerg. If the form-function complex of a feature changes in evolution, and if the change decreases the energy expenditure of an organism for its existence in a given environment, then the change has led to a higher degree of adaptation.

Although Bock & von Wahlert (1965, p.270) refer all their definitions and conclusions expressly to the organism as an individual, Peters & Gutmann (1971) unthinkingly transfer this approach to the level of species and supraspecific taxa. On the basis of the energy balance, they formulate a "**principle of economy**" which, in phylogenetic systematics, is supposedly the crucial "Lesrichtungs-Kriterium" (criterion of reading direction or polarity) for determining the alternatives of plesiomorphy and apomorphy. "The economy principle of phylogenetic systematics is based on determining the reading direction of phylogenetic series by reconstructing the process of adaptation" (translation of Peters & Gutmann, 1971, p.252).

In discussing this, I must logically begin at the level for which Bock & von Wahlert have produced their arguments. I therefore ask: Are the level of reproduction and of energy expenditure, in fact, two different things when working out the degree of adaptation of an individual?

Bock & von Wahlert (1965, p.287) do not omit to claim relative survival rate and relative numbers of offspring as measures for success in maintaining a niche (the latter being the total connection between an organism and its environment). In the last analysis, therefore, as with Stern (1970) and de Jong (1980), reproduction is the measure of adaptation even in Bock & von Wahlert's theoretical approach. And in that case, all attempts to introduce energy balance into phylogenetic systematics suffer from the defects which have already been encountered in applying the reproduction parameter at the level of species and supraspecific taxa.

Moreover, the assertion that decreasing an organism's energy requirement will improve its adaptation to particular environmental factors is without factual support, and nor is there any methodological procedure for recognizing such a decrease (Stern 1970, de Jong 1980). Bock's astounding sentence: "The use of energy as the measure of the degree of adaptation is possible, at least theoretically, for any adaptation ..." (1980, p.221) is followed by not one practical example. Rather, he makes the statement: "Certain problems exist in the use of energy requirements to measure the degree of adaptation because this measure does not work in all cases" (Bock 1979, p.44). Obviously there are some evolutionary transformations, recognized by everyone as adaptations, which take more energy, rather than less. "A simple example would be the length of muscle fibers within a muscle—longer fibers are needed if the muscle must shorten over a greater distance. Hence, if the selection force demands that the structure, for example, the tongue of a woodpecker, be moved over a longer distance, then a longer fibered muscle would be better adapted. However, a longer fibered muscle would require more energy, when it contracts and shortens" (Bock 1979, p.45). A different example seems more convincing to me in this respect: The selection pressure exerted by unstable littoral sands on their interstitial inhabitants may demand the evolution of effective anchoring mechanisms. If so, the adaptive evolution of complex, energy-expensive, "duo-gland adhesive systems" in the Plathelminthes (p.288), in the Gastrotricha and in a few annelid taxa has run contrary to the postulated decrease in energy. I agree that ... "Additional analysis is needed to develop a more comprehensive measure of the degree of adaptation" (Bock 1979, p.222; 1980 p.45).

But even if reliable methods for measuring the energy balances of different feature states did exist, such energy statistics would be pointless for recognizing the degree of adaptation of feature states, if evolutionary species and closed descent communities were being compared. Bock himself (1980) regards it as impermissible that his concept for determining the degree of adaptation should be applied to the theory of optimization (economy principle). Independent of his opinion, however, the result of my analysis is as follows: the "**economy principle**", proposed on the basis of energetics (Peters & Gutmann, 1971; Gutmann & Peters, 1973; Gutmann 1977), is **unusable** for formulating and justifying testable hypotheses on the direction of evolutionary changes in features. Cracraft (1981a) came to similar conclusions.

Having said all this, can anything be done to **use the phenomenon of adaptation in solving the problems of phylogenetic systematics**?
Something can, in fact, be done. A logically unobjectionable approach can be derived from the following considerations. The evolutionary species A and B are supposed to be the descendants of a stem species z common to them alone. After the speciation, species B entered a new environment and thus came automatically under the influence of new selective forces. In accordance with the synthetic theory of evolution, these are supposed to have led to adaptive transformations of particular features taken over from stem species z, or even to the adaptive evolution of new features. Sister species A remained, on the other hand, in the old environment under the selective forces which had already acted on stem species z. In species A there was therefore no reason to change any of the features which in species B, under new environmental conditions, underwent evolutionary transformation. In other words, the result is as follows: species B, in a new environment and under correspondingly changed selective pressure, acquired new apomorphous features or feature states. But species A, in the old environment with unchanged selective forces, retained in the corresponding features the plesiomorphous feature states of stem species z.
Where does this argument lead? It implies that judgements about evolutionary changes in individual features should be viewed in the light of the supposed change

in the feature-bearer's environment. With respect to the shared stem species, the conquest of a new environment is an apomorphy of descendent species B, while staying in the original environment is a plesiomorphy of descendent species A. If corresponding hypotheses could be concretely established from species to species and put at the beginning of the comparison, then obviously no special comparative procedure would be needed in evaluating single feature states, nor any measuring rod for their differing degrees of adaptation. Selection theory indicates the answer. For it predicts that a species entering a new environment will acquire apomorphous features under the pressure of new selective forces. By examining this species, therefore, these features can simply be read off.

How can these theoretical considerations be put into practice? When two or several species are compared, the formulation of a hypothesis about an evolutionary change in environment, or about the direction of change in a single feature, are basically one and the same process. Existence in a particular environment, like every bodily feature, is part of the feature pattern of an evolutionary species. Thus, in comparing evolutionary species, exactly the same methods apply when evaluating differences in environmental connections, as when formulating testable hypotheses about alternative features. If it is possible to interpret the **evironmental connections** of a particular species or species group as **an apomorphy** — e.g. by out-group comparison — then this hypothesis usually produces a whole series of statements about apomorphous feature states as being the results of adaptation to the the new environment.

A comparison between the adelphotaxa Crocodilia and Aves may seem trivial at this point. It is, nevertheless, an impressive example of the value of the above argument. An out-group comparison with other sub-taxa of the Sauropsida justifies the hypothesis that the stem lineage of the Crocodilia retained the original connection with the ground, while in the stem lineage of the birds a new environment was opened up by taking to the air. The evolutionary change in form and function of countless features of the birds is clearly correlated with the acquisition of flying ability, beginning with the evolution of feathers from dermal scales, the transformation of the fore limbs into the wings up to the transformation of the lungs with the evolution of parabronchi and air sacks (p.93). An abundance of evolutionary adaptations, which have originated in a new environment under the action of new selectional forces, are included among the apomorphies which make the Aves an unusually well validated monophyletic taxon.

Two other simple examples will show that adaptations associated with a change of environment may establish the monophyly of a group of animals, even when its sister group is not known.

Compared with the majority of Crustacea, the taxon Cirripedia is the only closed group of sessile crustaceans. Without considering which group of vagile crustaceans is the sister group of the Cirripedia (+ Ascothoracida), we can formulate a hypothesis by way of an out-group comparison with all other Crustacea. This hypothesis is that, in the stem lineage of the Cirripedia, a change occurred from primary free movement to sessility in the new environment of the hard substrate. Features clearly correlated with this change are, for example, the transformation of the carapace into a closed box, the transformation of the 1st. antennae into an attachment organ with adhesive glands, or the differentiation of the body appendages to form a complex filter apparatus. The validation of the monophylum Cirripedia is based on profound adaptive transformations which were governed by the selection pressures of a sessile mode of life.

Again, among the holometabolous insects, the Aphaniptera (fleas) are the only species-rich group of ectoparasites. The sister group of the fleas is unknown. Nevertheless, an out-group comparison justifies the hypothesis that the conquest of homoiothermous birds and mammals as a feeding space is an advanced mode of behaviour within the Holometabola. How should the loss of wings in the stem species of recent fleas be interpreted? Was it because they were a hindrance to locomotion in a dense covering of feathers or hair? Or was it for economy of material? For the purposes of phylogenetic systematics these questions can confidently be ignored, since, whatever the answer, the loss of superfluous wings is unmistakably correlated with the demands of a new feeding space. The same goes for the lateral compression of the flea's body or the adaptive transformation of the metathorax appendages into powerful jumping legs. Once again, a species-rich group of organisms is validated as a monophylum on the basis of a series of apomorphies which arose as evolutionary adaptations by the action of new selective forces in a new environment.

The same scheme of argument can obviously also begin from a single feature, and for the following general reasons. A particular complex of features (a form-function complex), which fulfils only one biological role in interaction with the environment, will very probably be an apomorphy as compared with the alternative which can cover a broad spectrum of different roles. So we can start with the *a priori* assumption that insect appendages such as the grabbing claws of the praying mantis (Mantodea), the digging shovels of the mole crickets (Gryllotalpidae, Ensifera) or the flattened rowing legs of water beetles (Gyrinidae, Coleoptera) are extreme apomorphies which, under the selection pressure of predatory feeding, or a subterranean or aquatic mode of life respectively, represent the end points of adaptive transformations. Similarly, in the mammalian taxon Edentata, the plantigrade walking hand of the armadillos can be seen as plesiomorphous, when compared with the digging hand of the anteaters or with the climbing-hook hand of the sloths with its huge sickle-shaped claws (p.170). The dentitions of herbivorous placentals are differentiated in various directions (rodents, lagomorphs, perissodactyls, artiodactyls, hyracoids, elephants etc.). They can all be interpreted, without hesitation, as the product of adaptive transformations which were started from a plesiomorphous heterodont dentition with the formula $\frac{3143}{3143}$, and occurred under a unidirectional selective pressure leading toward a better exploitation of plant food. Since, however, it is always a question of the interaction between an organism and its environment, it does not matter how we arrive, in the concrete individual case, at plausible explanations of a transformation that was obviously governed by selection.

The use of the possible **adaptive condition of feature states in interpreting the direction of evolutionary changes** can be summarized as follows: Suppose there is a clear correlation of the form of a feature with its function, and with the way its bearer interacts with particular circumstances of the environment. Then, in comparing alternative features, closer adjustment to particular special tasks will as a rule indicate the derived result of an adaptive change. In so far as particular feature states

133

càn be judged in this manner, they should be used in justifying hypotheses about the direction of evolutionary change, as a supplement to out-group comparison. Since, however, such judgements are themselves always empirically untestable, they can, on their own, never be the basis of objective hypotheses.

I stress, once again, that **phylogenetic systematics** is, in principle, **independent of hypotheses concerning the mechanisms of evolution**. Out-group comparison is the basic method for objective interpretations of alternative features. Without any constructional or functional explanation of changes in feature, and also without any interpretation of the adaptive state of the apomorphy, it leads to testable hypotheses about the direction of evolutionary transformations.

Finally, a striking example may demonstrate this fact. As is well known, a constitutive feature complex of the insect taxon Diptera is the combination of a pair of wings on the mesothorax with a pair of halteres on the metathorax. A simple out-group comparison, in which all other winged insects form the out-group, justifies a hypothesis which can be examined by everybody, and is always open to attempts at falsification. This hypothesis is as follows: The condition with halteres is the apomorphy, while the existence of a second pair of wings on the metathorax is the plesiomorphous alternative. In this hypothesis, in other words, a direction of evolution from wings to halteres is postulated. This result is reasonably certain. But it does not require any conception about the course of constructional or functional change from flying organs to tiny halteres, nor any interpretation of the adaptive significance of halteres or their role in the flight mechanism of the Diptera.

3. The information content of multiple series of features

General Considerations

Up to now, I have explained the elementary methodological arguments for interpreting the alternatives of plesiomorphy and apomorpy on the basis of features which, in the taxa compared, are expressed in only two states. I did this because such an approach best corresponds to the logical sequence of thought. When comparing several taxa, however, a particular feature may well appear in a whole series of different feature states.

An ordering, at first purely formal, of the different states according to the sequential degree of similarity leads to a **multiple feature series**. On the reasonable supposition that this sequence reflects the successive stages in the evolution of the feature, then the series of features becomes a transformation series or **morphocline** (Maslin 1952). If this supposition is justified, then it is still to be decided which end of the series represents the most derived state of the feature. When that has been done, the **polarity of the whole transformation series** has been established.

If transformation series are set up in this manner, the following questions must be faced: Are we not, at the outset, assuming, by a circular argument, exactly what we wish to discover by analysing and interpreting the features? Can the arrangement into a transformation series of feature states which have evolved to different extents, be undertaken other than by knowing the phylogenesis of the feature bearers, i.e. on the basis of already established hypotheses of relationship? This can be done, in fact, without preconceptions or circularity, but only by paying attention to the following strict conditions. Each **feature series**, at first arranged purely on similarities, **must step by step be broken up into pairs of features**. This produces alternatives of features once more, each with two states, and, for each of these pairs, probability judgements concerning the alternatives of plesiomorphy and apomorphy must be made, in correct sequence, by out-group comparison.

I shall begin with a schematic example, in which a particular feature 1 appears in different states in the four taxa A, B, C and D (Fig. 39a). A hypothesis that the four states form a transformation series: 1a ⟶ 1b ⟶ 1c ⟶ 1d can then only be justified under the following conditions: The sequence must be broken up into three pairs of features: 1a ⟶ 1b, 1b ⟶ 1c and 1c ⟶ 1d, and each pair of features must be examined to see whether the alphabetical sequence can be established as a transformation from plesiomorphy to apomorphy. If this is the case, then 1a becomes the most primitive state of the transformation series, and 1d the most strongly derived state, while, within the series, the evaluation of the states 1b and 1c necessarily changes in the direction of the arrow, from relatively plesiomorphous to relatively

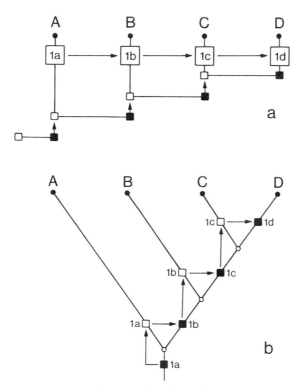

Fig. 39. The information content of multiple series of features for phylogenetic research. a) Evolutionary transformation series for the feature states 1a to 1d as developed in the four taxa A, B, C and D. The progressive change in value from plesiomorphy to apomorphy is shown for the individual feature states. b) Justification of the phylogenetic relationships of the four taxa A, B, C and D by way of the evolutionary transformation series of feature 1 (explanation in text).

apomorphous. It must further be examined, whether the feature state 1a is an apomorphy, relative to a further alternative outside the species group A + B + C + D.

Supposing that each change to a relatively more derived state was a unique evolutionary event (by the parsimony principle), then a feature series, evaluated in this way, justifies the phylogenetic relationship of the feature bearers in the following manner (Fig. 39b): The state 1a is an autapomorphy of the monophyletic taxon A + B + C + D, but at the next lower level of the hierarchy it becomes a meaningless plesiomorphy of the sub-taxon A. State 1b is an autapomorphy of the monophyletic taxon B + C + D, but at the next lower level of comparison it becomes a worthless plesiomorphy of sub-taxon B. State 1c is an autapomorphy of the monophyletic taxon C + D, but it necessarily becomes a plesiomorphy for sub-taxon C. And state 1d, at the end of the transformation series, is an autapomorphy of taxon D.

From this discussion, I wish to stress the following **law concerning the use of transformation series** in phylogenetic systematics: Apomorphous feature states are restricted to lower and lower levels of the systematic hierarchy to an extent which increases with the postulated direction of evolution. They characterize taxa that become step-by-step lower in rank as the degree of subordination increases.

Example: the proboscis of the Kalyptorhynchia (Plathelminthes)

I shall now apply these considerations to an example from the taxon Plathelminthes. Within the flatworms, a proboscis is the constitutive feature of the group Kalyptorhynchia (Fig. 40). The proboscis lies at the anterior end in a special proboscis sheath. It is extended terminally in catching prey. Among a great number of different states of this sheathed proboscis, four basic forms can be recognized[1]. I arrange them, on the basis of sequential similarity, in a purely formal sequence of feature states (Fig. 41). These feature states are characterized in the following list, which also records the number of taxa, of diverse categorial rank in which they occur among more than 200 species of Kalyptorhynchia given in Evdonin's monograph (1977):

— **Feature state 1a**

Cone-shaped proboscis ("conorhynch" Karling, 1961). Undivided bulb of muscle with retractable terminal cone—"Eukalyptorhynchia": 9 "families", 60 "genera", 120 species.

— **Feature state 1b**

Split proboscis ("schizorhynch" Karling, 1961). A bifurcated pincer apparatus consisting of two unarmed tongues of muscle—"family" Schizorhynchidae: 10 "genera", 60 species.

---

[1] The Schizorhynchia taxon *Nematorhynchus parvoacumine* Schilke is not considered in this general survey because its interpretation is uncertain (Schilke 1969, 1970).

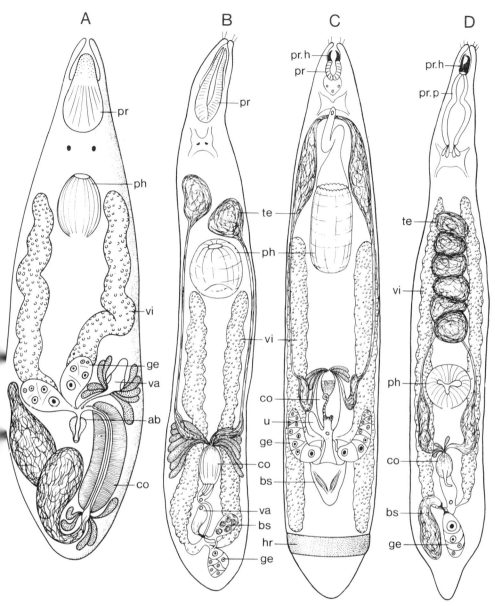

Fig. 40. Representatives of sand-dwelling Kalyptorhynchia (Plathelminthes) with a proboscis at the anterior end. A) *Cystiplex axi* Karling with cone-shaped proboscis—feature state 1a ("Eukalyptorhynchia", Cystiplanidae); B) *Prochizorhynchus gullmarensis* Karling with split proboscis—feature state 1b (Schizorhynchia, "Schizorhynchidae"); C) *Baltoplana magna* Karling with split proboscis—feature state 1c (Schizorhynchia, "Karkinorhynchidae"); D) *Diascorhynchus caligatus* Ax with split proboscis—feature state 1d (Schizorhynchia, Diascorhynchidae). Abbreviations: ab = atrial bulb, bs = bursa, co = copulatory organ, ge = germarium, hr = adhesive ring, ph = pharynx, pr = proboscis, prh = proboscis hooks, prp = proboscis pouch, te = testis, u = uterus, va = vagina, vi = yolk gland (from Ax, 1966).

137

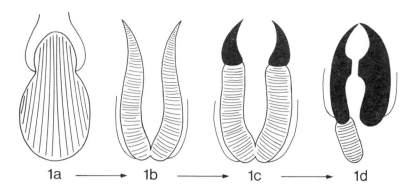

Fig. 41. The four basic forms of the proboscis in the Kalyptorhynchia (Plathelminthes), arranged in an evolutionary transformation series. Feature state 1a = cone-shaped proboscis with an undivided muscle bulb; feature state 1b = split proboscis with unarmed muscle tongues; feature state 1c = split proboscis with a pair of small symmetrical hooks; feature state 1d = split proboscis with a pair of large asymmetrical hooks.

— **Feature state 1c**

Split proboscis. Muscle tongues armed terminally with a pair of identically shaped grasping hooks which are relatively small compared with the tongues—"family" Karkinorhynchidae: 5 "genera", 19 species.

— **Feature state 1d**

Split proboscis. two large, dissimilarly shaped grasping hooks. A small muscle tongue attached to one hook only, or muscle tongues completely absent— "family" Diascorhynchidae: 2 "genera", 6 species.

In converting this multiple feature-state series into an evolutionary transformation series, I shall first discuss the principle of commonality again (p.121). In actual fact there is a remarkable correlation between the feature states of the proboscis in the sequence 1a ⟶ 1b ⟶ 1c ⟶ 1d and the step-wise decrease in the number of species from 120 to 60 to 19 to 6 in the same direction. It is certain that only part of the Kalyptorhynchia has been described up till now, but even so, the basic ratios are hardly likely to change. Assuming an identical rate of speciation in all lineages of the Kalyptorhynchia, the numerical ratios would thus give a clear-cut picture. The conorhynch of the Kalyptorhynchia would have to be the oldest and therefore most primitive state of the proboscis, while the schizorhynch with two unequal grasping hooks, as realized only in a few species of the taxon Diascorhynchidae, would represent the youngest, and thus the most derived, feature state.

Evolutionary theory, however, gives no legitimate grounds for assuming a constant rate of speciation. Apart from that, in this instance there is another explanation, even though it does not have the status of a testable hypothesis. The Eukalyptorhynchia with the cone-shaped proboscis are at home in all possible environments of the sea bed, while the Schizorhynchia with the split proboscis live almost nowhere but in the interstitial spaces of marine sands. Perhaps the rich speciation of the Eukalyptorhynchia is connected with ecological existence in a multitude of divergent

138

habitats. And perhaps the much lesser speciation of the Schizorhynchia is the result of limitation to a relatively very homogeneous habitat.

However that may be, the number of species must, in any case, be left out of account. And so I now divide the feature series of the proboscis of Kalyptorhynchia into **three pairs of features: 1a–1b, 1b–1c, and 1c–1d**. Having done so, the next task is to decide, by a three-taxon comparison for each pair of features, which alternative is plesiomorphous and which apomorphous (Fig. 42). I begin the argument at the supposedly most derived feature-state.

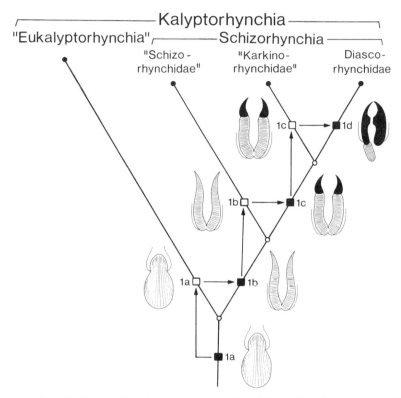

Fig. 42. A partial validation of the phylogenetic relationships of the Kalyptorhynchia by way of the evolutionary transformation of the proboscis. Explanation in text.

—Feature pair 1c–1d

The "family" Diascorhynchidae, with the proboscis of form 1d, includes only two "genera". Referring to the nearest "family", I now erect the three-taxon relationship: Karkinorhynchidae v. *Diascorhynchus* + *Diascorhynchides*. These three taxa agree in sharing the possession of grasping hooks but on the other hand there is a profound difference in the alternative feature states of: paired muscle tongues (1c) v. muscle tongue unpaired or completely lacking (1d). In an out-group comparison, paired muscle tongues without grasping hooks occur in the representatives of the "family" Schizorhynchidae. As concerns the pair of features 1c–1d, therefore, the hypothesis is justifiable that paired muscle tongues are the plesiomorphous alternative while the unpaired muscle tongue is the apomorphy.

—Feature pair 1b–1c

This brings me to the next three-taxon relationship of Schizorhynchidae v. Karkinorhynchidae + Diascorhynchidae. Here the crucial pair of alternatives is: grasping hooks absent (1b) v. grasping hooks developed (1c). An out-group comparison this time involves the whole of the Eukalyptorhynchia. The overwhelming majority of them have a muscular cone-shaped proboscis without grasping hooks [1]. This justifies the following hypothesis: in the feature pair 1b–1c, the lack of grasping hooks is the plesiomorphy and the existence of such hooks the apomorphy.

—Feature pair 1a–1b

The next task is to interpret the basic alternatives: conorhynch of the "Eukalyptorhynchia" (1a) v. schizorhynch of the Schizorhynchia (1b). Here an out-group comparison cannot be used because a corresponding proboscis does not exist outside the taxon Kalyptorhynchia (p.119). Also, because there are no fossils, the state of the proboscis in the stem lineage of the Kalyptorhynchia remains unknown.

Here, then, is a perfect example in which the arguments at present available do not allow any objective hypothesis about the direction of transformation, although an interpretation of the alternatives of the cone-shaped proboscis and the split proboscis is extremely important for discovering the phylogenetic relationships within the Kalyptorhynchia. In this situation it is legitimate to use a possible constructional and functional explanation of the change in the feature (p.126).

In a careful analysis, Karling (1961) derived the split proboscis of the Schizorhynchia from an undivided conorhynch. To mention only one detail, he regarded the radial muscle columns of the tongues of the split proboscis as resulting from the transformation of the retractor muscles of the terminal cone of the conorhynch. I shall therefore accept the plausible explanation of the evolutionary passage from the conorhynch to the schizorhynch, although this explanation is in no way a testable hypothesis. While stressing the provisional nature of this decision, I take the conorhynch (1a) as the plesiomorphous alternative, and the split proboscis (1b) as the apomorphy.

—Feature "sheathed proboscis"

The last step is to interpret the proboscis itself for the taxon Kalyptorhynchia. Here, out-group comparison can be used again, since the positive feature of the sheathed proboscis of the Kalyptorhynchia contrasts with the negative feature of the lack of such a proboscis in all other Plathelminthes. Probably nobody would seriously suppose that the stem species of the Plathelminthes had a sheathed proboscis which was lost in all Plathelminthes except the Kalyptorhynchia. On the contrary, out-group comparison allows only one justifiable conclusion: the absence of a sheathed proboscis within the plathelminths is a plesiomorphy while the existence of such a proboscis is an apomorphy of the Kalyptorhynchia.

This justifies the hypothesis that the four basic forms of the proboscis in the Kalyptorhynchia form a transformation series with the evolutionary direction 1a $\longrightarrow$ 1b

---

[1] I come back to an interesting exception in another connection (p.173).

140

$\longrightarrow$ 1c $\longrightarrow$ 1d. Assuming, on the basis of the principle of parsimony, that the changes in the series of features happened once only in evolution, then the following phylogenetic relationships between the feature bearers emerge (Fig. 42): The undivided cone-proboscis (feature state 1a), being an autapomorphy, establishes the monophyly of the taxon Kalyptorhynchia; at the next lower systematic level, it becomes a worthless plesiomorphy for the species included under the traditional name "Eukalyptorhynchia". The unarmed divided proboscis (feature state 1b) is an autapomorphy of the monophyletic taxon Schizorhynchia; at the same time it becomes a plesiomorphy for the species traditionally included in the "Schizorhynchidae". The divided proboscis with paired muscle tongues and equal grasping hooks (feature state 1c) is an autapomorphy for a monophyletic and still unnamed taxon "Karkinorhynchidae" + Diascorhynchidae; it becomes a plesiomorphy for the species united under the traditional name "Karkinorhynchidae". At the end of the transformation series, the divided proboscis with an unpaired muscle tongue and big asymmetrical hooks (feature state 1d) is an autapomorphy of the monophyletic taxon Diascorhynchidae.

At the risk of excessive repetition, I emphasis once again: On the basis of feature states of the proboscis apparatus, it is possible to establish as monophyletic taxa only the Kalyptorhynchia (by means of 1a), the Schizorhynchia (by means of 1b), the unity "Karkinorhynchidae" + Diascorhynchidae (by means of 1c) and the Diascorhynchidae (by means of 1d). On the other hand, the traditional classificatory units of the "Eukalyptorhynchia" (with 1a), the "Schizorhynchidae" (with 1b) and "Karkinorhynchidae" (with 1c), cannot be demonstrated as monophyletic taxa by way of the proboscis. This is simply because their representatives are held together, as regards this complex of features, only by plesiomorphous conditions.

So as to make the best possible use of all available arguments, I shall complete the evaluation of this feature series with a few thoughts on the constructional and functional transformation of the Kalyptorhynchia proboscis, as well as the possible adaptive values of the respective, relatively apomorphous states of the proboscis.

I have already discussed the **constructional change** from the cone-shaped proboscis to the divided proboscis. Subsequent to this change, there is, in fact, only one plausible explanation for the further transformation of the schizorhynch within the Schizorhynchia. Simple equal hooks were first added to the unarmed, divided proboscis as new elements. The later enlargement of the hooks was connected with a progressive decrease of the original muscle tongues. In the differentiation of large, asymmetrical grasping hooks, the muscle tongues were lost, at first on one side and finally also on the other. In their stead, a new system of hook-dilator muscles arose.

As to the **adaptive value** of individual feature states of the proboscis, I again begin with the alternatives of the conorhynch ("Eukalyptorhynchia") v. the divided proboscis (Schizorhynchia). The special habitat of the Schizorhynchia is the interstices between the sand grains of the sea bottom. The clear correlation of the divided proboscis with this habitat suggests that the grasping apparatus arose, under the influence of particular selective forces in the interstitial spaces, when the stem species of the Schizorhynchia invaded the sandy sea floor. This invasion may have necessitated the improvement of the organ of predation for the capture of mobile food objects in the interstitial system. According to this interpretation, the tweezer apparatus would have a higher adaptive value for the animal that carried it than the undivided cone-shaped proboscis. Pursuing this line of thought and applying it further to the Schizorhynchia, the evolution of grasping hooks, and their subsequent transformation into asymmetrical structures with new, effective dilator muscles, can be seen as a new, selectively controlled improvement in the mechanism for capture of prey. This improvement would have occurred, however, without any basic change in the habitat.

These explanations sound plausible, but they are no more than subjective judgements, burdened in several respects with uncertainty. Thus the sister group of the Schizorhynchia among the "Eukalyptorhynchians" is still unknown although these animals, with numerous sub-taxa, likewise inhabit the interstitial system of sand. If the sister group were already itself a taxon of interstitial organisms, then the evolution of the divided proboscis could not be seen as resulting from the conquest of a new habitat. Furthermore, the supposition that the divided proboscis is adaptively superior to the conorhynch remains an untestable assertion, however plausible it may seem. In any case, many sand-dwelling "Eukalyptorhynchia", complete with conorhynchs, lead a successful life in the interstitial habitat of the sandy sea bottom. Finally there is no method of determining the degree of adaptation (judged to be higher) of the divided proboscis with grasping hooks as compared with an unarmed divided proboscis.

IV.   The alternatives of synapomorphy and convergence

In deciding between the alternatives of plesiomorphy and apomorphy the out-group comparison, with an analysis of the distribution of differing feature states, is the basic, and at the same time universal, method. This procedure can be applied to the whole of living Nature and to all conceivable alternatives of features between organisms. In some cases it may be supplemented by reflecting on the constructional and functional transformation of features as well as by estimating the adaptive level of feature states.

A well grounded probability judgement about the alternatives of plesiomorphy and apomorphy is necessarily followed by the next question: Did the state hypothesized as an apomorphy, shared as it is between different taxa, arise only once, or was it evolved repeatedly? In other words, the next problem is to make a probability judgement between synapomorphy and convergence. In considering this problem more deeply, I wish to complete my earlier discussion of the matter (p.56). As already mentioned, in making the decision between synapomorphy and convergence, there is no proper empirical measuring rod. The formulation of objective hypotheses is dictated here by the principle of parsimony in scientific argument. This principle, also called Ockham's razor, starts from the plain assumption that "the simplest explanation is the best explanation", ... "an assumption, without which science would not be possible" (Ghiselin 1966b, p.214).

The shared existence of any apomorphy among several taxa can be more simply and parsimoniously explained by the hypothesis that it evolved once only in a stem lineage common to these taxa alone, than by the postulate of several, separate evolutionary origins. Nobody can reasonably deny this assertion. On the basis of the principle of parsimony, therefore, any apomorphous agreement between two species, or between supposedly monophyletic species groups, will in the first place be hypothesized as a synapomorphy[1]. I do not thereby assert that Nature must have followed the path which seems the simplest solution to our perceptual apparatus. But only Nature itself can force us to override this argument in particular points—it will do so if the pattern of distribution of apomorphous agreements causes conflicts and these can be rationally solved only by assuming convergent evolution in one or the other single case.

But phylogenetic systematics consistently applies the principle of parsimony even in such cases of "conflicting evidence"—that is to say, it considers the number and quality of those agreements which can compatibly be hypothesized as synapomorphies.

I have already expounded the numerical approach to evaluating apomorphous agreements as synapomorphies, and thus to deciding between rival hypotheses of relationship, by means of a schematic example (Fig. 15) and have illustrated it by using the mammals. The hypothesis of a sister-group relationship between the Marsupialia and the Placentalia is based on about 20 shared apomorphies which, with mutual compatibility, can be hypothesized as synapomorphies of the marsupials and placentals. This unequivocal result implies that the single agreement between the Monotremata and the Marsupialia which might at first sight seem to be a synapomorphy, must necessarily be taken as a convergence (p.85, Fig. 26).

To base a decision solely on the number of apomorphies in the different taxa under comparison would, of course, only be completely legitimate if basically identical weight could be assigned to the individual apomorphies. This is true whether it were a question of morphological structures, physiological processes or molecular patterns, and whether, also, the features were simple or complicated. "Weighting" of features is very difficult, or even impossible, to make objective and must therefore be viewed with great reserve (Eldredge & Cracraft, 1980; Patterson, 1982 in commenting on Hecht, 1976 and Hecht & Edwards 1977). Consequently, weighting differences in the quality of apomorphous features should be considered only if a probability decision between rival hypotheses of relationship cannot be made on the basis of the numerical distribution of apomorphous agreements alone. In such cases of conflict, it is justifiable to consider not only the number of apomorphies but also their quality, which again is decided by the principle of parsimony. In the first place, an agreement in a structure-poor apomorphy can more easily be explained as convergence than can the shared possession of a feature of high structural and functional differentiation. In the second place, positive apomorphous features must in principle be assigned a higher weight than negative apomorphous features. For, without regard to the degree of complexity, the repeated loss of a pre-existing form-function complex is easier to explain than the repeated convergent production of an identical end result.

Here I recall the example of the species A + B + C in which there is one apomorphous agreement in each case between A + B, A + C and B + C. This numerical equality would seem to make it impossible to decide between the three

---

[1] "Should this call for parsimony not be heeded, then nothing prevents one from postulating any phylogenetic hypothesis, whatsoever" (Szalay & Delson 1979, p.3).

rival hypotheses of relationship (p.58, Fig. 14). But if, perhaps, the apomorphous feature shared by B + C is a richly differentiated form-function complex, while the apomorphous agreements between the two other pairs of species concern very simple feature states, then it would be legitimate to see the first as a synapomorphy. If so, the other two shared features would necessarily be interpreted *a posteriori* as convergences. Numerical stalemates between rival hypotheses of relationship may be resolvable in the concrete individual case by the different weights of apomorphous agreements.

V. The relativity of the terms symplesiomorphy and synapomorphy

Plesiomorphy and apomorphy are terms used to characterize differently evolved features or feature states. Synplesiomorphy and synapomorphy are terms used to characterize agreements in the corresponding features or feature states when evolutionary species or closed descent communities are being compared (Hennig 1984).

No doubt it would be comfortable and convenient if all agreements between taxa that have evolved once only in evolution could be definitively described as symplesiomorphies or synapomorphies. If this were so, it would be possible, for example, to write the following sentences as statements which would always afterwards be true.

"If different vertebrates share three auditory ossicles in the middle ear, this agreement is a synapomorphy. If, on the other hand, species of vertebrate have one auditory ossicle in common, then they share a symplesiomorphy."

"Vertebrates with amnion and serosa are distinguished by a synapomorphous agreement in embryology. Vertebrates without amnion and serosa, by lacking embryonic membranes, share a symplesiomorphous agreement."

Unfortunately, however, this is not possible. The way a particular feature state is characterized within a particular transformation series changes from apomorphy to plesiomorphy according to the respective adjacent feature state (p.134). In the same way, whether a particular agreement in features could be characterized as a symplesiomorphy or synapomorphy varies in relation to different levels of the systematic hierarchy. This connection is logically simple, but often fundamentally misunderstood. I shall therefore discuss the feature states mentioned, in more detail, from this viewpoint.

Examples

"Alternative feature states—one or three auditory ossicles"

In the sister groups Monotremata and Theria of the taxon Mammalia, the feature state of three auditory ossicles in the middle ear occurs throughout. In the Sauropsida, the alternative feature state of one auditory ossicle in the middle ear is likewise universal.

An out-group comparison with other land vertebrates shows that the feature state with **three auditory ossicles is an apomorphy**. The next logical decision is between

144

synapomorphy and convergence and, according to the principle of parsimony, it is postulated that this feature state evolved once only in the stem lineage of the Mammalia. On the basis of this hypothesis, the characterization of the feature state and of its agreement can be stated as follows for the three consecutive levels of the systematic hierarchy (i.e. for the levels of super-, co- and sub-ordination) (Fig. 43).

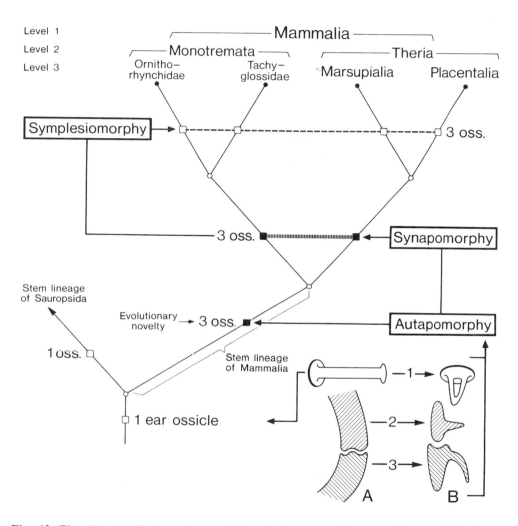

Fig. 43. The characterization of a particular feature state (three auditory ossicles in the middle ear) at three consecutive levels of the phylogenetic system of the mammals.
    Autapomorphy at level 1 = taxon Mammalia.
    Synapomorphy at level 2 = the adelphotaxa Monotremata and Theria.
    Symplesiomorphy at level 3 = adelphotaxa Ornithorhynchidae and Tachyglossidae; adelphotaxa Marsupialia and Placentalia.
Diagrams: A. Plesiomorphous condition of the Tetrapoda with one auditory ossicle (1 = columella auris) and the primary jaw articulation (2 = quadrate, 3 = articular).
B. Apomorphous condition of the Mammalia with three auditory ossicles (1 = columella auris, 2 = incus, 3 = malleus). After Wake (1979, Fig. 8.9).

- **Autapomorphy**. At the uppermost relevant level, that of superordination, the three auditory ossicles constitute an autapomorphy of the taxon Mammalia (level 1).
- **Synapomorphy**. The next level is that of coordination, at which the existence of three auditory ossicles is a synapomorphy of the sister groups Monotremata and Theria (level 2).
- **Symplesiomorphy**. At the level of subordination, the possession of three auditory ossicles becomes a symplesiomorphy, worthless for working out phylogenetic relationships (level 3). This statement holds for the comparison of the Ornithorhynchidae with the Tachyglossidae, of the Marsupialia with the Placentalia, and also, obviously, for the comparison with each other of any sub-taxa of the Monotremata or the Theria.

When it comes to working out the phylogenetic relationships between the Monotremata, Marsupialia and Placentalia, and all species groups subordinate to these, then the general feature state of the hammer, anvil and stirrup in the middle ear is no longer of any use.

This brings me to the plesiomorphous alternative of "**one auditory ossicle in the middle ear**". As concerns this feature state, and its agreements in different vertebrates, there must be corresponding characterizations, but these must logically refer to other, and indeed higher, levels of the systematic hierarchy.

- **Autapomorphy**. The middle ear with one auditory ossicle arose only once, as an evolutionary novelty, in the stem lineage of the land vertebrates (tetrapods). Under this supposition, which is dictated by the principle of parsimony, the feature state becomes an autapomorphy of the taxon Tetrapoda.
- **Synapomorphy**. If the level of the Tetrapoda, as just discussed, is taken as the level of superordination, then the condition of the middle ear with one auditory ossicle is a synapomorphy of the coordinate adelphotaxa Amphibia and Amniota placed within the Tetrapoda. This statement holds although an evolutionary transformation of the feature has taken place within the taxon Amniota.
- **Symplesiomorphy**. Only at the following level—that of subordination—where the subtaxa of Sauropsida are compared with each other, and likewise also for the sub-taxa of the Amphibia, does the feature state with one auditory ossicle become a symplesiomorphy of no value for working out phylogenetic relationships.

"Alternative feature states—existence or absence of the embryonic membranes amnion and serosa"

On the basis of the parsimony principle, I postulate a single evolutionary origin of the embryonic membranes of the amnion and serosa. This hypothesis correspondingly gives the following characterizations for the positive feature.

- **Autapomorphy**. The existence of the embryonic membranes of the amnion and serosa is an autapomorphy of the taxon Amniota.
- **Synapomorphy**. At the next lower systematic level, the embryonic membranes become a synapomorphy for the coordinate adelphotaxa of the Sauropsida and the Mammalia.

146

— **Symplesiomorphy**. Passing in the direction of subordination into the Sauropsida and into the Mammalia, then the amnion and serosa constitute a symplesiomorphy between the Lepidosauria and Archosauromorpha, between the Monotremata and the Theria and between all further, subordinate sub-taxa within these.

With the negative feature of the absence of embryonic membranes, there are logically no correspondingly different characterizations. Shared primary absence is always a symplesiomorphy, whether in comparing the Amphibia with each other or with any group of "fishes". Shared primary negative features are the only ones which can definitively be characterized as symplesiomorphous agreements for all systematic levels whatever.

Result

Leaving the primarily negative feature on one side, the connections that I have just illustrated can be summarized in the following **general propositions**. A single agreement in features between different real unities in Nature can be characterized as a synapomorphy only at a particular level, fixed by rule, in the systematic hierarchy. Merely by descending to the next lower level, the valuable synapomorphy becomes in-principle-worthless symplesiomorphy. While at the next higher level, the synapomorphous agreement becomes an autapomorphous feature state of the superordinate taxon. When once determined as an autapomorphy, the feature is of no further use for the study of relationships.

VI. The ground pattern of the closed descent community

General survey

Erecting or constituting a monophylum means uniting evolutionary species to form a systematic unity, which latter is equivalent to a closed descent community in Nature. Just as every evolutionary species is characterized by a species-specific pattern of features, so must every monophyletic species group possess a single group-specific pattern of features. This composition of features can be called the ground pattern of the closed descent community in Nature and likewise of the monophylum in the system.

Hennig spoke of the "ground plan" of a monophyletic group but I purposely choose the term "ground pattern", for the following reasons:

1) The word "plan" often implies intentionality and this has no place in the theory of evolution.
2) A pattern is defined as a relatively invariant structure in the diverseness of reality, its elements being features (Löther 1972, p.108). With this meaning, the word "pattern" is extremely well suited for characterizing the mosaic of features of real unities in Nature. "Pattern" has been used similarly by Eldredge & Cracraft

(1980, p.1) for: "aspects of the apparent orderliness of life".

This brings me to the **relationship between the pattern of features of evolutionary stem species and the ground patterns of the descent communities** which have arisen from those stem species. I refer back here to the following connection: every closed descent community has its own stem lineage in which the evolutionary novelties of its ground pattern arose (p.44,202). If this stem lineage coincided with the life span of a single evolutionary species, then logically there was only a single stem species. The ground pattern of the descent community corresponds in this case to the pattern of features of the stem species at the end of its period of existence, before it had split up into descendent species. On the other hand, it is possible that the stem lineage was formed by more than one evolutionary species. If so, the ground pattern of the closed descent community is identical with the feature pattern of the latest common stem species of the extant members of the descent community, i.e. the latest stem species shared by the highest ranking constituent adelphotaxa which have recent representatives.

This, however, has already brought me to the next point—the fact that features comprised in the ground pattern are not all equally important in the study of phylogenetic relationships. In the **composition of the ground pattern**, there are always two groups of elements:

1) Novelties evolved within the stem lineage, i.e. **autapomorphies** of the monophylum. These existed in the single stem species—or respectively in the latest common stem species of the highest ranking sub-taxa with recent representatives, but did not exist in the stem species of the adelphotaxon. Following Hennig (1969), the autapomorpies can also be referred to as the "constitutive features" of a taxon or of its ground pattern, because they are crucial when monophyletic systematic groups are being constituted (p.153).
2) **Plesiomorphies** which, being shared with other unities of the phylogenetic system, did not originate in the stem lineage of the monophylum. In any ground pattern, a whole series of primitive features has been taken over from successive stem species of increasing age. Among these primitive features, at least the agreements with the sister group should always be mentioned when the ground pattern of a closed descent community is being characterized.

I come back, once again, to the **autapomorphies of the ground pattern**. The sheathed proboscis mentioned above is the constitutive feature in the ground pattern of the taxon Kalyptorhynchia (Plathelminthes). Two pairs of wings are the crucial ground-pattern feature in constituting the Pterygota as a monophyletic taxon of insects. But I have already described four "basic forms" of the proboscis of Kalyptorhynchia and, naturally, every biologist knows the great diversity of different conditions of wing within the insects. Which of these conditions, therefore, corresponds to the ground pattern of the Kalyptorhynchia or of the Insecta? For both these ground patterns represent the feature mosaic of a particular evolutionary stem species which could have possessed a single condition only.

The answer is that **the ground pattern of each closed descent community always includes the most primitive expression of the constitutive features of that**

148

community[1]. With the taxon Kalyptorhynchia, the undivided cone-shaped proboscis must be assigned to the ground pattern as being the most primitive condition of the autapomorphy. With the taxon Pterygota, the stiff-winged condition of the may flies and dragon flies must be assigned to the ground pattern, whereas the ability to turn the wings rearward and bring them in contact over the back is already a fundamental modification of the ground-pattern feature.

Ground pattern of the Mammalia

The fully developed ground pattern of the taxon Mammalia corresponds to the mosaic of features of the latest common stem species of the adelphotaxa Monotremata and Theria. This was the evolutionary species that contained the complete series of about **60 autapomorphies** which had evolved in the stem lineage of the Mammalia as novelties. This list runs from hair and milk glands to the secondary jaw articulation with the three auditory ossicles and extends into the cellular organisation with the enucleate erythrocytes (p.87–88).

Among the **plesiomorphies**, I shall choose with care. **Young plesiomorphies** in the ground pattern of the Mammalia are all those features which the mammals share solely with the adelphotaxon Sauropsida. They include, for example, the epidermis with a thick stratum corneum, and the embryonic membranes of the amnion and serosa (p.92). These were developed as evolutionary novelties in the stem lineage of the Amniota, were present as autapomorphies in the latest common stem species of the recent Sauropsida and Mammalia, and correspondingly form synapomorphies in comparing these two taxa together. In the ground pattern of the Mammalia (and obviously also in that of the Sauropsida) they logically become plesiomorphies.

In understanding the ground pattern of the Mammalia, **extremely old plesiomorphies** are also of interest. These, such as ovipary, the coracoid of the shoulder girdle or the cloaca, were already present in the stem species of the Tetrapoda. Even though they now occur only in three recent species of Monotremata, they are completely specific ground pattern features of the taxon Mammalia. The widespread alternative conditions —whether vivipary, the absence of a coracoid or the separation of the urogenital system from the anus—first belong to the ground pattern of the descent community Theria (Marsupialia + Placentalia).

In characterizing the ground pattern of the Mammalia, it is scarcely necessary to go beyond the vertebrates. However, the embryonic notochord goes back to the ground pattern of the Chordata, bilateral symmetry to that of the Bilateria and spermatozoa to the stem species of the Metazoa. If we wished to follow all plesiomorphies back to their presumed evolutionary origins, then cellular elements like the cilia with the 9 + 2 pattern or the mitochondria would land us back at the stem species of all eucaryotes.

---

[1] Hennig (1984, p.46, 54, 63) suggested the term "archapomorphy" for an apomorphous feature of the ground plan (ground pattern). This suggestion is logical but, in my opinion, it could easily lead to a problematical overloading of the terminology of phylogenetic systematics. If we keep to the accepted term "autapomorphy" (cf. Hennig, 1980, p.20), then the intended meaning can be expressed clearly and sufficiently by speaking of the most primitive condition of the autapomorphy or the most primitive condition of the constitutive feature of the descent community.

Ground pattern of the Onychophora

All zoologists know the "living fossil" *Peripatus*. Along with a few other animals, the Onychophora have become famous by way of this doubtful label and, perhaps for this exact reason, their interesting mosaic of features has often been basically misunderstood.

According to Kaestner's "Lehrbuch der speziellen Zoologie" (1969), *Peripatus* and its relatives are animals "whose construction is somewhat intermediate between the annelids and the arthropods". They possess "important annelid features" but at the same time also show "features which are highly reminiscent of the arthropods". "Peculiarities of their construction which cannot be seen as preliminary stages of arthropod organisation" suggest, however, that they should not be regarded as, "so to speak, a stage on the evolutionary path from the annelids to the arthropods". As a result of these considerations, a special phylum is erected, and thus any attempt to solve the phylogenetic relationships of the Onychophora is totally obstructed[1].

In working out the ground pattern of the Onychophora, I intentionally reverse Kaestner's sequence of the groups of features (Fig. 44).

1) "Peculiarities" are the **autapomorphies** or constitutive features, of the ground pattern on the basis of which the Onychophora are erected as a monophylum. Obviously they do not, in any sense, speak against systematizing the Onychophora as a subordinate sub-taxon of the Arthropoda. On the contrary,

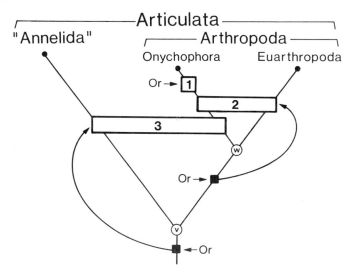

Fig. 44. Diagram of phylogenetic relationships of the Articulata to show the ground-pattern features of the Onychophora. Feature block 1 = autapomorphies of the Onychophora. Feature block 2 = synapomorphies between the Onychophora and the Euarthropoda which become plesiomorphies in the ground patterns of the Onychophora and of the Euarthropoda. Feature block 3 = symplesiomorphies with the Annelida which among Arthropoda are retained only by the Onychophora. Or = origin of evolutionary novelties.

---

[1] In the new edition of the text book (1982) the passages quoted are repeated, either verbatim or with the same sense.

it would be most remarkable if no peculiar features had arisen in the stem lineage of the Onychophora, for the common stem species w shared with the Euarthropoda split in the Precambrian. Among such features peculiar to the Onychophora are the oral hooks, the oral papillae with defensive glands, the tuft-like tracheae and the presence of numerous commissures of the nervous system in each segment (p.196).

2) "Features highly reminiscent of the arthropods" are those features in the ground pattern of the Onychophora which are shared with the sister group Euarthropoda. They were evolved in the stem lineage of the Arthropoda and were present as autapomorphies in the latest common stem species w of the Onychophora and Euarthropoda. In the ground pattern of the Onychophora (and obviously also that of the Euarthropoda) these synapomorphies of the two taxa became **plesiomorphies**. I refer the reader to the series of features mentioned below (p.196), running from the chitinous cuticle to the nephridia with sacculi.

3) "Annelid features" of the ground pattern, finally, are **symplesiomorphies with the annelids**. These were taken over into the taxon Arthropoda from the constitutive features of the stem species v of the Articulata but, among arthropods, are retained only in the Onychophora. Thus, in the gigantic army of the arthropods, they persist in only about 70 species. As examples I mention the dermal muscle layer and the segmentally arranged nephridia. Because of these parts of the ground pattern, the Onychophora can be regarded as relatively the most primitive taxon of the Arthropoda, just as the Monotremata, because of the plesiomorphies of ovipary, the coracoid and the cloaca, can be seen as relatively the most primitive taxon of Mammalia.

Relatively most primitive taxon

The "relatively most primitive part" of a particular taxon is always the unity with the greatest number of plesiomorphous ground-pattern features. On the other hand, the often-heard assertion that a species or species group is simply the plesiomorphous unit within a larger taxon is false. For the terms plesiomorphy and apomorphy were coined for features and not for feature bearers. On terminological grounds, therefore, it is not permissible to characterize any taxon whatever as plesiomorphous or apomorphous. But also, as a question of fact, it would be mistaken to consider the Onychophora as the plesiomorphous sister group of the Euarthropoda or the Monotremata as the plesiomorphous adelphotaxon of the Theria. For, in addition to a great number of plesiomorphies, both taxa possess various apomorphies, by means of which they are themselves validated as monophyla.

Ground pattern and type

Finally, I now have to confront the precisely defined term "ground pattern of the closed descent community", on the one hand, with the pretty-looking term "type", on the other.

The word "type" has entered into biology with a confusing diversity of meanings. Among these, only Remane's "systematic type" approaches the concept of the ground pattern. The systematic type was not viewed by him as an idea or as a conceptual abstraction "but as a being which once, in fact, existed as a species" (Remane 1952, p.142). The type, or bauplan, is an "expression for the basic morphological structures of a natural systematic group" (Remane, Storch & Welsch 1973, p.44).

In the modern literature of research into phylogenetic relationships, however, the term "systematic type" has almost no noticeable role. This may be because of the ineradicable historical loading of the word "type", closely connected as it is with the vague terms of ancestral form and stem form (*Urform*, *Stammform*), as well as the lack of any precise expositions or exemplifications of the feature pattern of systematic types.

Serious reasons, however, for totally excluding the term "type" from phylogenetic systematics lie in the fact that logically invalid conceptions of the type have been emphatically defended even up to the present day. According to Schindewolf (1969), the type concept is absolutely indispensable for biology, and for throwing light on the sequence of phylogenetic development. In this connection, he sets out his conception of the "type" as a biological reality, using the example of the insects, with the following formulation:

"We speak of the type or bauplan of the insects and understand by it the special complex of features which characterizes the largest known group of organisms, holds it together like a rigid, impassable frame and distinguishes it from other, comparable groups of animals. An essential characteristic of this bauplan is the division of the body into: 1) a head, with one pair of antennae and three well developed pairs of mouth appendages; 2) the thorax with three pairs of legs and mostly with two pairs of wings; and 3) an abdomen without typical extremities. This type complex has existed unchanged for 370 million years, from the moment when the insects first appeared in the history of the earth up to the present day. It is made incarnate by the reality of more than 800,000 species of insect." (Schindewolf 1969, p.15).

This description of the "insect type", in fact, contains one important ground-pattern apomorphy by which the Insecta are constituted as a monophylum within the group of the arthropods—I refer to the division of the body into the tagmata of the head, the thorax with three pairs of locomotory appendages, and the abdomen. But then, even when characterizing the insect head, the description entirely misses the point. For a pair of antennae and three pairs of mouth parts are in no way "typical" in the sense of being constitutive for the taxon Insecta. Rather they are plesiomorphies, differing greatly in age, taken over into the ground pattern of the insects from the stem lineages of other arthropod taxa. The evolution of the first pair of antennae, as present in insects, was probably completed in the common stem lineage of all arthropods, or, in any case, at latest in the stem lineage of the Euarthropoda. The loss of the next pair of head appendages occurred in the stem lineage of the Tracheata and the transformation of a pair of limb gnathobases into the mouth parts known as mandibles happened in the stem lineage of the Mandibulata. Equally misleading, on the other hand, is the reference

152

to "mostly with two pairs of wings" for the type of the insects. This statement would only be valid for the ground pattern of insects if the absence of wings within the taxon were always due to loss (and then, of course, the word "mostly" would have to be deleted). This, however, is not so. The wings are not typical, or constitutive, for the taxon Insecta—they first developed within the insect taxon Ectognatha and belong to the ground pattern of the subordinate monophylum Pterygota (p.185).

I shall avoid any further discussion as to whether, and in what fields of biology, various conceptions of the type may be of use. For whole books have been devoted to this problem (Jürgens & Vogel 1965) and another one, of corresponding size, would be needed to deal with it. In any case, in the study of phylogenetic relationships the term "type" has caused so much damage that I shall hold it at proper distance.

What is true of the word "type", understandably also holds for the word "typology". In modern discussions between rival tendencies in systematics, it has become unkindly fashionable for each side to damn the other for its old-fashioned typological mode of thinking incompatible with the theory of evolution. Attempts to impose order that are made on the basis of subjectively chosen features characterized as essential, and which, even now, consciously permit non-monophyletic groups, can certainly be referred to as typology or typological classifications. I have already explained this in the introduction. However, quarrels about words whose meanings are very differently interpreted will not bring us much farther forward in factual questions. If all attempts to impose order which are not compatible with the bases and aims of a logically consistent phylogenetic systematics are referred to neutrally as "traditional classifications", then perhaps these quarrels can be avoided ... But I leave that question on one side.

VII. Constitutive features (= autapomorphies) and diagnostic features

Traditional classifications, as is well known, contain a diagnosis for each taxon, of whatever rank. This diagnosis delimits the taxon from the supposedly "closest related" taxa on the basis of "taxonomically relevant" features. In the framework of the Linnaean classification, diagnostic features thus serve, for example, to separate one genus from other genera within a family or a particular order from other orders within a class.

However, the description and delimitation of a supraspecific taxon by means of diagnostic features in traditional classifications is a fundamentally different procedure from the establishment of a monophyletic unity in phylogenetic systematics. Monophyletic taxa, being portrayals of closed descent communities in Nature, can be erected or constituted only by means of derived features evolved once only. For this reason Hennig (1969) made a sharp terminological distinction between the diagnostic features of a taxon and its constitutive features, i.e. its evolutionary novelties or autapomorphies. I shall analyse the relationships of these two sets of features using three groups of animals.

Mammalia

Every diagnosis of the Mammalia will concentrate on those features which are unique within the vertebrates, tetrapods and amniotes. Diagnoses of mammals thus include hair, the sweat- and milk-glands, the secondary jaw articulation formed on the squamosal and the dental, the middle ear with three auditory ossicles, heterodont dentition with diphyodont tooth replacement, seven cervical vertebrae, the enucleate red blood corpuscles, etc. Homoiothermy would, no doubt, also make part of the diagnosis of the Mammalia, of course with the reservation that this feature is not unique to the mammals, for a constant warm body temperature occurs a second time in the Amniota, in the birds.

What position does phylogenetic systematics adopt to these diagnostic features of mammals? It interprets them as evolutionary novelties acquired once only in the stem lineage of the Mammalia and correspondingly requires them to have existed in the latest common stem species of the sister groups Monotremata and Theria. The diagnostic features of a traditional characterization of the mammals thus agree, point by point, with the constitutive features in the ground pattern of the Mammalia by which this taxon can be justified as equivalent to a closed descent community in Nature (p.87–88).

What then becomes of the seemingly superfluous talk of a basic difference between the diagnoses of traditional classifications and the constituting of monophyletic taxa in phylogenetic systematics by means of autapomorphies? But let us proceed!

Crustacea

From a subtle analysis of the problem of monophyly of the taxon Crustacea (Lauterbach 1983a), I shall take up the simple question: what is a crustacean? Probably every zoologist will believe that he can recognize the crustaceans with certainty and can delimit them from millipedes, insects and spiders. And indeed, in every textbook of systematic zoology there will be found an impressive series of diagnostic features, by means of which the position of the Crustacea within the arthropods seems to be determined without any doubt. In separating the Crustacea from the Chelicerata, Myriapoda and Insecta by means of a diagnosis, traditional classifications would certainly be content with the following survey.

The Crustacea are:
1) primarily aquatic arthropods; with
2) respiration adapted to this mode of life by gill-breathing through the epipodites of the limbs of the cephalothorax;
3) the existence of two pairs of antennae on the head is a unique character of the Crustacea by comparison with all other sub-groups of recent Arthropoda—indeed this striking "taxonomic feature" is emphasized in many classifications by using the group name "Diantennata";
4) except for the first antennae, the limbs of crustaceans are primarily developed as biramous appendages with exopodite and endopodite;
5) the ontogeny of Crustacea commonly passes through a nauplius stage—a larva

with few segments and only three pairs of limbs (1st. antenna, 2nd. antenna, mandible);

6) many crustaceans have a carapace—a large duplication of the epidermis with cuticle at the posterior margin of the head;
7) the nauplius-eye occurs in the head—this is a median, unpaired photosensory organ which on superficial examination seems to be a unitary structure (this is additional to paired compound eyes);
8) two pairs of segmental nephridia occur as the excretory organs specific to the group—these are the antennal glands in the segment of the second antenna and the maxillary glands in the segment of the second maxilla.

Although not all the named features are unique to the Crustacea (*Limulus* has gills and there are various aquatic mites and insects) and although in different groups of Crustacea this or that feature may be lacking (as with the biramous limbs, nauplius larva, nauplius eye, carapace and nephridia), this summary provides in every case an unmistakable combination of diagnostic features. By means of the complete or at least partial realization of these features, any crustacean could doubtless be established as a representative of the Crustacea.

Why then do I not accept these diagnostic features, as with the mammals, and simply adopt them in constituting a taxon Crustacea in the phylogenetic system of the Arthropoda? Phylogenetic systematics cannot do this because, on closer examination, most of the diagnostic features are of no use at all in establishing the Crustacea as a monophylum within the arthropods.

Thus points 1 to 5 describe plesiomorphous ground-pattern features of the crustaceans. These were first evolved not in the stem lineage of the Crustacea but in much earlier phases of the phylogenesis of the arthropods. An aquatic mode of life in the sea is an extremely ancient feature of the stem species of the Mandibulata (Crustacea + Tracheata), of the Euarthropoda (Chelicerata + Mandibulata), of the Arthropoda and, beyond even these, of the stem species of the Articulata. Epipodial gills and a pair of premandibular appendages (= 2nd antennae) already belong in the ground pattern of the Mandibulata. In the Tracheata, which are the adelphotaxon of the Crustacea, the premandibular head segment occurs along with its ganglia and even with appendages for a short time in ontogeny. The crustaceans acquired the second antennae as a plesiomorphous feature which was lost in the stem lineage of the Tracheata at the transition to a terrestial mode of life. As shown by the extinct Trilobites, from the stem lineage of the Chelicerata (p.212), biramous limbs existed already in the latest common stem species of the Chelicerata and the Mandibulata. Finally, according to Lauterbach, even the nauplius, as a larva with few segments, must be assigned to the ground pattern of the Mandibulata.

On the other hand, after the discovery of the Cephalocarida, the carapace can no longer count as an apomorphous ground-pattern feature of the crustaceans—this structure obviously arose later, within the crustaceans (Lauterbach 1974).

In this situation, what remains to suggest that the Crustacea of traditional classifications are a valid taxon of the phylogenetic system? In this connection, I shall now focus more sharply on diagnostic features 7 and 8, i.e. the nauplius eye and the nephridial organs.

A nauplius eye made up of four ocelli is once again a plesiomorphy which was taken over into the stem lineage of the Crustacea from the ground pattern of the Euarthropoda. Within the Chelicerata, the Pantopoda have an equivalent number of four median ocelli and the same is true of the Collembola within the Insecta (Paulus 1979). Unique to the Crustacea, however, is the joining-together of the four ocelli to form an almost unitary structure. In all other Euarthropoda, in so far as ocelli are present, they are always far apart from each other. By means of an out-group comparison, therefore, the following hypothesis can be justified: the moving together of four primarily separate median ocelli has led, in the stem lineage of the Crustacea, to the evolution of the nauplius eye. According to Lauterbach (1983a), this particular phenomenon represents an autapomorphy of the Crustacea.

The segmental organs (nephridia) are of comparable interest for this problem. Nephridia in the posterior four head segments and the first two trunk segments can be assigned to the latest common stem species of the Euarthropoda. This ground pattern of the nephridia has then been altered in different directions in particular stem lineages of the Euarthropoda. In the stem lineage of the Crustacea, the reductive process limited the nephridia to the segments of the second antenna and second maxilla. There is no parallel to this in other taxa of the Euarthropoda. The peculiar distribution of the nephridia can be regarded as one autapomorphy in the ground pattern of the Crustacea (Lauterbach 1983a)—whether or not the antennal glands, the maxillary glands, or even both pairs of nephridia, were later lost in particular sub-taxa of the Crustacea.

Consequently, among the eight chosen diagnostic features of the Crustacea there are only two relatively inconspicuous characteristics which allow the Crustacea to be constituted as a monophyletic taxon. That is very little. It remains the task of phylogenetic systematics to search steadfastly for further possible autapomorphies, so that the basis for the monophyly of the Crustacea, at present so weakly grounded, shall be improved.

Plathelminthes

Every zoologist knows the division of the Plathelminthes into the three "classes" of the Turbellaria, the Trematoda and the Cestoda. And in every diagnosis there are two essential features by which the turbellarians are distinguished from the flukes and tapeworms. Turbellarians are free-living aquatic plathelminths with a ciliated epidermis. Trematodes and cestodes, on the other hand, are parasites—their epidermis is fundamentally free of cilia in the adult stages.

What judgement is given by phylogenetic systematics on the two essential diagnostic features of the turbellarians? The features "free-living" and "ciliated epidermis" are plesiomorphies which simply must have existed in the ground pattern of the common stem species of all plathelminths. As such, however, they cannot logically be used to justify the taxon "Turbellaria" as a monophylum within the Plathelminthes. Moreover both features are not even constitutive for the taxon Plathelminthes.

156

Rather were they taken over from the ground pattern of the Bilateria, and of all Metazoa in general, into the stem lineage of the Plathelminthes.

These diagnostic features, being ancient plesiomorphies, are worthless and, equally, the search for a constitutive feature for the traditional group "Turbellaria" is vain. There is not even a single autapomorphy for the turbellarians. Being a non-monophyletic grouping of species, they must disappear from the phylogenetic system of the Plathelminthes (p.281).

Result

These examples of the relationship between diagnostic and constitutive features give the following general insights. Diagnostic features may be useful in practice, when it is a question of formulating identification keys and of recognizing individual taxa quickly, by a particular combination of features, as representatives of more inclusive systematic units. In so far as the diagnostic features of traditional classifications are autapomorphies, they coincide with the constitutive ground-pattern features of monophyletic species groups. Diagnostic features of a taxon which represent plesiomorphies, however, are worthless for phylogenetic systematics. When the diagnoses of traditional classificatory units are based solely on plesiomorphies, and no constitutive features can be shown to exist, then the corresponding species groups must be rejected from the phylogenetic system as artefacts without any equivalent in Nature.

H. The Relationship of Evolutionary Agreements to the Concepts of "Homology" and "Non-homology"

The definitions of the words "symplesiomorphy", "synapomorphy" and "convergence" supply a terminological apparatus which covers, completely and without valid objections, all the forms of agreement which are possible in principle between different evolutionary species. The meanings of the words "homology", and of its supposed partner "analogy", on the other hand, are insufficient for the purposes of phylogenetic systematics. I shall justify these statements by a precise exposition of the relationship between the three possible forms of evolutionary agreement, on the one hand, and the concepts of homology and non-homology, on the other.

The word "**homology**" was coined in the middle of the previous century by Owen and, together with the word "**analogy**", was defined for the first time in his "Lectures on the Comparative Anatomy and Physiology of the Invertebrate Animals" (1843). [1] Owen's definitions were as follows:

"Homologue. The same organ in different animals under every variety of form and function."

"Analogue. A part or organ in an animal which has the same function as another part or organ in a different animal."

Thus, according to Owen, one and the same organ is homologous in different animals under all possible transformations of its form and function. On the other hand, totally different organs or parts of organisms are analogous if they have the same function in different animals. In this confrontation of "morphological equivalence" against "functional comparability", homology and analogy are still generally regarded as antonyms (Remane 1952; Voigt 1973).

A causal interpretation of the "equivalence" of homologous organs results from the theory of evolution. Homologous organs, or to speak more generally, homologous features, exist between different organisms because they were already present in their common stem species and were taken over from this species. The consequences were obvious. Homologies in the mosaic of features of different species were considered to be the crucial elements in working out the phylogenetic relationships of their bearers. But joy over this insight has not, of course, been totally unclouded. Even today, the use of homologies for the study of phylogenetic relationships has been attacked by means of the following argument: homologies between different organisms are supposed to indicate phylogenetic relationship; but the precondition for the establishment of homology is knowledge of the phylogeny of those very same organisms. Thus "Phylogenetic homologies can only be stated under an accepted phylogenetic hypothesis, i.e. deductively" (Rieppel 1980b).

If the "recognition" of homologies were, in fact, only possible in the context just

---

[1] Compare Boyden (1973, p.80)

postulated, then indeed we would be trapped in a completely circular argument. I shall consider this problem by strictly separating the theoretical definitions of the words homology and non-homology, on the one hand, from the possible ways of recognizing their meaning in practice, on the other.

I. Theoretical definitions

"Homologous features are features in two or more evolutionary species which go back to one and the same feature of a common stem species. They may have been taken over from the stem species unchanged or else with evolutionary transformation".

In the great diversity of different definitions of the word "**homology**" (e.g. Bock 1973; Osche 1973, 1975, 1982; Riedl 1975, 1980; Holmes 1980; Patterson 1982a; Mayr 1982c) this **definition** is, in fact and logic, the widest possible formulation within the framework of evolutionary theory[1]. The meaning corresponds to that definition which, according to Hennig (1984, p.37), is the only one valid for phylogenetic systematics. At the same time, it is with this meaning that the word "homology" is most widely used in biology.

I shall apply the word in this meaning to a simple schematic example (Fig. 45). Evolutionary species A and B are supposed to possess a particular feature 1 which was present in their common stem species z. Species A and B received feature 1 from this stem species; it was either transferred to them unchanged, was altered in the lineage of one descendent species, or even altered in the lineage of both descendants.

The theoretical definition of the word "homology" makes no statement about the similarities or agreements between the homologous features of different species (Bock 1963, 1969, 1973), and therefore no form of similarity can possibly form its logical alternative.

The **antonym** of "**homology**" is not "analogy" but simply the word "**non-homology**" with the following theoretical definition:

"Non-homologous features in two or more evolutionary species are features which were not present in the common stem species; they were evolved independent of each other".

In the schematic example, species A and B each have a further comparable feature. This feature 2 was not present in the stem species z. It was developed independently in the lineages of the species A and B. For the definition of non-homology,

---

[1] Genetic inheritance, as the transmission of genetic information, and tradition, as the transmission of learned (acquired) information are two fundamentally different mechanisms for the transfer of characteristics or features between organisms. The strict definition of the word "homology" to mean the common possession of features which are transferred to the descendent species by way of the genome, results in a corresponding strict rejection of the term "traditional homology" (Wickler 1965, 1967; Meissner 1976; Immelmann 1979). The unqualified widening of the term "homology" to include agreements in learned modes of behaviour has no scientific value. In the study of phylogenetic relationships it can only lead to confusion.

The transmission of genetic information in the process of reproduction is a phenomenon unique to living organisms. Correspondingly, there can fundamentally be no definition of the word "homology" which is equally binding for biology and for other sciences. In the comparison of organisms, therefore, it is logically necessary to reject Voigt's (1973) attempted general definition, which was as follows: "Homology is the structural identification of systems or parts of systems which can be made on historical grounds".

159

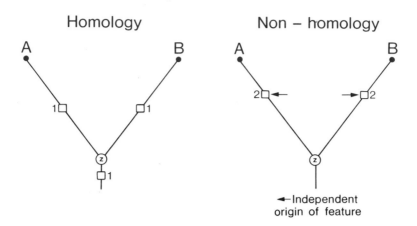

Fig. 45. The meaning of the terms "homology" and "non-homology". The species A and B have a homologous feature 1 which was taken over from the common stem species z. They also have a non-homologous feature 2 which was not present in the common stem species, but was only evolved, independently, in the separate lineages of A and B.

it is fundamentally irrelevant under what aspect independently evolved features are compared or to what extent similarities, or even agreements, exist between non-homologous organs.

II. The probability judgement between the alternatives of homology and non-homology

This brings me to the possible ways of recognizing the meaning just defined. Is it possible to formulate objective hypotheses on the homology or non-homology of particular features in different evolutionary species without knowledge of the phylogenesis of the feature-bearers?

A. Remane (1952, 1955) set up several "criteria of homology". These have since been widely accepted and extensively treated in text books, though they have also sometimes been in part rejected. The three "principal criteria" of Remane are as follows:

1) Criterion of position. Homology is revealed when a feature has the same position in comparable feature-complexes.
2) Criterion of special quality. Similar structures can be homologized, even without reference to similar position, when they agree in numerous particular features. Certainty increases with the degree of complication and the degree of agreement of the structures compared.
3) Criterion of constancy or continuity. Even dissimilar structures, and structures of different position, can be homologized if transitional forms between them can be found, such that conditions 1 or 2 (or both) are fulfilled when two adjacent forms are considered. The intermediate forms can be taken from the ontogenesis of the structures or may be true systematic intermediate forms.

160

From this catalogue, I first derive the following basic insight: In formulating independent hypotheses about the existence of homologous features in different organisms, there is only one source of data (well known though it be) — this is similarity or agreement between features in the taxa compared.

Of course, I reject the word "criterion" in the present connection, just as when deciding between the alternatives of plesiomorphy and apomorphy. In recognizing homologies, the "criteria of homology" are neither decisive identifiers nor infallible touchstones. They merely suggest how to search in a directed way for the similarities and agreements which will be the basis for suspecting homologous connections between features. Moreover, the proposal to bring "true systematic intermediate forms" into the comparison, as under point 3, must be deleted. Whether or not, in the concrete individual case, a recent or fossil organism is drawn into the comparison, the "systematic intermediate form" must always be recognized as a representative of a particular monophyletic taxon. In that case, however, the attempt to use it in recognizing homologies inevitably leads into the circular argument already described (Bock 1969; Dohle 1976). For the intermediate condition in a particular feature, as possessed by a "systematic intermediate form", is supposed to allow us to homologize strongly divergent states in other species of the taxon. But a precondition is that the "intermediate form" be systematized as belonging to just this group of species or, in other words, that its phylogenetic relationship be recognized *a priori*.

I shall illustrate the situation by using an example cited by Remane himself. This is the homologization of the endostyle of the Acrania, on the one hand, with the thyroid gland of the vertebrates on the other, completely different though this is in construction and function. The homologization is made by way of the Cyclostomata, labelled as an "intermediate stage". But in this connection I must first make clear that in a phylogenetic system there are no "systematic intermediate stages" nor "intermediate forms". The closest possible phylogenetic relationship between two taxa is the adelphotaxon relationship. In the present example this relationship takes the following form: the Cyclostomata[1] are the adelphotaxon of all the other vertebrates and these are united under the name of Gnathostomata. At the next higher level of the system the Acrania probably constitute the sister group of the Vertebrata (Fig. 46). Obviously this phylogenetic systematization is totally independent of the feature complex of the endostyle and thyroid. Rather, and conversely, in dependence on the previously assumed hypothesis of relationship, the structure of the pouch-shaped endostyle of the ammocoete larva of the Cyclostomata is brought in *a posteriori* for the formulation of a hypothesis of homology. On the basis of our insights into the phylogenetic relationships between the Acrania, Cyclostomata and Gnathostomata, the homologization of the endostyle of amphioxus with the thyroid of gnathostome vertebrates is made *a posteriori* by way of the intermediate condition of the feature complex in the lamprey. Note that this is a logically unassailable and factually completely legitimate process of gaining knowledge by way of deduction. It has nothing whatever to do with the problem of recognizing homologies as a precondition for erecting hypotheses of phylogenetic relationships.

---

[1] Convincing evidence for the monophyly of the Cyclostomata was recently given by Yalden (1985).

I shall therefore stipulate more exactly the **empirical basis for the formulation** of such **independent hypotheses of homology**. If, in the mosaic of features in different organisms, there are great similarities or agreements in the spatial and temporal structure of particular features, then homologous relationships must *a priori* be assumed for them. The term "spatial structure" subsumes the spatial position in the organism's pattern of features and the macroscopic lay-out of features as well as their light-microscopic and ultra-structural details and molecular pattern. The term "temporal structure" denotes the ontogenetic development of features and all the further phases of differentiation in the life cycle of the individual. Exactly corresponding considerations are logically valid in judging identical physiological processes or accordant elements of behaviour in different species.

Basically, the empirical situation does not allow more than the formulation of suppositions or hypotheses about the homology of features. In deciding between the alternatives of homology and non-homology, there is no special measuring rod for use in testing, such as, for example, the out-group comparison provides in judging between plesiomorphy and apomorphy. This, of course, does not at all mean that homologies can only be "established" deductively on the basis of previously accepted hypotheses of relationship. Rather, the logic of the probability judgement between homology and non-homology corresponds exactly to the procedure for deciding between synapomorphy and convergence (p.56,142). Given a very similar or identical spatial and/or temporal structure, the parsimony principle demands that, feature by feature, we assert the hypothesis of a homologous connection so long as this does not conflict with the distribution of features in the organisms under comparison. In the concrete individual case, only imcompatibility can force us to adopt a hypothesis of non-homology for similar or accordant features.

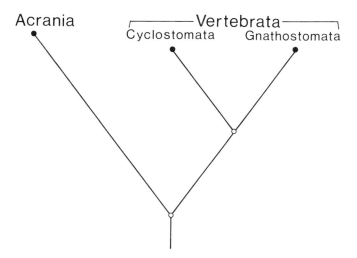

Fig. 46. The phylogenetic relationships of the Acrania. Cyclostomata and Gnathostomata.

Let me illustrate this line of reasoning by means of the remarkable distribution of the **arthropod "Malpighian tubules"**.

Within the Arthropoda, the terrestrial Arachnida and Tracheata possess comparable organs of excretion, equally called "Malpighian tubules" in both taxa. These structures, situated at the limit between the mid-gut and the end-gut, are so similar in function and spatial structure that the parsimony principle at first demands a hypothesis of homology. This is true even though differences in the temporal structure emerge, for the tubules are endodermal in origin in the Arachnida and ectodermal in the Tracheata. Corresponding excretory organs are lacking in the aquatic Pantopoda, Xiphosura and Crustacea. If we do not assume that they have been secondarily lost in these groups, then on the basis of a hypothesis of homology we must demand a common stem species, equipped with Malpighian tubules, for the Arachnida and the Tracheata. Such a hypothesis, however, conflicts with the distribution of features which validates the phylogenetic system in which the Xiphosura and Arachnida, on the one hand, and the Crustacea and Tracheata, on the other, form adelphotaxa (p.197). The fact that the homology hypothesis is incompatible with this systematization forces us, *a posteriori* to the following conclusion (Fig. 47): the aquatic stem species of the Euarthropoda (x), of the Chelicerata (y) and of the Mandibulata (z) had no Malpighian tubules. New excretory organs in the form of evaginations at the junction of the mid- and end-gut arose twice, independent of each other, in the stem

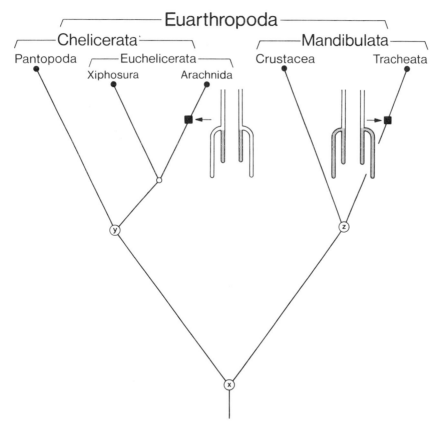

Fig. 47. Diagram of phylogenetic relationship of the Euarthropoda to show the independent evolution of non-homologous Malpighian tubules in the stem lineages of the Arachnida and the Tracheata.

lineages of the Arachnida and of the Tracheata—obviously in connection with the conquest of terrestrial environments. In other words, in comparing these two taxa the existence of Malpighian tubules is a non-homology.

As already said, the probability decision between the alternatives of synapomorphy and convergence corresponds to that between homology and non-homology. Nevertheless, there is a **fundamental difference between synapomorphy and homology**. For the term "synapomorphy" refers to the possession of apomorphous features or feature states, which the compared adelphotaxa inherited only from the latest stem species common to them alone. The term "homology", on the other hand, extends to cover the shared possession of apomorphous and of plesiomorphous features, and the latter may go back to indefinitely remote stem species. As I show in the following survey, hypotheses of homology are valueless for the study of phylogenetic relationships if the alternatives of plesiomorphy and apomorphy have not been evaluated.

III. The possible forms of homologous relationship between features

Given two evolutionary species A and B which arose from a common stem species, there are four possibilities for the expression of homologous features (Fig. 48).

1) After the splitting of stem species z, feature 1 passed unchanged to the descendants A and B.

Homologous relationships of features

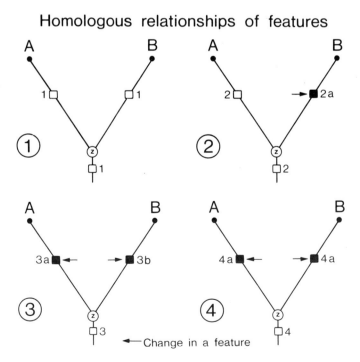

Fig. 48. The possible forms for the expression of homologous features. Explanation in text.

2) Feature 2 was preserved unchanged in species A but in the lineage of species B, on the other hand, was changed to a new condition 2a.
3) Feature 3 was altered in the lineages both of descendent species A and B and was transformed to two mutually different states 3a and 3b.
4) Feature 4 was likewise altered in both lineages but this time changed, both in species A and B, to a similar or identical condition 4a.

I shall illustrate these four possible forms of homologous relationship between features by means of examples and show in each case the relations to synapomorphy and symplesiomorphy. Instead of A and B, I shall use actual evolutionary species or monophyletic species groups, the latter in each case being represented by the latest common stem species of their highest ranking subtaxa with Recent representatives.

1. Homology with agreement

The grasping hand of the primates (Fig. 49)

The taxon *Pan* includes two species, the chimpanzee *Pan troglodytes* and the dwarf chimpanzee *Pan paniscus*. Both have a pentadactyl grasping hand with an opposable thumb. Nobody would doubt the assertion that their common stem species had a corresponding grasping hand which has been transferred unchanged to the two recent descendent species. Although the agreement is complete, this homologous feature at the lowest level of the systematic hierarchy tells us nothing. The identical grasping hand is totally worthless in validating the sister-group relationship between *Pan troglodytes* and *Pan paniscus*. Why? Because a grasping hand with opposition between the rotatable thumb and the four fingers does not occur in the two chimpanzee species alone, but also in all other Hominoidea (*Homo, Gorilla, Pongo, Hylobates, Symphalangus*) as well as in the Old World Cercopithecoidea.

Only if we push the comparison farther, beyond the Catarrhini (= Hominoidea + Cercopithecoidea), do we meet with an alternative. In the South American Platyrrhini and in the paraphyletic group of the "Prosimiae" the thumb is not fully, but only partially, opposable—it can be abducted but always moves in the plane of the other phalanges. The grasping hand of the primates thus occurs in two basic expressions: 1) the condition of partial opposability in which the thumb is not rotatable and 2) that of full opposability with a rotatable thumb (Napier 1962; Starck 1974).

It is now necessary to decide between plesiomorphy and apomorphy, but this cannot be done by out-group comparison since a grasping hand is absent outside the taxon Primates. Progress is possible by way of the following considerations. Full opposability of the thumb is attained by the evolution of a saddle joint between the metacarpal of the thumb and the trapezium bone of the carpus. This saddle joint was certainly not yet present in the common stem species of all primates but first arose within the primate descent community by modifying a grasping hand with an abductable but not rotatable thumb. According to the parsimony principle, the apomorphous condition of the grasping hand with the thumb fully opposable can be interpreted as a synapomorphy of the Cercopithecoidea and Hominoidea.

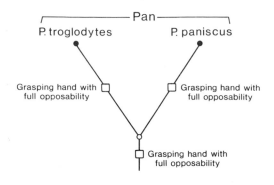

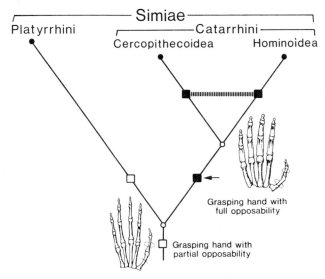

Fig. 49. The phylogenetic significance of the primate grasping hand with full opposability. Above: as a symplesiomorphy worthless for validating the adelphotaxon relationship between the chimpanzee species *Pan paniscus* and *Pan troglodytes*. Below: as a synapomorphous agreement between the Cercopithecoidea and the Hominoidea, helping to validate the monophyly of a taxon Catarrhini to include the Old World monkeys, the great apes and Man. Figures of the hand after Napier (1962).

This implies, in other words, its unique evolution in the stem lineage of the taxon Catarrhini. And this train of thought fixes the exact level at which the homologous relationship of the feature, from which we began in comparing the two species of chimpanzee, is of value in phylogenetic systematization. The grasping hand with full opposability is among the autapomorphies, or constitutive features, by which the Catarrhini are established as a monophyletic taxon within the Primates.

The ciliated creeping sole of free-living plathelminths (Fig. 50)

In the taxon *Philosyrtis* (Seriata), *P. sanjuanensis* from the American Pacific Coast and *P. santacruzensis* from the Galapagos Islands are sister species (Ax & Ax 1974; Ax 1977). In both species the ciliation of the body is restricted to the ventral surface.

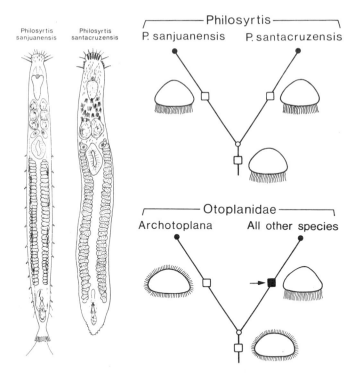

Fig. 50. The significance of the ciliated creeping sole for working out phylogenetic relationships in the plathelminths. Above: symplesiomorphy between the species *Philosyrtis sanjuanensis* and *P. santacruzensis*, worthless for establishing their sister-group relationship. Below: autapomorphy of an unnamed sub-taxon of the "family" Otoplanidae to which all the species belong, except for representatives of the taxon *Archotoplana*. The ciliation is shown by transverse sections of the body.

This condition is known as the ciliated creeping sole. Without doubt this is a homologous feature which has been taken over unchanged from a shared stem species. But as in the case of the identical grasping hand, the homologous creeping sole is completely worthless for validating the sister-group relationship between *P. sanjuanensis* and *P. santacruzensis*. For a corresponding ciliated creeping sole is found in all other species of *Philosyrtis* and almost all other representatives of the proseriate taxon Otoplanidae with more than one hundred species. Only in the species included in the taxon *Archotoplana* is the whole body ciliated. Only at this point does the state of ciliation become relevant for the study of phylogenetic relationships, for only here are there alternative states—as between the ventral ciliated creeping sole, on the one hand, and holotrichous ciliation on the other. What is plesiomorphous, in this case, and what apomorphous? The question can be decided by an out-group comparison. All other "families" of Proseriata, and the overwhelming majority of all free-living plathelminths have a completely ciliated body. The ventral creeping sole in the taxon Otoplanidae can therefore only represent the apomorphous alternative. In accordance with the principle of parsimony, I postulate that this feature evolved once only within the Otoplanidae. In this way an effective locomotory organ was produced, by loss of the dorsal and lateral cilia and lengthening of the ventral cilia, for movement in the interstices of marine sands. The autapomorphy of the ventral creeping sole validates the monophyly of a large, but still unnamed, sub-group of the Otoplanidae which excludes only the species of the taxon *Archotoplana*. The plesiomorphous alternative of complete ciliation has no value whatever for establishing phylogenetic relationships within the free-living plathelminths.

Summing up

Even when they show complete agreement between two or more evolutionary species, homologous features do not in themselves give the slightest insight into phylogenetic relationships. To ascertain the level of hierarchy at which such features can be applied for validating monophyletic taxa, the following procedure must be followed: 1) the basis of comparison must be widened to more and more taxa until some alternative feature is found; 2) a probability judgement must be made between the alternatives of plesiomorphy and apomorphy; and 3) the agreement in the apomorphous feature state must be interpretable, without contradictions, as a synapomorphy.

2. Homology with alteration on one side only

In discussing the homologous relationships of features in the antennae and mouth parts of the highest ranking taxa of insects, I refer back to previously mentioned examples and begin immediately with the relevant alternative features (Fig. 51).

Segmented antenna v. annulated antenna

The segmented antenna of the Entognatha consists of several uniform segments. The antennal musculature extends to the last segment but one. The annulated antenna is divided into a proximal scape, a pedicel with Johnston's sense organ, and a distal slender many-jointed flagellum. The antennal musculature supplies only the proximal segment (the scape).

Ectognathy v. entognathy

In the Ectognatha, the mandibles, the first maxillae and the labium (second maxillae) are inserted free on the head capsule. In the Entognatha a head pouch is developed which involves the labium and in which the mandibles and maxillae are deeply inserted.

On the basis of extensive similarities in the spatial structure, temporal structure and function, I uphold the hypothesis that the antennae and mouth parts were inherited from the common stem species of insects and passed into the stem lineages of the Entognatha and Ectognatha. But since the stem species can only have had one particular condition in each case, the homologous features must have suffered

168

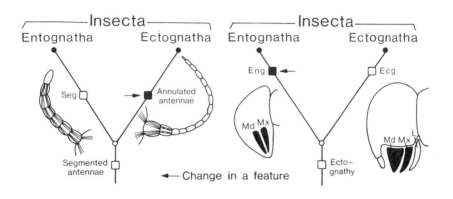

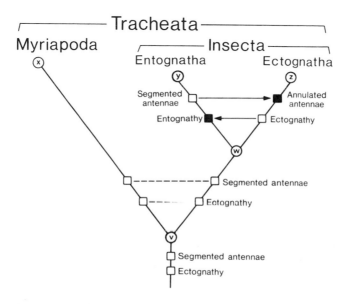

Fig. 51. The significance of differing states of the homologous antennae and mouth parts for the phylogenetic systematization of the Insecta. Above: evolution of the apomorphous annulated antenna in the stem lineage of the Ectognatha and evolution of apomorphous entognathy in the stem lineage of the Entognatha. Below: diagram of phylogenetic relationships for the Tracheata with indications of two plesiomorphous alternatives (the segmented antenna and ectognathy) for the latest common stem species of the recent Tracheata (v), of the recent Myriapoda (x) and of the recent Insecta (w). L = labium; Md = mandible; Mx = 1st maxilla.

evolutionary changes, respectively, in at least one of the two stem lineages. In outgroup comparison, segmented antennae and ectognathous mouth parts are found in the Myriapoda and, beyond these, in the Crustacea. These feature states must therefore have existed in the stem species v of the Tracheata and in the stem species w of the Insecta. Within the Insecta, therefore, they must be interpreted as plesiomorphies. Annulated antennae and entognathous mouth parts represent the apomorphous alternatives of the homologous features.

169

Result

These statements give the result of the analysis. Mandibles and first maxillae sank into the head capsule in the stem lineage of the Entognatha, i.e. the autapomorphy of entognathy came into existence at latest in stem species y and it validates the mono-phyly of the taxon Entognatha. But the segmented antenna was transformed into the annulated antenna in the stem lineage of the Ectognatha, i.e. the autapomorphy of the annulated antenna came into existence at latest in stem species z and it validates the monophyly of the taxon Ectognatha. On the other hand, the plesiomorphous al-ternative states of the homologous organs are totally worthless for the phylogenetic systematization of the Insecta, Myriapoda, Tracheata and Mandibulata.

3. Homology with divergent alteration in two lineages

In the example of the differing states of the homologous antennae and mouth parts, out-group comparison showed, in both cases, that only one of the two conditions is an apomorphy while the other state represents the unaltered plesiomorphy as it existed in the stem species of all insects. This corresponds to a widespread situation, but I now pass to the phenomenon already mentioned where both observed alternatives are apomorphies (p.115). Once again, a particular homologous feature is supposed to occur in two differing states. On out-group comparison, however, neither of these two states is found. Instead there is a third condition, different again. In such case there is a suspicion that both of the states observed in the two taxa represent evolutionary transformations compared with the plesiomorphous condition of their latest common stem species.

The climbing-hook hand of the sloths v. the digging hand of the anteaters (Fig. 52)

In the mammalian taxon Edentata (Xenarthra), the Bradypodidae (sloths) and the Myrmecophagidae (anteaters) are united, under the name Pilosa, as presumed sister groups. Opposite this group, with the same rank, are the Dasypodidae (armadillos) under the name Cingulata.

The hands of the sloths are grasping and anchoring devices. They have at most three fingers with huge sickle-shaped claws. The hand of the anteaters is a hitting organ with pickaxe-like claws for opening the nests of termites and ants. In walking, the primitively pentadactyl hand is used with the outer edge downwards.

Would the common stem species z, of the Bradypodidae and the Myrmecophagi-dae, have possessed one of these two states? In the adelphotaxon Cingulata the armadillos have predominantly a plantigrade walking hand. This must be seen as a plesiomorphy, not only compared with the climbing-hook hand of the sloths, but also compared with the digging hand of the anteaters, since the plantigrade condition is widespread within the Mammalia. This probability decision suggests the following interpretation: the plantigrade hand of the stem species x of the Edentata passed, on the one side, into the stem lineage of the Cingulata but, on the other side, was also present in the latest common stem species z of the Pilosa. From this plesiomorphous

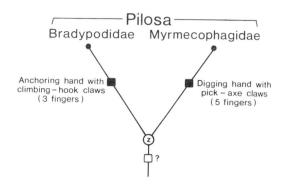

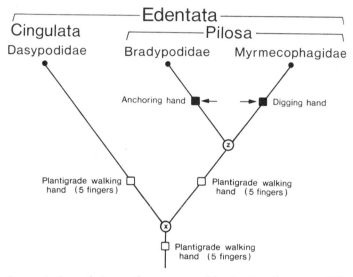

Fig. 52. Above: the evolution of alternative apomorphies by the divergent differentiation of homologous fore limbs to give the climbing-hook hand of the sloths and the digging hand of the anteaters. Below: justification of the argument by way of a diagram of the phylogenetic relationships of the Edentata.

feature state, the divergent differentiations of the climbing-hook hand and digging hand were evolved only in the stem lineages of the sloths and of the anteaters. The climbing-hook hand validates as an autapomorphy the monophyly of the Bradypodidae while, at the same time, the digging hand, as an alternative autapomorphy, validates the monophyly of the Myrmecophagidae.

Summing up

If, in the stem lineages of different taxa, a homologous feature has been transformed in divergent directions compared with the feature state in the latest common stem species, then both states can be used, as alternative apomorphies, to establish monophyletic species groups of the phylogenetic system.

171

4. Homology with alteration in the same direction in two lineages

Alternative apomorphies do not result from divergent transformations only. For logically they may also arise by independent alterations of a homologous feature in the same direction to produce approximately similar states in separate stem lineages.

The fore limbs of the Artiodactyla and Perissodactyla

The Artiodactyla (even-toed hoofed animals) and the Perissodactyla (odd-toed hoofed animals) are monophyletic subtaxa of the Ungulata (hoofed animals) but, according to McKenna (1975), they do not represent sister groups within the taxon Ungulata.

The plesiomorphous state of the fore limbs is extraordinarily similar in the Artiodactyla and the Perissodactyla. The artiodactyl *Hippopotamus* and the perissodactyl *Tapirus* both have four-toed, unguligrade fore limbs in which the first digit is lacking and the third digit is the strongest.

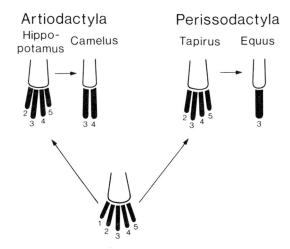

Fig. 53. The evolution of alternative apomorphies by alteration in the same direction of homologous fore limbs to produce a four-toed basic form in the artiodactyls (*Hippopotamus*) and the perissodactyls (*Tapirus*). The diagram does not imply that the Artiodactyla and the Perissodactyla arose from a stem species shared by them alone. It is only intended to show how the identical fore limbs arose from a plesiomorphous limb with five digits.

Since the pre-accepted phylogenetic systematization of the Artiodactyla and the Perissodactyla does not place them as adelphotaxa, the identical fore limbs of the hippopotamus and tapir must have arisen independently within the Ungulata from a plesiomorphous limb with five digits. The four-toed unguligrade fore limb thus becomes an autapomorphy validating the monophyly of the Artiodactyla and, at the same time, an autapomorphy of the ground pattern of the Perissodactyla justifying the monophyly of this taxon.

Later evolutionary transformation of the four-toed ground pattern then happened very differently in the artiodactyls and the perissodactyls, though this, of course, does not affect the present problem. In the artiodactyls, the reduction of further phalanges concerned toes 2 and 5. These have been completely lost in the two-toed limb of the Camelidae (*Camelus, Lama*) and of the Giraffidae (*Giraffa*). Within the perissodactyls, on the other hand, the digits 2, 4 and 5 were reduced. The end product here is the one-toed limb of the taxon *Equus*.

The proboscis hooks of the Kalyptorhynchia (Plathelminthes) (Fig. 54)

For a second example of change in homologous features in the same direction, I refer back to the proboscis apparatus of the Kalyptorhynchia (p.136).

In kalyptorhynchians from the interstitial system of marine sands, hard grasping hooks are found in the undivided "conorhynch" of the "Eukalyptorhynchia" and also in combination with the divided proboscis of the Schizorhynchia. Grasping hooks for the capture of prey have arisen twice, independently, by alterations in the same direction of the homologous proboscis organ. The apomorphy of "grasping hooks on the proboscis" validates a monophyletic taxon Gnathorhynchidae among the eukalyptorhynchians. And, at the same time, it serves to constitute a monophyletic group Karkinorhynchidae + Diascorhynchidae within the Schizorhynchia.

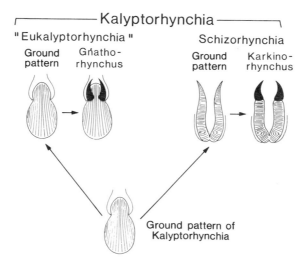

Fig. 54. Alteration in the same direction of the homologous proboscis organ of the Kalyptorhynchia by independent evolution of grasping hooks on the plesiomorphous conorhynch of the "Eukalyptorhynchia" and on the apomorphous schizorhynch of the Schizorhynchia.

Summing up

Alternative apomorphies of a homologous feature can be used to justify monophyletic taxa if, compared with the plesiomorphous condition of the common stem species, they have been changed independent of each other in an identical manner. Of course, the outcome of such evolutionary transformations hides the danger of serious mistakes in interpretation. In accordance with the principle of parsimony, the agreements must *a priori* be seen as synapomorphies. Only *a posteriori*, on the basis of previously accepted hypotheses concerning the phylogenetic system of the feature bearers, can they be recognized as independent alterations of a homologous feature.

IV. Negative features

Homology and non-homology, in my understanding of language, extend only to features that are present. Nothing whatever can be hypothesized as homologous or non-homologous if it does not even exist[1]. As already explained, however, phylogenetic systematics uses information about negative features as logically and consistently as information about positive features. In its terminology, the three possible forms of shared negative features in different taxa are appropriately covered by the terms "symplesiomorphy", "synapomorphy" and "convergence". I shall illustrate this matter by the absence of wings in the taxon Insecta (cf. Hennig 1966, 1980; Patterson 1982a).

Absence as plesiomorphy

In the Diplura, Ellipura, Archaeognatha and Zygentoma[2] there is not the slightest empirical indication (such as wing vestiges), nor any argument from pre-accepted hypotheses of relationship, to suggest descent from wing-bearing stem species. The lack of wings is, correspondingly, interpreted as a primary, plesiomorphous phenomenon and the shared lack of wings in these apterygote taxa is referred to as a symplesiomorphy. In comparison with this, the existence of two pairs of wings on the meso- and metathorax is the apomorphous alternative. Assuming, in accordance with the parsimony principle, that wings evolved only once, they become a constitutive feature for erecting the monophylum Pterygota within the Insecta.

Absence as apomorphy

Bed bugs, lice and fleas belong respectively, within the phylogenetic system of insects, to more inclusive monophyletic units which primarily bear wings—the Cimicidae in the taxon Hemiptera, the Phthiraptera in the taxon Psocodea and the Aphaniptera in the taxon Holometabola. From the pre-accepted systematization it

---

[1] In an attempt to subsume negative features under the term "homology", Hennig (1984, pp.38,39) discusses an extension of the meaning of the term "homologous features" (which includes positive features only), to cover "homologous positions of features" in the total structure of the organism. From this point of view, as concerns the negative feature of the absence of wings in primarily and secondarily wingless insects, there exist, on the meso- and metathorax, the "homologous positions" of the positive feature "two pairs of wings".

This widening of meaning, however, remains problematic, because it cannot be applied universally to negative features. For example, in the holoblastic ontogeny of the Amphibia, it would be vain to search for the "homologous position" of the embryonic organs of the amnion, allantois and serosa of the adelphotaxon Amniota.

[2] Compare p.185.

necessarily follows that the lack of wings is the outcome of evolutionary reductions. In relation to the existence of two pairs of wings, which is now a plesiomorphous feature, the winglessness of bed bugs, lice and fleas is, in each case, an apomorphy.

Absence as synapomorphy or convergence

The same example can be logically extended. Since wings are lacking in all bed bugs, all lice and all fleas, a single loss of the organs of flight can be postulated in each case, following the principle of parsimony, i.e. once in the stem lineage of the Cimicidae, of the Phthiraptera and of the Aphaniptera. The negative feature thus becomes a synapomorphy of the highest ranking adelphotaxa within the bed bugs, the lice and the fleas.

It remains to interpret the agreement between the Cimicidae, Phthiraptera and Aphaniptera in the negative feature of winglessness. The phylogenetic systematization of the Pterygota gives convincing grounds for the assertion that the stem species of the bed bugs, of the lice and of the fleas became secondarily wingless independent of each other. And thus, at this level of comparison, the identical evolutionary outcome counts as a convergence.

V.   Comparative overview

As a result of this analysis, I first stress two aspects.

1) There is no congruence between the distribution of homologous features among organisms and the relevance of those features for discovering closed descent communities. Homologies between different species can basically exist as two forms of evolutionary agreement — symplesiomorphy and synapomorphy. But, of these, only synapomorphies signify for erecting monophyletic taxa as equivalents of closed descent communities in Nature.
2) The terms "homology" and "non-homology" do not extend, in my usage, to negative features. The shared primary absence of a feature is a symplesiomorphy, but not a homology. The secondary absence, evolved once only, of a feature in a pair of adelphotaxa is a synapomorphy, but equally is not a homology. The independently evolved loss of a feature is a convergence, but probably can scarcely be called a non-homology.

These statements contradict a view pointedly advocated by Patterson (1982a), in which "homology" and "synapomorphy" are seen as synonyms. The centre of his argument is the fact that homologous agreements change from symplesiomorphies to synapomorphies in passing from lower to higher levels of hierarchy: "Symplesiomorphies are also synapomorphies. They are simply homologies that circumscribe a group whose generality is wider than, and therefore irrelevant to, the problem under study" (Patterson 1982a, p.33). I have already discussed in detail this obligatory change in value (p.144). But, firstly, Patterson's statement only partly applies to negative features, for a shared primary absence remains a symplesiomorphy at all levels of hierarchy. And, secondly, it is undeniably **necessary** that **evolution-**

175

ary agreements between species and closed descent communities be concretely and unmistakably characterized at every level of comparison. The grasping hands of *Pan troglodytes* and *Pan paniscus* are evolutionary agreements which, in both recent species of chimpanzee, have been derived from a common stem species. Thus they are homologous organs by definition. However, in comparing the two chimpanzee species together, and at many other levels in ascending the hierarchy of the phylogenetic system of primates, this homology is a symplesiomorphy. The homologous agreement of a grasping hand with full opposability becomes a synapomorphy only at the level of the sister groups Cercopithecoidea and Hominoidea.

Bock (1969, 1973, 1977b) has attempted to give more precision to the concept of homology by introducing a conditional phrase which describes the nature of the feature in the common ancestor: "Thus it is erroneous to say ... that the humerus of the gorilla is homologous to the humerus of the chimpanzee. Rather, one must say that the wing of birds and the wing of bats are homologous as the forelimbs of tetrapods" (Bock 1973, p.387).

I shall state my position on this. The first sentence is in no way mistaken or meaningless. Rather, it is the completely correct formulation of a hypothesis of homology. But just as when homologizing the grasping hands of the species of chimpanzee, the interpretation of a particular bone in the upper arm of the gorilla and chimpanzee as a homologous feature is, at this low level of the hierarchy, a worthless symplesiomorphy in working out phylogenetic relationships. The adjoining conditional phrase merely states, long-windedly, the hierarchical level at which a homology of features is relevant, as a synapomorphy, to the process of systematization (or, in other words, when the feature behind the homology is relevant as an autapomorphy). A pentadactyl fore limb, from which the wing of birds and also the wing of bats is derived, is a synapomorphy of the Amphibia and Amniota or an autapomorphy of the taxon Tetrapoda which unites them.

These considerations establish precisely where the terms of "homology" and "non-homology" fit into the three possible forms of evolutionary agreement. With reference to the definitions of the words synapomorphy, symplesiomorphy and convergence (p.52–53) I summarize the situation in the following survey.

1. Synapomorphies

a) Positive features

These evolved once only in the shared stem lineage of the adelphotaxa under comparison and were first present as autapomorphies in the stem species that was common to these adelphotaxa alone.

Synapomorphous agreements in positive features are always homologies.

Examples are:
— the grasping hand with full opposability, being a synapomorphy of the adelphotaxa Cercopithecoidea and Hominoidea and an autapomorphy of the taxon Catarrhini;

176

— the annulated antenna, being a synapomorphy of the adelphotaxa Archaeognatha and Dicondylia and an autapomorphy of the taxon Ectognatha.

b) Negative features

These were lost once only in the shared stem lineage of the taxa under comparison and as autapomorphies were no longer present in the stem species that was common to these taxa alone.

Synapomorphous agreements in negative features are not covered by the term "homology".

Examples are:
— absence of limbs on the premandibular head segment, being a synapomorphy of the adelphotaxa Myriapoda and Insecta and an autapomorphy of the taxon Tracheata;
— absence of wings within the Pterygota, being an autapomorphy of the taxon Cimicidae, of the taxon Phthiraptera and of the taxon Aphaniptera.

2. Symplesiomorphies

a) Positive features

These did not evolve in the stem lineage of the taxa under comparison but were inherited from more remote stem species.

Symplesiomorphous agreements in positive features are always homologies.

Examples are:
— the grasping hand with full opposability, being a symplesiomorphy of the chimpanzee species *Pan paniscus* and *Pan troglodytes* because already present in the stem species of the taxon Catarrhini;
— the segmented antenna, being a symplesiomorphy of the Diplura and Ellipura (Protura + Collembola) because inherited from the stem species of Insecta and, still more remotely, from the stem species of the Mandibulata and of the Euarthropoda.

b) Negative features

These are primarily absent in the taxa under comparison.

Symplesiomorphous agreements in negative features are not included under the term "homology".

Examples are:
— lack of wings within the Insecta, being a symplesiomorphy of the taxa Diplura, Ellipura, Archaeognatha and Zygentoma;
— the lack of embryonic membranes in the Vertebrata, being a symplesiomorphy between fishes (a collective term for primarily aquatic vertebrates) and the taxon Amphibia.

3) Convergences

a) Positive features

These were not present in the latest common stem species but first evolved, independently, in the separate stem lineages of the taxa under comparison.

Convergent agreements in positive features are always non-homologies.

Examples of evolutionary novelties are:
— independent evolution of Malpighian tubules at the junction of mid- and hind gut in the stem lineages of the Arachnida and of the Tracheata;
— independent evolution of homoiothermy in the stem lineages of the Aves and of the Mammalia.

Examples of evolutionary alterations are:
— independent evolution of a four-toed fore limb in the stem lineages of the Artiodactyla and of the Perissodactyla, from a five-toed limb;
— independent evolution of grasping hooks on the conorhynch of the "Eukalyptorhynchia" and on the divided proboscis of the taxon Schizorhynchia (Plathelminthes).

b) Negative features

These existed in the latest common stem species but were then lost, independently, in the separate stem lineages of the taxa under comparison.

Convergent agreements in negative features are not covered by the term "non-homology".

Examples are:
— lack of wings, in comparing the insect taxa Cimicidae, Phthiraptera and Aphaniptera;
— lack of limbs in comparing the vertebrate taxa Gymnophiona (caecilians) and Serpentes (snakes).

I. The Artificial Character of Non-monophyletic Groupings

I. Definitions, recognition and characterization

In combining evolutionary species to form supraspecific unities, phylogenetic systematics accepts none but monophyletic taxa as valid representations of the closed descent communities in Nature. Accordingly, every form of non-monophyletic group must be strictly avoided and those of traditional classifications must consistently be eliminated.

What, then, are non-monophyletic groupings of species? The logical antonym of "**monophyly**" is not "polyphyly", coined by Haeckel in the previous century, but rather the term "**non-monophyly**". In the case of "monophyly" and "non-monophyly" I again distinguish between their theoretical definitions and the practical possibilities of recognizing or characterizing what they mean. This gives rise to the following distinctions.

1. Monophyla

Definition

Monophyletic taxa are groups of species which derive from a stem species common to them alone. They comprise this stem species and all its descendent species (p.21).

Recognition

Monophyletic taxa are representations of real unities in Nature. A monophylum is recognized on the basis of those agreements between its members which arose once only as evolutionary novelties in the stem lineage of the taxon. In this methodology monophyla are based, in other words, on evolutionary agreements which can be hypothesized as synapomorphies.

I recapitulate the facts of the case in two diagrams of phylogenetic relationship (Fig. 55, upper pair). In the left three-taxon statement, the species C + D, together with the stem species z common to them alone, form a monophylum. The taxon is recognized on the basis of synapomorphy 2 which arose as an evolutionary novelty in the time span of the species z. In the diagram on the right, because of an additional species split, there is an additional monophylum which comprises the species A + B and the stem species y shared by them alone. This extra monophylum is based methodologically on synapomorphy 1 which was evolved by species y as an apomorphy.

179

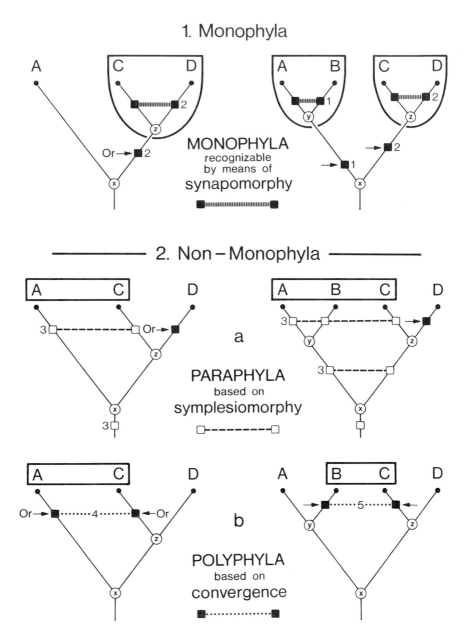

Fig. 55. The fundamental difference between monophyletic taxa and non-monophyletic groupings of species and the methodological possibilities of recognizing them or characterizing them. Explanation in text. Or = origin of evolutionary novelties.

2. Non-monophyla

Definition

Non-monophyletic taxa are groups of species which do not derive from a stem species common to them alone. They include only part of the descendants of their latest common stem species and, correspondingly, cannot include this stem species.

Characterization

Non-monophyletic groupings are artificial creations of mankind without equivalence to the closed descent communities in Nature. Correspondingly, they cannot be discovered in Nature—it is only possible to characterize the sort of mistake which has led to the erection of a non-monophyletic unit. On this methodological basis, Hennig (1965, 1966, 1974) undertook to distinguish between paraphyletic and polyphyletic groups.

a) Paraphyla

The erection of paraphyletic groups depends on the unification of species on the basis of symplesiomorphies.

In the middle pair of diagrams in Fig. 55, A + C on the left and A + B + C on the right are respectively combined together to form a taxon. But neither A + C nor A + B + C possess a stem species common to them alone, for D also is a descendant of stem species x. And thus the legitimation is lacking for including this stem species in the groups.

The amalgamation of species is based, in both cases, on the shared possession of a primitive feature 3 which had already been evolved in the latest common stem species x from which it was taken over unchanged. An apomorphous alternative developed only in the lineage of the species D.

b) Polyphyla

The erection of polyphyletic groups depends on the unification of species on the basis of convergences.

In the bottom pair of diagrams in Fig. 55, A + C have again been united to form a species group on the left, whereas B + C have been united on the right. A + C, and also B + C, possess no stem species common to them alone. Once again, only part of the descendants of the latest common stem species x has been considered, and the stem species itself is not included.

This mistaken grouping, however, is based on features which were not present at all in the latest common stem species x. Rather they were evolved in the lineages of species A and C (feature 4) or of species B and C (feature 5) independent of each other and in approximately similar form.

Discussion

The chaotic quarrels about this straightforward situation and about how to characterize it terminologically are of very limited interest. I refer to various surveys (Platnick 1977a; Ashlock 1979; Holmes 1980; Wiley 1981; Charig 1982) and go into a few points only.

Platnick (1977a) finds it inconsistent that the word "monophyly" is defined on the sort of phylogenetic relationship, whereas the terms "paraphyly" and "polyphyly", on the other hand, are based on agreements in particular features. All definitions, however, which attempt to show differences in the phylogenetic relationships between paraphyletic and polyphyletic groups (Ashlock 1971, 1979; Nelson 1971; Farris 1974; Wiley 1981; Table 3.1) are completely irrelevant to the aims of phylogenetic systematics. With the best will in the world, when groupings do not exist at all in Nature as unities with reality and individuality, but are simply Man's inventions, I have neither the inclination nor the energy to disentangle the phylogenetic relationships between the species placed in them. The only thing that matters is consistently to distinguish between monophyla and non-monophyla (Eldredge & Cracraft 1980, p.210; Patterson 1981a, p.206). Since phylogenetic systematics sternly rejects non-monophyletic taxa, however they are constituted, then the nature of phylogenetic relationships in such groupings is a matter of complete indifference. The only requirement, of course, is to know the methodological errors which led to the setting-up of those taxa and which, in our own work, we wish to eradicate or avoid—and these are simply groupings on the basis of symplesiomorphies or of convergences.

But on the methodological level, also, Platnick (1977a) argues against the terminological distinction into paraphyla and polyphyla by pointing out that a particular non-monophyletic grouping may be, at one and the same time, both paraphyletic and polyphyletic. This happens, namely, when its representatives show combinations of plesiomorphous and convergent agreements. Such a situation is in fact possible. In Fig. 55, I have shown a paraphylum A + C and a polyphylum A + C with precisely identical phylogenetic relationships. Supposing that A and C simultaneously have a symplesiomorphy (feature 3) and a convergence (feature 4) in common, then these taxa could be exactly replaced by the Monotremata and the Marsupialia. I have already rejected, as a non-monophyletic grouping, the unification of these under the name Marsupionta. If the shared possession of a marsupial bone is stressed, then the Marsupionta are a paraphylum on the basis of a symplesiomorphy. As regards the existence of a marsupium, on the other hand, the group Marsupionta is a polyphylum based on a convergence. In other words, this single example demonstrates a double mistake, which shows with great clarity that for phylogenetic systematics nothing else counts but recognition of artificial groupings as non-monophyla. The Marsupionta are to be eradicated—but for the progress of work on a phylogenetic system it is totally unimportant whether they fail as a paraphylum, as a polyphylum or as a *mixtum compositum* of both.

According to Platnick (1979) and Patterson (1982a), paraphyletic groups are not characterized at all on the basis of features, but merely on the lack of features, for "the absence of a character is not a character" (Nelson 1978, p.340; Patterson 1982a p.33). I contest this view, not only because of what I have said concerning the value of negative features, but also by simply mentioning a few classical paraphyla whose erection is based partly on plesiomorphous negative features and partly on plesiomorphous positive features.

The "Apterygota" among the insects and the "Anamnia" among the vertebrates are probably the widest known examples of non-monophyletic artefacts based on the primary absence of features. Evaluating the plesiomorphous absence of wings as a symplesiomorphy of particular insects, or the primary absence of amnion and serosa as a symplesiomorphy of particular vertebrates, is open to objection. For wings are likewise primarily lacking in the other arthropod taxa and in all other invertebrates. And embryonic membranes are likewise lacking in the acraniates and tunicates and in the other deuterostomes etc. These facts are undeniable, but equally undeniable is the assertion that counting the primary absence of features as symplesiomorphies only makes sense within the limits of that taxon in which the symplesiomorphy contrasts with an apomorphous alternative. Thus, the primary absence of wings counts as a symplesiomorphy only within the limits of the taxon Insecta while the primary absence of amnion and serosa counts as such only within the limits of the monophylum Vertebrata.

This brings me to the characterization of paraphyletic groupings by means of primitive positive features. In the taxon Plathelminthomorpha, the monociliary epidermis of the Gnathostomulida is simply the plesiomorphous condition, whereas the multiciliary condition of the epidermis of the sister group Plathelminthes is the apomorphous alternative (p.258). Within the taxon Plathelminthes, the multiciliary epidermis then becomes a symplesiomorphy for the sub-groups united under the name "Turbellaria". Only in the stem lineage of the parasitic Neodermata (Trematoda + Cercomeromorpha) was it completely replaced by the evolution of an unciliated epidermis made of stem cells (p.296)—the original epidermis is only retained in the larval phase (miracidium, coracidium). Within the limits of the superordinate taxon Plathelminthomorpha, therefore, it is only the paraphyletic "turbellarians" which are characterized by the plesiomorphous positive feature of a multiciliary epidermis throughout the life cycle.

I finish by referring back to the plathelminth taxon Kalyptorhynchia (p.136. Figs. 40–42). The solid cone-shaped proboscis of the "Eukalyptorhynchia" represents the plesiomorphous condition while the divided proboscis, with two tongues, is the apomorphous state of this prey-catching apparatus. Up till now, no evolutionary agreement which can be interpreted as an apomorphy is known for the "Eukalyptorhynchia" though these are equipped with the primitive cone-shaped proboscis. In other words, the "Eukalyptorhynchia" form a paraphylum whose members are held together only by a symplesiomorphous agreement in the structure of the proboscis. They are a striking example of the erroneous setting-up of an artificial systematic unit on the basis of a positive feature which occurs solely in members of the unit. Despite the common possession of the "conorhynch" symplesiomorphy, the "Eukalyptorhynchia" had no stem species shared by them alone. This is because the stem species for which the evolutionary novelty of the cone-shaped proboscis can be postulated is also ancestral to the Schizorhynchia.

II.   The formation of groups in the Systema naturae of Linnaeus

To illustrate the historical burden carried by work towards a phylogenetic system, a glance at the famous 10th. edition (1758) of the Systema naturae is of use. In it, Linnaeus divided the "Regnum Animale" into the six classes of Mammalia, Aves, Amphibia, Pisces, Insecta and Vermes. In so doing, he exemplified all the possible forms in which valid or artificial taxa can be created.

Thus, in the first place, with the Mammalia and Aves Linnaeus specified two supraspecific taxa which today can be accepted, incontrovertibly, as monophyla. In both cases the constitutive features of the mammals and the birds were so obvious that the Systema naturae, as an example of a pre-phylogenetic classification of animals, already contained valid equivalents of closed descent communities in Nature.

The Pisces and the Amphibia of Linnaeus are two paraphyletic artefacts. The former results from bringing together the primarily aquatic fishes while the latter combines the amphibians, in the modern sense, with the reptiles as relatively primitive terrestrial vertebrates.

The Vermes form a heterogeneous polyphylum of species from the most diverse taxa such as the cnidarians, nemathelminths, annelids, molluscs and echinoderms.

And finally the order Aptera of the class Insecta is of particular interest. Like the Marsupionta discussed above (p.182), it too is a *mixtum compositum* brought together on the basis of symplesiomorphous and convergent absence of wings. Thus in the Linnaean group Aptera there are representatives of the Arachnida, Crustacea and Myriapoda, as well as primarily wingless insects such as the silver fish *Lepisma* and the collembolan *Podura*, in addition to secondarily wingless insects like the termite *Termes*, the louse *Pediculus* and the flea *Pulex*.

III. Paraphyletic groupings

Paraphyletic taxa are extensively represented in all traditional classifications of animals. Even today, they are not merely tolerated as provisional measures which will successively be broken up as knowledge advances. To the contrary, the advocates of evolutionary classification approve of paraphyla and expressly defend them.

In traditional classifications, paraphyletic groupings are frequently formed by the splitting of a particular, often monophyletic taxon into two sub-units which are given equal rank. The first of these subunits, based on plesiomorphous agreements, is the "more primitive" subtaxon and represents a paraphylum while the second subunit, as a rule, is a monophylum based on autapomorphies. I shall now mention a few examples that have been discussed above or are generally known. I place the paraphyla in quotation marks.

Kalyptorhynchia:	"Eukalyptorhynchia"—Schizorhynchia
Annelida:	"Polychaeta"—Clitellata
Clitellata:	"Oligochaeta"—Hirudinea
Insecta:	"Apterygota"—Pterygota
Hymenoptera:	"Symphyta"—Apocrita
Vertebrata:	"Anamnia"—Amniota
Sauropsida:	"Reptilia"—Aves
Primates:	"Prosimiae"—Simiae

With all traditional classifications that show dichotomies of this sort, a critical examination is necessary to discover whether the subtaxa seen as primitive are, in fact, valid systematic units, or whether they must be thrown out as the artificial products of human imagination.

184

But even in classifications where several subtaxa are arranged next to each other with identical rank, paraphyletic groupings are widespread. This holds, for example, for the traditional subdivision of the flat worms into three classes of equal rank, for the traditional subdivision of the vertebrates into five classes or the subdivision of the man-like apes and Man into three taxa with the rank of families.

Plathelminthes: "Turbellaria"—Trematoda—Cestoda
Vertebrata: "Pisces"—Amphibia—"Reptilia"—Aves—Mammalia
Hominoidea: Hylobatidae (*Hylobates, Symphalangus*) —"Pongidae" (*Pongo, Pan, Gorilla*)—Hominidae (*Homo*)

A consistent phylogenetic system is the only possible reference system in biology which allows the correct storage and the unambiguous retrieval of information on the phylogenetic relationships between evolutionary species and closed descent communities in Nature. I made this claim at the beginning of the book and I shall now substantiate it with examples from traditional classifications where the presence, beside each other, of paraphyletic and monophyletic taxa is the source of fundamentally false conclusions. These examples emphasize that, whatever form of paraphyletic group is being dealt with, phylogenetic systematics requires that all combinations of species be rejected which cannot be derived from a stem species common to these species alone.

The "Apterygota" in the system of the Insecta

The traditional subdivision of the insects into the paraphylum "Apterygota" and the monophylum Pterygota can be contrasted with a well founded phylogenetic system in which the highest ranking adelphotaxa are the Entognatha and the Ectognatha (Hennig 1953, 1969, 1981; Kristensen 1975, 1981; Boudreaux 1979).

I shall begin with the phylogenetic systematization and give evidence in the first place for the first two levels of subordination (Fig. 56). The monophyly of the individual taxa is supported by selected autapomorphies (A). The plesiomorphous alternative feature (P) of the sister group is mentioned afterwards in each case.

Insecta: A1 = heteronomous subdivision of the trunk into the thorax, consisting of three segments, each with a pair of walking limbs, and an abdomen of eleven segments without locomotory limbs. (Myriapoda: P = homonomous trunk with a larger number of pairs of locomotory limbs.)

Entognatha: A2 = sinking of the mandibles and maxillae into the head capsule. (Ectognatha: P = free articulation of the mouth parts on the head capsule).

Diplura: A4 = secondary absence of ocelli and compound eyes. (Ellipura: P = photosensory organs present.)

Ellipura: A5 = reduction of the antenna to four segments at most; reduction of the cerci. (Diplura: P = antennae with a large number of segments; cerci present.)

Ectognatha: A3 = annulated antenna with scape, pedicel and many-jointed flagellum; musculature in the scape only. (Entognatha: P = segmented antenna consisting of several uniform segments; musculature extends to the last segment but one.)

185

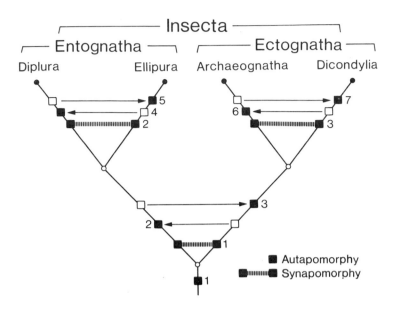

Fig. 56. Phylogenetic systematization of the monophylum Insecta for the first two levels of subordination. The autapomorphies of the individual taxa (= synapomorphies of the highest ranking sister groups within them) are given in the text.

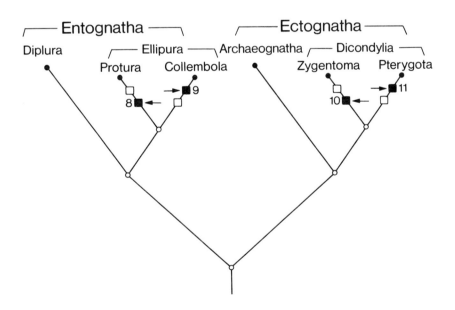

Fig. 57. The phylogenetic system of the Insecta continued on the next level of subordination by dividing the Ellipura and Dicondylia into adelphotaxa.

186

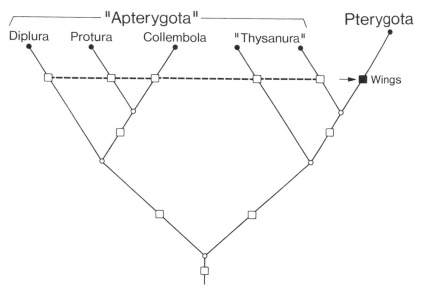

Fig. 58. Fitting the phylogenetic system of the Insecta into the traditional apterygote-pterygote classification by: 1) raising the Protura and the Collembola to the same rank as the Diplura; 2) uniting the Archaeognatha and the Zygentoma to form the "Thysanura"; and 3) raising the Pterygota to the hierarchical level of the "Apterygota".

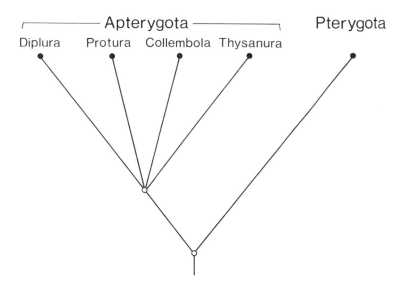

Fig. 59. Demonstration of the mistakes which result from transforming the artificial apterygote-pterygote classification into a diagram of phylogenetic relationships. Explanation in text.

187

Archaeognatha: A6 = compound eyes pushed together dorsally into the mid-line of the head. (Dicondylia: P = compound eyes separate from each other.)

Dicondylia: A7 = mandible connected to the head capsule by two condyles. (Archaeognatha: P = mandible with only one condyle.)

To complete the comparison which I wish to make between the traditional classification and the phylogenetic system, I need, in the case of the Ellipura and Dicondylia, to characterize the adelphotaxa at the next level of hierarchy (Fig. 57.)

Ellipura

Protura: A8 = antennae secondarily absent. (Collembola: P = antennae present.)

Collembola: A9 = ventral tube, retinaculum and furca on the abdomen. (Protura: P = corresponding limb derivatives not developed.)

Dicondylia

Zygentoma: A10 = reduction of the compound eyes to a few ommatidia. (Pterygota: P = complex eyes developed normally.)

Pterygota: A11 = two pairs of wings on the meso- and metathorax. (Zygentoma: P = primarily without wings.)

This brings me to the traditional classification of the insects with the dichotomy between the "Apterygota" and the Pterygota (Fig. 58). It divides the primarily wingless species into four units which are considered to have equal rank and together form the paraphylum Apterygota. Along with this, the combination of the Archaeognatha (bristle-tails *Machilis*) and the Zygentoma (silver fish *Lepisma*) to form a class "Thysanura" leads to a new mistake, for this is a paraphyletic grouping within the paraphylum "Apterygota".

I shall now tabulate the alternative schemes of subdivision and ask a precise question: Can a traditional classification, which allows monophyletic and paraphyletic groups alongside each other, store information on the phylogenetic relationships of its supraspecific units?

Traditional classification	**Phylogenetic systematization**
1. Apterygota	1. Entognatha
1.1 Diplura	1.1 Diplura
1.2 Protura	1.2 Ellipura
1.3 Collembola	1.2.1 Protura
1.4 Thysanura	1.2.2 Collembola
2. Pterygota	2. Ectognatha
	2.1 Archaeognatha
	2.2 Dicondylia
	2.2.1 Zygentoma
	2.2.2 Pterygota

The answer is an absolute "No". In classifying the Insecta into a paraphylum "Apterygota" and a monophylum Pterygota, the content of phylogenetic information equals nil. The subdivision of the primarily wingless insects into four classes with identical categorial rank obscures any information concerning the mutual phylogenetic relationships of the Diplura, Protura, Collembola and "Thysanura" to each other. Moreover, from this classification it is impossible to decide whether a particular group of the paraphylum Apterygota, and if so which one, is closer related to the Pterygota than to the other groups of primarily wingless insects. However, there is worse to follow. If the apterygote-pterygote classification is transformed into an equivalent diagram of relationship, as is required for every phylogenetic systematization, then the following pieces of basic misinformation result (Fig. 59): 1) The Diplura, Protura, Collembola and Thysanura together arose from the multiple splitting of a stem species shared by them alone. 2) The four classes of primarily wingless insects together form the sister group of the Pterygota. 3) Accordingly, no "class" of the "Apterygota" can be closer related to the Pterygota than to any other "class" of primarily wingless species.

This discussion, to my mind, sufficiently shows the worthlessness of conventional classifications which permit paraphyletic groupings on the basis of plesiomorphous agreements. In the example just treated, the traditional classification contains not a single piece of information as to the five adelphotaxon relationships which exist in Nature between the ten descent communities of insects that comprise this part of the system. In the corresponding phylogenetic system, on the other hand, all this information is stored precisely, level by level, by means of a logically consistent subordination of groups.

The "Reptilia" in the system of the Amniota

The Amniota are undeniably a monophyletic group of land-living vertebrates. As with the Insecta, I give the traditional classification and contrast it with the logically consistent phylogenetic systematization. The traditional classification divides the Amniota into three classes of equal rank, i.e. the reptiles, birds, and mammals. For the phylogenetic system, I have based the tabulation of this part of the system on the hypothesis of phylogenetic relationship which, as discussed above (p.91), seems relatively best founded.

Traditional classification	**Phylogenetic systematization**
1. Reptilia	1. Sauropsida
1.1 Chelonia	1.1 Lepidosauria
1.2 Rhynchocephalia	1.1.1 Rhynchocephalia
1.3 Squamata	1.1.2 Squamata
1.4 Crocodilia	1.2 Archosauromorpha
2. Aves	1.2.1 Chelonia
3. Mammalia	1.2.2 Archosauria
	1.2.2.1 Crocodilia
	1.2.2.2 Aves
	2. Mammalia

Whatever diagnosis of the vertebrate "class" "Reptilia" is studied, it is vain to look for an evolutionary novelty common only to the rhynchocephalians, lizards + snakes, turtles and crocodiles. As an example, I shall quote the "Lehrbuch der Zoologie" of Claus, Grobben & Kühn (1932, p.957), a work which is taken as standard. It diagnoses the reptiles as follows: "Scaled or armoured vertebrates, of inconstant body temperature, limbs provided with feet, respiration exclusively by lungs, the heart with two auricles and two ventricles but usually incompletely divided, embryos with amnion and allantois."

The diagnostic features listed for this ostensibly valid grouping of vertebrates belong, without exception, to the ground pattern of the closed descent community Amniota. Some of them are original tetrapod features. The others are features which, like the thick epidermis with scales or the embryonic amnion, serosa and allantois, evolved in the stem lineage of the Amniota and which, at latest, already existed in the latest common stem species of the living representatives of the Amniota. In other words, all the agreements between the Rhynchocephalia, Squamata, Chelonia and Crocodilia represent symplesiomorphies which these taxa have taken over, feature by feature, from the ground pattern of the Amniota.

Like the paraphylum "Apterygota" in the insects, the paraphylum "Reptilia" obscures for the Amniota every scrap of information about phylogenetic relationships. The arrangement of Rhynchocephalia, Squamata, Chelonia and Crocodilia into four "orders" of equal rank can say nothing whatever about the mutual phylogenetic relationships of these taxa with each other nor about the relationship of the individual units to the Aves or to the Mammalia. In connection with the traditional classification it is customary to say that the Aves and the Mammalia arose phylogenetically from the reptiles. This statement must be rejected because supraspecific taxa can never be the ancestors of closed descent communities (p.32).

The reptiles are probably the best known paraphyletic grouping among animals. As such they are suited to show, with the requisite clarity, yet another erroneous aspect involved in the creation and advocacy of paraphyletic classificatory units.

Why do the defenders of an evolutionary classification not allow the taxon "Reptilia" to be excluded from the system of the vertebrates, even though they recognize that the throwing together of Chelonia, Rhynchocephalia, Squamata and Crocodilia is based on plesiomorphous agreements? The reason is that they cannot do without a "class" Reptilia if the birds are to be given "a suitable position" in the system of the vertebrates with the rank of class alongside the reptiles and mammals. Their excitement, in fact, is all about the degradation of the honourable "class" of Aves to be a subordinate subtaxon of the Archosauria, as phylogenetic systematics proposes. How can anybody dare to demote the birds to one and the same low level as the crocodiles? After all, the birds, in the conquest of the air, have undergone profound evolutionary changes in nearly every system of organs, whereas the changes in the crocodiles, compared with other reptiles, seem very slight. In the face of this huge difference in phenetic divergence, the close phylogenetic relationship between the Crocodilia and the Aves becomes of secondary importance in classification. Thus E. Mayr (1969, p.70) writes, with reference to the example of the birds and crocodiles: " ... if one of the lines is exposed to severe selection pressures and as a result diverges dramatically from its genealogically nearest relatives[1], it may become genetically so different that it would be a biological absurdity to continue calling them near relatives."

---

[1] i.e. the phylogenetic relative of the first degree.

What a "biological absurdity" may be, I leave on one side. Contrary to Mayr's astounding assertion, the argument that relationship depends only on descent from common ancestors, and on nothing else, can be understood, in my experience of academic teaching, on the basis of the following comparison. A simplified diagram of the phylogenetic relationships of the Sauropsida (Fig. 60) is squeezed to fit into the traditional classification of the paraphylum "Reptilia" with Aves arranged on the same hierarchical level. Next to this diagram of relationship, I place an example of an intraspecific genealogical relationship, i.e. a family tree with three generations of men (Fig. 61).

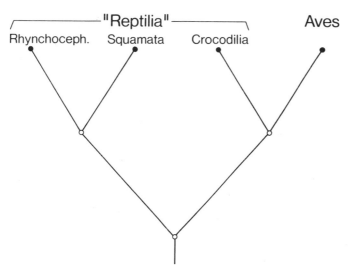

Fig. 60. Simplified diagram of the phylogenetic relationships of the Sauropsida to demonstrate the artificial grouping "Reptilia" (Chelonia omitted, see Figs. 29–31).

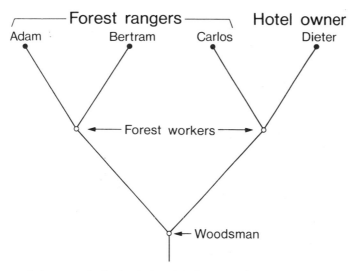

Fig. 61. Diagram of the genealogical relationship between three generations of men, considering the male sex only.

Obviously, this comparison between the phylogenetic relationships of supraspecific taxa and particular genealogical relationships within the species *Homo sapiens* is partly lame.

With reference to the structure of the relationships, it would only be correct if an example with uniparental reproduction could be used (p.31). This, however, is basically impossible with the male sex and, in the case of *Homo sapiens* is equally impossible with the female sex. The comparison achieves its end, however, and this, for once, justifies the means.

The stem species of the Sauropsida is supposed to correspond to a poor woodsman as "father of the clan". His two sons, like himself, also became forestry workers, and earned their keep by the sweat of their brow. Both of them, likewise, had two sons who, as still living offspring in the woodsman's family tree, occupy the positions taken by the Rhynchocephalia, Squamata, Crocodilia and Aves in the diagram of phylogenetic relationships of the Sauropsida. Three of the four men in the new generation retained their love of the open-air life—the brothers Adam and Bertram and their cousin Carlos. All three worked in the forest as rangers and remained extraordinarily like each other in general appearance, life habits and the clothes they wore. From time to time they played a game of cards with each other. Only Dieter went to town and there became the owner of a hotel. His only remaining connection with his family was the wild game delivered from his relatives' game reserves to the hotel kitchen. Apart from that, because of his fundamentally different mode of life, the hotelier's appearance changed so much compared with the forest rangers that no outsider would guess them to be kin. Nevertheless, nobody has yet contradicted me when I say that the forest ranger Carlos and the hotelier Dieter, being brothers, are the closest possible genealogical relatives on their level of hierarchy, however much Carlos may resemble his cousins Adam and Bertram in his phenotype.

In exactly analogous manner, the Crocodilia and Aves remain phylogenetic relatives of the first degree whatever evolutionary changes may have occurred in the stem lineage of the Aves when the ancestors of the recent birds changed to a new environment. Correspondingly, in a phylogenetic system they obviously belong together and, as adelphotaxa which arose simultaneously from a stem species common to them alone, they take one and the same rank. The Zygentoma and the Pterygota stand forever in the first degree of phylogenetic relationship with each other. This is true although the former remain a numerically insignificant group of insects, deceptively similar to the Archaeognatha, while the Pterygota, by the evolutionary acquisition of wings, were able to exploit a new environment and thus developed into a gigantic descent community with almost 800,000 species. Despite the extreme dissimilarity between a silver fish and a dragonfly, the Zygentoma and the Pterygota, as sister groups within the taxon Dicondylia, logically take an identical rank in the phylogenetic system of insects.

IV.   True and ostensible polyphyla

The paraphyletic taxa of evolutionary classifications are, even now, obstinately defended. But, on the other hand, there is at least verbal unanimity in the rejection of supraspecific units based on convergent agreements.

Combinations of species based on independently evolved common features do not portray closed descent communities in Nature. In so far as mistakes of this sort are recognized, the invalid taxa must be broken up and their names must disappear. Famous polyphyla of pre-phylogenetic classifications which have suffered this fate include, for example, the Vermes of the "Systema naturae" of Linnaeus (1758) or the Radiata from "Le règne animal" of Cuvier (1817).

The word "phylogenesis" refers to the origin of closed descent communities and thus to complete natural entities. It does not refer to the evolution of features in populations of species (p.29). Correspondingly, the paired terms of monophyly and non-monophyly (= paraphyly + polyphyly) cover the phylogenetic origin of groups of species, and not the evolutionary origin of features in groups of species. It is, therefore, a grave etymological mistake to refer to the evolution once only of a feature as being its monophyletic origin. And it is equally wrong and disastrous to label the repeated independent evolution of a particular feature as being its polyphyletic origin. Like Patterson (1978), I regard the impermissible extension to features of the terms "monophyly" and "polyphyly" as essentially a source of confusion. It has led to some strange assertions, such as the "polyphyletic origin" of the Tetrapoda (Jarvik), of the Arthropoda (Manton) or of the Mammalia (Simpson)—assertions that these taxa are non-monophyletic though each is characterized by a whole series of apomorphous features common to it alone.

Mammalia

Every modern treatment of the phylogenesis of the mammals, up to the very latest monograph on the Mesozoic Mammalia (Lillegraven, Kielan-Jaworowska & Clemens 1979), discusses the polyphyly problem as raised by Simpson, though usually without revealing the logical errors in the hypothesis of a "polyphyletic origin of mammals". These errors are firstly caused by deriving the mammals from the paraphyletic "reptiles"—an artificial creation of mankind without real equivalent in Nature, and, secondly, by arbitrarily fixing a limit between these "reptiles" and the Mammalia.

This limit, set for practical reasons, states that all fossil Amniota with a synapsid skull and a primary (articular–quadrate) jaw articulation should be placed in the "reptiles" while all fossil amniotes with this sort of skull and a secondary (squamosal-dentary) jaw articulation belong to the mammals (Simpson 1959). This limit may seem "appropriate" and it may be that all authorities in the study of mammals agree upon it. I, however, must reject it with emphasis for it runs clean through the middle of a closed descent community in Nature. This descent community extends from the fossil "Pelycosauria" and "Therapsida" of the late Palaeozoic up to the recent Monotremata and Theria and thus includes all those amniotes, whether with primary or secondary jaw articulation, which derive from a stem species, common to them alone, with a synapsid skull (p.217).

In the course of argument, it is interesting to note the frequent assertion that the artificial "reptile"–mammal limit was passed several times independently in the Mesozoic by parallel transformations of the primary into the secondary jaw articulation. This is supposed to have happened at least four times, and perhaps in as many as nine separate lines (Simpson 1959). Among these were the stem lineages of the Monotremata and the Theria which are supposed to have originated, along-side each other, from the †"Therapsida". If we concentrate on these two taxa alone, then it necessarily follows that the middle-ear complex of the mammals, with the three auditory ossicles of the malleus, incus and stapes, would have arisen at least twice independently—in the stem lineage of the Monotremata and in that of the Theria. Why? Because the quadrate and articular could only be drawn as sound-conducting elements into the middle ear after they had been freed, separately in the Monotremata and Theria, from their primary task as the bones of the jaw articulation. The convergent evolution of the complex mammalian middle ear is indeed postulated with different degrees of explicitness or, at least, held to be possible (Hopson 1966; Thenius 1979)—a more than astonishing position for which no shadow of an argument exists. So long as there are no such arguments, and so long as there are no conflicts with the interpretation of other apomorphous features, the principle of parsimony dictates a single plausible hypothesis—the secondary jaw articulation, and the middle-ear complex with three sound-conducting bones and a tympanic bone, have arisen once only and are synapomorphies of the Monotremata and Theria.

Nevertheless, in a logical assessment of the polyphyly problem let us accept, temporarily and for the purposes of argument, that these features were evolved convergently, several times, in the Mesozoic Amniota. Would this be evidence, as is widely believed, for a "polyphyletic origin" of the mammals? No. Not at all! In addition to the jaw articulation and the sound-conducting apparatus of three bones, any number of the approximately 60 apomorphous agreements might have arisen as evolutionary novelties separately from each other in the stem lineages of the Monotremata and the Theria, but this would not decisively contradict the derivation of both taxa from a stem species common to both of them alone—not even with the total of all conceivable convergences. Why not? Because, whatever evolutionary changes may have happened after the splitting of the single stem species into the lineages of the descendent taxa, they are logically irrelevant to the question of the phylogenetic relationship of these taxa. To validate the monophyly of the Mammalia, it would theoretically be enough to show that one single agreement could be interpreted incontrovertibly as a synapomorphy of the Monotremata and Theria.

Under what conditions, then, could the Mammalia, or—to speak generally—any other taxon, be shown to be a non-monophyletic combination of species made on the basis of convergent agreements? Only if it could be shown as probable that both of the highest ranking sub-groups within the combination had a totally different taxon as its sister group. Thus, for the mammals, it would be necessary to show that the Monotremata were the sister group of a particular unit of non-mammals among the Amniota (say the Squamata) and that the Theria were the adelphotaxon of yet another group (such as the Chelonia or the Archosauria). In that case, the Monotremata and the Theria would never have had a stem species common to them alone. All apomorphous agreements between them, in other words, would need, without exception, to be convergences. If, under these conditions, shared apomorphous features of the Monotremata and the Theria were called upon in constituting a taxon Mammalia, then, and only then, would a polyphylum have been erected on the basis of convergent agreements. Turning to the Arthropoda, this imagined situation would be comparable to uniting the Arachnida (sister group of the Xiphosura) with the Tracheata (sister group of the Crustacea) to form a systematic unit on the basis of a convergence i.e. the common possession of Malpighian tubules (p.162).

This brings me to a last remark. Even if the mammals, in the way just outlined, were shown to be a polyphylum based on convergences, it would be necessary to reject any talk of the polyphyly of the Mammalia or of the polyphyletic origin of the mammals. "It should be logically impossible to speak of the polyphyletic origin of any taxonomic groups" (Reed 1960, p.319). For it is plainly and simply nonsense to think about the origin, several times independently, of a unity invented in the brain of Man which does not exist at all in Nature as a descent community. What ought, instead, to happen in this case is easy to say. The group of mammals would have to be eliminated, as an invalid taxon, from the phylogenetic system of the Amniota and the name rejected as a designation for a monophyletic unit. Of course, I have been able to show that there is not, in fact, the slightest occasion for such a procedure.

Arthropoda

Phylogenetic systematics passes a basically similar verdict on the notion of a "polyphyletic origin" of the Arthropoda, as summarized in the latest publications of its leading protagonists (Manton 1977, 1979; Manton & Anderson 1979; Anderson 1973, 1979).

Compared with the mammalian example, the starting situation is somewhat different in that various arthropod lineages, independent of each other, are said to be derived, not from a particular "stem group" of fossil organisms, but from unknown ancestors that are not more precisely described. This is true, in the first place, for the proposed separate derivation of three "arthropod phyla" — the Uniramia, Chelicerata and Crustacea — from soft-skinned ancestors with numerous unjointed appendages. But even the "phylum Uniramia", made up of ten sub-units, is supposed, in turn, to have had a "polyphyletic origin". Manton postulated a separate origin of the Onychophora, of the four "myriapod classes" accepted by her (Pauropoda, Diplopoda, Symphyla, Chilopoda) and of the five "hexapod classes" (Collembola, Protura, Diplura, Thysanura, Pterygota). All these are supposed to have arisen separately from ancestors with lobopodial limbs.

The essential reasons for supposing a dozen or so independent "arthropodizations" are, in the first place, seemingly unbridgeable differences between the gnathobases of the Chelicerata and the mandibles of Crustacea, Myriapoda and Insecta. Secondly, in the myriapods and insects the articulations of the limb coxae with the body are said to be so different from taxon to taxon that, in Manton's opinion, not one of the "classes" of myriapods, and none of the insect "classes", could have given rise to any other "class of Uniramia".

By contrast with these views based on functional morphology, phylogenetic systematics bases its answer on an abundance of apomorphous agreements which can be interpreted, stepwise, as synapomorphies of subordinate adelphotaxa in the arthropods. By using them, I shall validate, in the following survey, the monophyly of the

195

arthropods and also determine the adelphotaxon relationships of the monophyletic subgroups of the Arthropoda for the four highest levels of subordination. The corresponding part of the phylogenetic system of the arthropods[1] is as follows:

Arthropoda
  1. Onychophora
  2. Euarthropoda
    2.1. Chelicerata
    2.2. Mandibulata
      2.2.1. Crustacea
      2.2.2. Tracheata
        2.2.2.1. Myriapoda
        2.2.2.2. Insecta
          2.2.2.2.1. Entognatha
          2.2.2.2.2. Ectognatha

I have summarized some selected autapomorphies of the first nine taxa in nine blocks of features and the relative sequence of times of origin for these blocks is shown in the appended diagram of relationship (Fig. 62). The systematization of the Insecta has already been justified in the preceding chapter. All the named autapomorphies refer to the pattern of features present in the latest common stem species of the living members of the respective taxon.

1) ■ **Autapomorphies of the Arthropoda**

= **synapomorphies of the adelphotaxa Onychophora and Euarthropoda**
  — Chitino-proteinaceous exoskeleton which can be moulted
  — Mixocoel (haemocoel)
  — Heart with ostia
  — Pericardium with pericardial septum
  — "Open" blood-vascular system
  — Complex brain comprising the ganglia of the acron and of the three following metameres
  — Nephridia with sacculi

2) ■ **Autapomorphies of the Onychophora**

= **synapomorphies of the highest ranking contained adelphotaxa**
  — 2nd. pair of appendages = oral hooks
  — 3rd. pair of appendages = oral papillae
  — Oral papillae with defensive glands producing a sticky secretion
  — Salivary glands as the modified nephridia of the oral-papillar segment
  — Tuft-like tracheae with the stigmata irregularly scattered over the body
  — Nervous system with widely separated nerve cords and nine to ten commissures per segment

---

[1] The phylogenetic relationship of the Tardigrada has not been determined. If their suspected adelphotaxon relationship with the Onychophora proves to be valid, then the Onychophora + Tardigrada should be placed in a monophylum Pararthropoda alongside its sister group Euarthropoda and with the same rank (Lauterbach 1978).

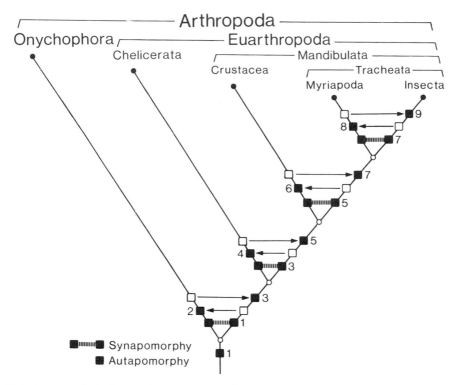

Fig. 62. Diagram of phylogenetic relationship for the Arthropoda down to the hierarchical level of the adelphotaxa Myriapoda and Insecta. The apomorphies corresponding to feature blocks 1–9 are explained in the text.

3) ■ **Autapomorphies of the Euarthropoda**

= **synapomorphies of the adelphotaxa Chelicerata and Mandibulata**
— Formation of a cephalon with a pair of pre-oral segmented antennae and four pairs of post-oral biramous appendages
— Trunk with jointed appendages, probably developed as biramous limbs with endopodite and exopodite (Lauterbach 1978)
— One pair of lateral compound eyes
— Four median eyes
— Nephridia in the four posterior head segments and in the first two post-cephalic trunk segments (Lauterbach 1983a)

4) ■ **Autapomorphies of the Chelicerata**

= **synapomorphies of the adelphotaxa Pantopoda and Euchelicerata[1] (Xipho-sura + Arachnida)**
— Division of the body into prosoma (head of the Euarthropoda + two trunk segments) and opisthosoma

---
[1] Cf. p.212

197

- Secondary absence of the pre-oral segmented antennae
- 2nd. pair of head appendages differentiated to form three-jointed chelicerae

5) ■ **Autapomorphies of the Mandibulata**

= **Synapomorphies of the adelphotaxa Crustacea and Tracheata**
- Mandibles developed as differentiations of the coxal endites (gnathobases) of the 3rd. pair of head appendages
- Ommatidia comprising a cornea, with two corneal cells, four Semper cells and a quadripartite, euconic crystalline cone; retinula formed of eight cells

6) ■ **Autapomorphies of the Crustacea**

= **synapomorphies of the highest ranking contained adelphotaxa**
- Combination of the four median eyes to form a unified nauplius eye
- Restriction of the segmental organs to one pair of antennal nephridia in the segment of the 2nd. antenna and one pair of maxillar nephridia in the segment of the 2nd. maxillae (p.00–00)

7) ■ **Autapomorphies of the Tracheata** (= Antennata)

= **Synapomorphies of the adelphotaxa Myriapoda and Insecta**
- Secondary absence of the 2nd. pair of head appendages (= 2nd. antennae of the Crustacea)
- Loss of the palp (endopodite) of the mandibles
- Air-breathing organs in the form of tracheae with paired segmental stigmata
- Malpighian tubules at the limit between the mid- and hind-gut and ectodermal in origin
- Indirect transmission of sperm by means of spermatophores

8) ■ **Autapomorphies of the Myriapoda**

= **synapomorphies of the adelphotaxa Progoneata and Chilopoda**
- Secondary loss of the median eyes

Further possible autapomorphies have been compiled by Boudreaux (1979). Obviously, the monophyly of the taxon Myriapoda is established only weakly by this loss. However, it is in no way valid to devalue the shared absence of median eyes in Myriapoda by referring to their convergent loss in other groups of arthropods (Dohle 1980, Spies 1981). However often the median eyes have otherwise been lost, their shared absence in the Progoneata and the Chilopoda remains a fact. And it is decidedly simpler to explain this fact by a single loss within a stem lineage common to all myriapods, than to assume, without any compelling reasons, that it has been lost twice or more in this group. No such compelling reasons exist—there are no known agreements, neither between the Progoneata and the Insecta, nor between the Chilopoda and the Insecta, which can convincingly be interpreted as synapomorphies.

9) ■ **Autapomorphies of the Insecta**

= **synapomorphies of the Adelphotaxa Entognatha and Ectognatha**
- subdivision of the body into head, thorax (three segments) and abdomen (11 segments and telson)
- three pairs of locomotory limbs on the thorax
- limb derivatives on abdominal segments 1–9 and 11; the latter form the cerci

The speculations about a polyphyletic origin of the arthropods have been criticized in detail (Siewing 1960; Lauterbach 1972; Patterson 1978; Weygoldt 1979; Paulus 1979; Boudreaux 1979). I shall now confront them with the phylogenetic systematization just presented. I do this so as to show the root-causes for such deep misunderstandings in working out phylogenetic relationships.

In the first place, I share Manton's view that no "class" of myriapods, nor of insects (nor of Crustacea nor Chelicerata . . .) can be the ancestral group of any other class of arthropods. My reasons for holding this view, however, are diametrically opposed to hers. For she considered that allegedly unbridgeable differences in the construction and function of particular features forbad the "derivation" of one class from another class. Whereas I hold that the concept of the supraspecific taxon as the ancestor of any other group of species is an artefact incompatible with the theory of evolution (p.32). Referring to the particular example, there cannot ever be a derivation of the Euarthropoda from the Onychophora, of the Tracheata from the Crustacea nor of the various "classes" of Insecta one from another. It is irrelevant how large and unbridgeable the differences between them may seem. In my view, the only thing that counts is that, allowing for all conceivable divergences, apomorphous common features are demonstrable between certain taxa and these features can, with mutual compatibility, be interpreted as synapomorphies. In so far as this is the case, it is legitimate to trace these taxa back to a stem species common to them alone—a stem species which would have possessed, as evolutionary novelties, all those features which occur in its descendants as derived agreements. I have validated the derivation of the Onychophora and Euarthropoda, of the Crustacea and Tracheata, of the Entognatha and Ectognatha etc. respectively from a stem species common to each pair of groups alone and I have performed the phylogenetic systematization, group by group, by means of a series of apomorphous agreements. In such case it is obvious that "unbridgeable" divergences in feature between the descendent taxa never need to be derived from each other. They are nothing else than the product of evolutionary changes in the stem lineages of the descendent taxa—evolutionary transformations of features which were concretely realized in the latest common stem species, each with a particular form and function.

In the arthropod example, as with the Mammalia, I must again stress that it is those who advocate "polyphyly" who must bring evidence to support their views. When they postulate the repeated separate origin of apomorphous agreements between taxa, they must establish, feature by feature and without exception, that all these agreements result from convergent evolution. There is, however, no methodological yardstick, nor any logically legitimate reason, for concluding *a priori* that convergent evolution has occurred. The interpretation of an apomorphous agreement as a convergence can happen only *a posteriori*, when a phylogenetic systematization favoured by other apomorphies demands it (p.56,142). The interpretation of the arthropods as a polyphylum would only be a valid topic for discussion if the Onychophora, for example, were the sister group of the Mollusca while the Euarthropoda could be established as the adelphotaxon of some other group of animals—say the Sipunculida. In such case it would be necessary to regard all apomorphous agreements between the Onychophora and the Euarthropoda as convergences. Even if this were so, however, the thoughtless use of the word "polyphyly" should be rejected. I repeat again that it would be nonsense, under such circumstances, to speak of the polyphyly of a taxon Arthropoda. The only course would be to break up the group Arthropoda—much like Cuvier's group Radiata—and to forget the group name. For this procedure, however, as with the Mammalia, there is not the slightest occasion.

J. The Phylogenetic Systematization of the Fossil Record

I. The nature of the problem

Present-day organisms are the starting point for the study of phylogenetic relationships. Without considering any fossil record whatever, phylogenetic systematics can formulate objective hypotheses about the phylogenetic relationships of recent species and species groups and can then translate the hypotheses of relationship favoured by testing into a system with hierarchical structure.

The aims of phylogenetic systematics, however, are larger than this. It seeks to discover and display the phylogenetic relationships of all organisms with each other — whether they exist today as evolutionary species or are known only as fossil fragments from the geological past.

This brings me to the problem of the phylogenetic systematization of fossil organisms in relation to the fauna and flora of the present day. I shall begin my discussion with some famous fossils. The "proto-bird" *Archaeopteryx lithographica* from the Jurassic lithographic limestones of Solnhofen in southern Germany is placed in the taxon Aves on the basis of a series of detailed agreements with recent birds. In a similar way, the king crab *Mesolimulus walchi*, likewise from the Jurassic, is assigned to the chelicerate taxon Xiphosura. And the Pleistocene mammoth *Mammonthus primigenius* is, with no doubt whatever, correctly placed in the mammalian taxon Proboscidea on account of extensive agreements with the recent species of elephant. What, in the first place, do these fossils tell us? They say that, if the taxa Aves, Xiphosura and Proboscidea can be justified as monophyletic units of the system — and for such a conclusion there are good arguments in each case — then the following statement is logically implied: closed descent communities in Nature may comprise recent organisms and also extinct organisms known only as fossils. Being self-evident, this statement may seem banal and superfluous. I make it, nevertheless, because from it there logically follows an extremely important requirement which is not at all generally accepted. It follows, namely, that all organisms, whether recent or known as fossils, are to be united together in a single phylogenetic system. Why is this requirement inevitable? Because the monophyletic taxa of the system that we aim to build must always include all descendants of their respective stem species if they are to be valid representations of the closed descent communities in Nature.

The next basic consideration is that our insights into the **composition of living Nature are to be extended into the past from the time-horizon of the present**. This way of looking at the problem results in the following incontrovertible hypothesis: after the origin of eucaryote organisms with bisexual reproduction, animals and plants have existed in all geological periods in the form of evolutionary species which were unities with individuality and reality. Logically there can be no grounds whatever, in the theory of phylogenetic systematics, for treating extinct organisms differently from recent ones. The systematization of both must in principle be subject to one and the same procedure.

Plain though this statement is, it cannot, of course, remove the obvious difficulties which hinder the logically consistent phylogenetic systematization of fossils. It is an unalterable fact that the fossilizable elements of past organisms preserve only a fraction of the information which can be got for phylogenetic research, in a great number of different ways, by a comparative analysis of the features of recent organisms.

The problems begin even with the understanding and delimitation of evolutionary species in the fossil record. Fundamentally, there is no objectifiable yardstick by which different individual fossils can be interpreted as members of a single evolutionary species, or, on the other hand, as representatives of reproductively separate species. Obviously, identity or great agreement in all fossilizable structures may be taken to indicate that different individuals belong to one and the same species. It is equally obvious that significant differences in the pattern of features of different fossils can be seen as indicating a former reproductive isolation, suggesting that the individuals in question belonged to different evolutionary species. However, with rare exceptions —such as the preservation of viviparous ichthyosaurs with the young animal inside the mother's body— the reproductive connection between individuals preserved as fossils is obscure. As a result the decisions which can and must be made in this field are in the nature of subjective estimates, even when extremely similar fossil fragments are preserved from a particular place and from one and the same time horizon. The high degree of error which must be taken into account in judging the number of extinct species, however, in no way decreases the validity of the evolutionary species concept nor the legitimacy of extending it to fossils. The necessary limitations of the analysis and comparison to fossilizable structural features, and the resulting inadequacy in our systems of observation and method (p.20), do not in any way justify the notion that palaeontology should operate with a special concept of species "usable" in practice. I emphasize as strongly as possible: just as every recent species of animal or plant exists today in the form of single individuals, so also in the past every fossilized organism belonged to a particular evolutionary species which, at one time, existed as a real unity in Nature.

But the central problem of phylogenetic systematics lies, of course, at a different level. For phylogenetic systematics must indicate a path, acceptable in logic and practicable in methodology, by which extinct and recent organisms can be united, as postulated, in a single system. I shall indicate this path in the next section.

II.   The conditions for systematization—the concept of the stem lineage

Thus we seek to establish objective hypotheses about the phylogenetic relationship of extinct organisms to particular recent species or monophyletic species groups. Formulating the conditions under which such hypotheses can be formed is linked with the following incontrovertible requirement. In a phylogenetic system which comprises both past and present, only the equivalents of real unities in Nature can be considered, even in the field of extinct organisms. That is to say, only taxa which can be interpreted as evolutionary species or as portrayals of closed descent communities. Without regard to the question of how well, or badly, the unities created by Nature can be recognized in the concrete single case, the fossil record in principle offers, on these assumptions, three possibilities as legitimate starting points for phylogenetic systematization:

1) A **single fossil specimen** which, as an individual or the remains of an individual, is the representative of a particular evolutionary species. An example is the headless body of *Xenusion auerswaldae* which, as the only known specimen of an evolutionary species, may be a representative of the Onychophora from the Early Cambrian or late Pre-Cambrian.
2) A **group of fossil individuals** which are judged to be the representatives of one and the same **evolutionary species**. An example is the six known specimens of "proto-birds" from the lithographic limestone of Solnhofen which, in the most parsimonious interpretation, are seen as individuals of a single species *Archaeopteryx lithographica* (De Beer 1954a; Ostrom 1972, 1976; Wellnhofer 1974; Tarsitano & Hecht 1980).
3) A **group of extinct evolutionary species**—each represented by one specimen or several individuals—in so far as the hypothesis can be justified that the units thought to be species together form a **closed descent community** in Nature. An example is the well known taxon Trilobita characterized as a monophylum by Lauterbach (1980b).

If fossil taxa of these sorts are compared with recent organisms, the next step is to consider the empirical basis of systematization, i.e. the similarities and agreements between them. As in the case of recent species, we are not interested here in all agreements of any sort whatever. Rather we are concerned, once again, only with those shared features which can be interpreted as synapomorphies. What, however, are synapomorphies between fossil and recent taxa? Every closed descent community in Nature has its own stem lineage within which its characteristic evolutionary novelties (autapomorphies) originated (p.44). Every such constitutive feature which can be recognized in a particular fossil is automatically a synapomorphy between the fossil taxon and the recent representatives of the closed descent community.

A **fossil taxon**—whether it is a single species or a monophyletic group of species—must show at least **one autapomorphy of the ground pattern of a recent taxon** in order to be considered at all as part of the phylogenetic system of organisms. As with recent taxa, I stress again: the more agreements which can compatibly be interpreted as synapomorphies, the better. The more derived features that a fossil taxon possesses, which agree with the autapomorphies of a closed descent community with recent representatives, the more certain its phylogenetic systematization will be. *Archaeopteryx lithographica* can be placed in the Aves with a probability approaching certainty because of a series of agreements with recent representatives of the taxon. These go from the feathers, and the construction of the fore limb as an organ of flight, to the legs with opposable hallux (p.93). These agreements only make sense if each is assumed to have evolved, once only, in the stem lineage of a descent community in Nature which includes all birds. As we shall soon see, it is also true that differences in the number of agreements with the recent representatives are crucial when several fossil taxa are to be systematized within the stem lineage of a particular closed descent community.

Perhaps, on the other hand, not one agreement in the preserved feature pattern is recognizable which can be hypothesized as a synapomorphy with a recent species or species group. If so, the extinct taxon cannot be integrated into the phylogenetic system. This is true even though, for theoretical reasons, every fossil must belong to a group that exists in the recent fauna and flora (Hennig 1969).

This statement can be exemplified by the **Conodonta**—a diverse group of fossils whose phylogenetic relationships are indeterminate. The record of this "phylum", in the form of millimetre-sized tooth-like structures of calcium phosphate, extends from the Cambrian to the Trias and comprises about 4000 "species". Only a single specimen from the Carboniferous of Scotland can be regarded, for good reasons, as the soft body of a conodont-animal. In this specimen a grouping of "teeth" (a conodont assemblage) is found in the head region of a slender, eel-shaped organism about 4cm. long (Briggs, Clarkson & Aldridge 1983; Higgins 1983). The individual has been compared in detail with the Chordata and the Chaetognatha but, even with this specimen, no structures are recognizable which can be seen as synapomorphies with any element in the feature pattern of any taxon of recent animals. Phylogenetic systematization of the Conodonta is therefore impossible with the material at present available.

These considerations provide a clear and sufficient basis for the position that I formulated at the beginning. They are reason enough to claim, namely, that the fauna and flora of the present day "are the solid starting point for phylogenetic research" (Hennig 1969, p.17). From this point of view, I shall now specify the conditions for integrating fossil taxa into the phylogenetic system. I do this using the schematic example with recent species A, B and C. In the **stem-lineage concept** that I am about to present, the fossil units are basically treated as terminal taxa with no descendants. This will be justified later (p.225).

Case 1

In a monophyletic unit with three recent species it is supposed that taxon I (with the single recent species A) and taxon II (with the species B and C) are sister groups of each other (Fig. 63). But now fossils of an extinct species M are found which share a particular apomorphous agreement with the recent species B and C. If interpreted as synapomorphy 1, the corresponding feature 1 must have arisen in the stem lineage of taxon II and have been present in the stem species v—the splitting of which has produced the fossil species. This fossil species M, however, lacks feature 2. For this first arose as an evolutionary novelty in the stem lineage of taxon II after the origin of species M and then became synapomorphy 2 of the recent species B and C. Under these conditions, species M can be placed, without any objections, in taxon II with the following very exact formulation: the extinct species M is a representative of the stem lineage of taxon II.

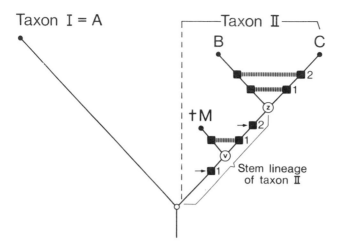

Fig. 63. A diagram of phylogenetic relationship for the three Recent species A, B and C and the extinct species M. The latter is placed in the stem lineage of taxon II (with the Recent representatives B and C).

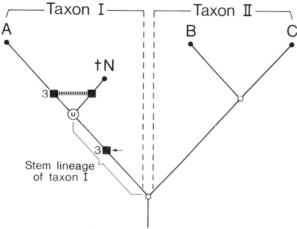

Fig. 64. A diagram of phylogenetic relationship for the three recent species A, B and C and the extinct species N. The latter is placed in taxon I at the end of the stem lineage.

Case 2

Another fossil species N is discovered and in it an apomorphy is recognized shared with the recent species A (Fig. 64). If, once again, this agreement is interpreted as a synapomorphy, then taxon I, which up till now has been monotypic, automatically becomes a supraspecific unit with one recent and one extinct species and a stem lineage common to them alone. In this stem lineage, feature 3 must have arisen as an evolutionary novelty. Fossil species N can be interpreted as a representative of taxon I. It stands at the end of the stem lineage, this end being marked by the splitting of stem species u into the descendent species A and N.

205

Case 3

For the third schematic example, I shall consider taxa whose recent representatives are characterized by an abundance of autapomorphies. As I have already argued, the origin of approximately 50 autapomorphies for the birds, or approximately 60 autapomorphies for the mammals (Hennig 1983) must be seen as a slow process with the successive addition of evolutionary novelties (p.45). This process could scarcely have occurred within the limits of a single evolutionary species but must, as a rule, have been connected with numerous speciation events in the stem lineages of these closed descent communities. If the conditions for fossilization are favourable, this would lead us to expect an abundant record of fossil units from the stem lineages. In the Mammalia, indeed, such is the case.

This brings me to the next question: what is the correct approach to the systematization of fossils when several extinct units from the stem lineage of a particular taxon are known?

In the light of what has just been said, the answer is easy. If several extinct units—whether evolutionary species or monophyletic taxa—can be seen as representatives of the stem lineage of a recent descent community, then they must be systematized within this stem lineage, in an ascending series, according to the increasing number of agreements with the ground-pattern apomorphies of the recent representatives. In the appropriate schematic example (Fig. 65) I specify again, as clearly as possible, the limits of the stem lineage of a supraspecific taxon (cf. p.44). The stem lineage of monophylum II, with the adelphotaxa B and C (whether species or supraspecific units), begins with the lineage of the stem species v and ends with the splitting of the stem species z—this latter being the latest common stem species of the units B and C which are recent species or have, as supraspecific taxa, recent representatives.

In case 3, taxa B and C are supposed to share a series of fossilizable features (1–4), the common possession of each of which can be interpreted as a synapomorphy. I also assume good preservation of the fossils of the extinct species M, P, R and S, for which reason the named features are in principle recognizable. In this situation, the following arrangement of fossil units within the stem lineage of taxon II would result.

The species M possesses apomorphy 1, which was the first fossilizable novelty evolved in the stem lineage of the taxon and existed in the stem species v. Being the relatively most plesiomorphous taxon, species M takes the lowest position. It is followed in the stem lineage by species P. This has a stem species x, later in time than v, in common with the recent representatives B + C, and in the lineage (life-span) of this stem species apomorphy 2 was added as an evolutionary novelty. The species R and S possess the apomorphous features 1–3. Being relatively the most strongly apomorphous fossils, they take a position in the stem lineage nearest to the recent taxa. They lack apomorphy 4, however, for this arose in the stem lineage of taxon II only after the splitting of stem species y.

This example also assumes that R and S together form an extinct monophylum. Obviously, in fossil organisms as in recent ones, such a hypothesis can only be justified if at least one agreement can be established which can be seen as a synapomorphy. For the fossil species R and S, this justification comes from apomorphy 5 which evolved, after the splitting of stem species Y, in the stem lineage of the extinct monophylum γ.

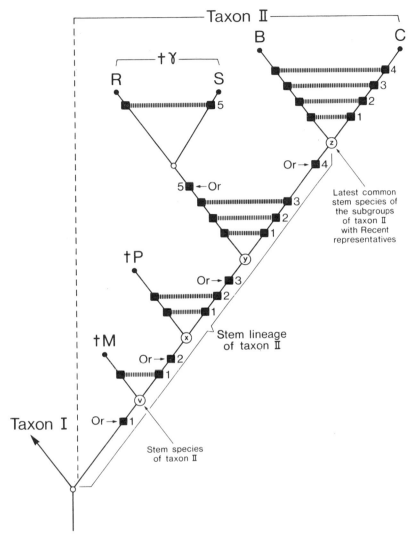

Fig. 65. Arrangement of several extinct units—the evolutionary species M and P and the monophylum γ (R + S)—in the stem lineage of a monophyletic taxon II containing the recent taxa B + C. Or = origin of evolutionary novelties.

III. Labelling the adelphotaxon relationships—the plesion concept

Having specified the conditions for the systematization of extinct organisms, I necessarily come to a basic problem of arranging the fossil record in practice.

I have defined adelphotaxa (p.36) as unities showing the first degree of phylogenetic relationship. Accordingly, in the examples just discussed, there are clearly the following adelphotaxon relationships:

In case 1, the extinct species M is the adelphotaxon of the recent unit B + C. In case 2, the extinct species N is the sister species of the recent species A. And in case 3, with several fossil taxa in a stem lineage, firstly, the species M is the adelphotaxon of all the other units; then the species P is the adelphotaxon of R + S + B + C; and finally, the extinct taxon γ (i.e. R + S) is the sister group of the recent unit B + C.

At this point, I recall the previous discussion of how recent adelphotaxa should be treated. They arose at one and the same time in the past and, accordingly, they take in principle the same rank. Equally, they are united at the next higher level of superordination to form a monophylum with its own name. If this procedure is applied consistently to part of the system with fossil as well as recent taxa, the following tabulation will result in case 3 (cf. Fig. 65).

1. Taxon I
2. Taxon II
   2.1. Taxon with the single species M (same rank as 2.2.)
   2.2. Taxon comprising P + R + S + B + C with a new proper name
      2.2.1. Taxon with the single species P (same rank as 2.2.2.)
      2.2.2. Taxon comprising R + S + B + C with a new proper name
         2.2.2.1. Taxon γ (same rank as 2.2.2.2.)
            2.2.2.1.1. Taxon R
            2.2.2.1.2. Taxon S
         2.2.2.2. Taxon comprising B + C with a new proper name
            2.2.2.2.1. Taxon B
            2.2.2.2.2. Taxon C

In this example, therefore, an adequate labelling of the adelphotaxon relationships would require three new taxa, each of which would have to be provided with its own name. Moreover, the fossil species M and P would have to be raised in rank to equal the supraspecific taxa coordinate with them.

Such a procedure might seem tolerable for units where the number of fossil representatives of the stem lineage was small. It is, for example, generally accepted as regards the taxon Aves. For ever since Haeckel (1866) first put *Archaeopteryx* into the system of birds, the "class" Aves has mostly been subdivided into two "subclasses" of equal rank. Haeckel called these the Saururae and the Ornithurae. But nowadays *Archaeopteryx* is usually put in a taxon Archaeornithes and this is contrasted with a taxon Neornithes, of equal rank, uniting all other birds. This is true even of attempts to construct a consistently phylogenetic system for the birds (Cracraft 1981b).

If, however, many fossils are known from the stem lineage of a taxon, this procedure involves worrying complications. McKenna's systematization of the mammals (1975) is a striking illustration of this. For it does not even include the diverse stem lineage of the Mammalia—it concentrates solely on the stem lineages of the Monotremata (Prototheria) and Theria. But, even so, it leads to an abundance of new taxa, each with its own name.

208

However, phylogenetic systematics has been attacked even when it sets up the necessary new taxa for recent organisms in a consistently subdivided systematic hierarchy (p.243). In this situation, the **plesion concept** formulated by Patterson & Rosen (1977) offers a logically incontrovertible and very practical basis for the combined systematization of fossil and recent taxa.

After fully considering the principles of phylogenetic systematics, Patterson & Rosen propose that each fossil taxon be referred to as a "**plesion**". They then suggest that: 1) a plesion should not be given the same rank as its adelphotaxon; and 2) no superordinate taxa for the respective sister groups should be erected and named. Instead, the plesia in an ascending stem lineage are simply arranged in a sequence corresponding to the increasing number of apomorphous agreements with the recent units of the taxon and listed in series, one after the other, in the equivalent tabulation of the respective part of the system. Obviously, only real unities in Nature can be accepted as plesia. Thus they must be either evolutionary species or supraspecific taxa, in so far as these can be characterized as monophyla. In this connection they take no definite rank, but are simply called by the names given them in the literature. In subdividing a fossil monophylum, just as with recent monophyla, a hierarchical structure with subordination according to the adelphotaxon relationships is obviously necessary. In the schematic example of case 3, use of the plesion concept produces the following simple tabulation:

1. Taxon I
2. Taxon II
       Plesion † species M
       Plesion † species P
       Plesion † monophylum γ
              1. † species R
              2. † species S
    2.1. Taxon B
    2.2. Taxon C

Because of the consistently sequential arrangement, the sister-group relationships formulated above can be read off without difficulty from the tabulation. Plesion M is the adelphotaxon of all the taxa that follow it. Plesion P is, in turn, the adelphotaxon of all the taxa that follow *it*. And the monophylum γ is the sister group of the recent unit B + C. New fossils can easily be integrated into the sequence of fossil taxa and the sequence can be re-arranged, at any time, to suit improved knowledge of the fossil record. And this re-arrangement will not, in the slightest degree, affect the known first-degree phylogenetic relationships between the recent taxa—in this instance between taxa I and II. In general, adelphotaxon relationships between the recent taxa, once they have been hypothesized, will not at all be changed by placing fossils in the stem lineages of these taxa.

The plesion concept was accepted by Eldredge & Cracraft (1980) and it has been applied in detail especially by Wiley (1976, 1979c, 1981). From this point of view, I shall now illustrate the phylogenetic systematization of taxa that have fossil and recent representatives.

IV. Examples of systematization

Hominini

My first example concerns the phylogenetic system of the Homininae. Into the relevant part of the system for the recent taxa (p.76, Fig. 24) I shall now integrate the fossil units which arose in the lineage leading to *Homo sapiens* after the split into the adelphotaxa Panini and Hominini. As in the schematic case 2 above, the recent monotypic taxon Hominini becomes a supraspecific taxon. This comprises the recent species *Homo sapiens*, the fossil representatives and a stem lineage.

For the purposes of this very basic discussion, the taxa *Sivapithecus* and *Ramapithecus* need not be considered. Likewise, the problem of delimiting evolutionary species in the fossil record is unimportant for present purposes. I shall concentrate on those fossil Homininae which can be placed with good grounds in the stem lineage of the taxon Hominini because of the proveable autapomorphy of bipedal locomotion. These are the taxon *Australopithecus* with the three species *A. africanus*, *A. afarensis* and *A. robustus* as well as two species placed in the taxon *Homo*, i.e. *H. habilis* and *H. erectus* (Eldredge & Tattersall 1975; Delson, Eldredge & Tattersall 1977; Tattersall & Eldredge 1977; Schwarz, Eldredge & Tattersall 1978; Szalay & Delson 1979; Ciochon & Corrucini 1983; Ciochon 1983).

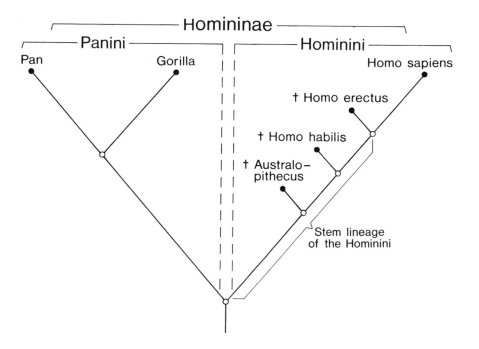

Fig. 66. Arrangement of the fossil taxa *Australopithecus, Homo habilis* and *Homo erectus* as plesia in the stem lineage of the Hominini.

The sequence of plesia in the stem lineage of the Hominini is logically derived from the successive increase of apomorphous agreements shared with *Homo sapiens*. For a detailed justification, I refer the reader to Eldredge *et al.* (1975) and Delson *et al.* (1977). In these papers the evolutionary novelties evolved, one after the other, in the stem lineage of the Hominini are listed, stem species by stem species.

The resulting diagram of phylogenetic relationships (Fig. 66) can be translated into a hierarchical tabulation as follows:

Homininae
    1. Panini
        1.1. *Pan*
        1.2. *Gorilla*
    2. Hominini
        Plesion †*Australopithecus*
        Plesion †*Homo habilis*
        Plesion †*Homo erectus*
        Homo sapiens

The plesion *Australopithecus* forms the fossil adelphotaxon of all the Hominini that follow it. The plesion *Homo habilis* forms the adelphotaxon of the species group *Homo erectus* + *Homo sapiens*. And, finally, *Homo erectus* is the fossil sister species of *Homo sapiens*.

Two general statements may be made in connection with the systematization just suggested:

1) The adelphotaxon relationship between the Hominini and the Panini, hypothesized for the present day on the basis of their recent representatives, is not affected by placing fossil taxa in the phylogenetic system. Obviously this adelphotaxon relationship would still remain unchanged if new fossils were inserted in the stem lineage of the taxon Hominini or if improved knowledge altered the position of known fossil taxa.
2) Conversely, the arrangement of fossils in the lineage leading to *Homo sapiens* would remain unaltered if the hypothesis here presented—of an adelphotaxon relationship between Panini (*Pan* + *Gorilla* and Hominini (*Homo sapiens*)— were replaced by some other, better validated hypothesis on the relationships between the recent taxa *Pongo, Pan, Gorilla* and *Homo* (cf. Ax 1985a).

Chelicerata

For my second illustration of the stem-lineage concept, I apply it to the Chelicerata. I shall arrange some well enough analysed fossils in the stem lineage of the taxon and thus characterize certain steps in the evolution of the chelicerate ground pattern (Lauterbach 1973, 1980a,b, 1983b; Weygoldt & Paulus 1979a,b).

I again present the relevant part of the system in two mutually equivalent ways—as a diagram of phylogenetic relationships (Fig. 67) and as a tabulation, as follows:

211

Chelicerata
    Plesion †Trilobita
            1.   Emuellida
            2.   Eutrilobita
    Plesion †"Olenellinae"
    1. Taxon Euchelicerata
            1.1. Xiphosura
            1.2. Arachnida
    2. Taxon Pantopoda

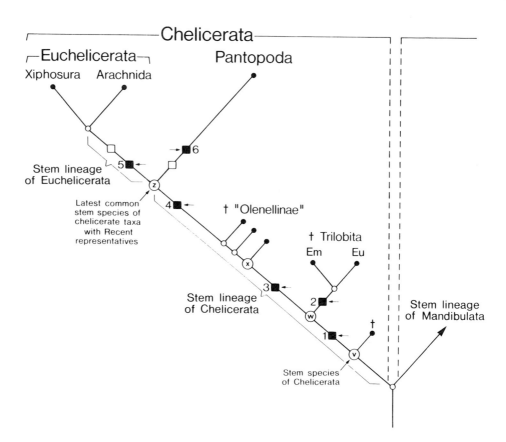

Fig. 67. Diagram of phylogenetic relationship for the Chelicerata with fossil units arranged in the stem lineage of the taxon. The blocks of apomorphies are explained in the text. Em = Emuellida: Eu = Eutrilobita.

I shall first validate the **monophyly of the Chelicerata** by way of some basic autapomorphies postulated as belonging to the differentiated ground pattern of the taxon. In other words, they belong to the mosaic of features of the latest common stem species z of the chelicerate taxa that have recent representatives. These autapomorphies are as follows:

— Division of the body into two tagmata: a) the prosoma, produced by fusion of the head with two thoracic segments (segments 1 + 2) and bearing six pairs of appendages; and b) the opisthosoma, comprising twelve complete thoracic segments (segments 3–14) and a long tergal spine which is the vestige of a 13th trunk segment (segment 15).
— 1st antennae of the Euarthropoda lost.
— 2nd prosomal appendage (2nd antenna of the Crustacea) differentiated to form a three-jointed chelicera.

On this basis I now turn to the stem lineage of the Chelicerata. There are thousands of trilobite-like fossils, but Lauterbach recognizes as representatives of a **monophyletic taxon Trilobita** only those species characterized by a pair of facial sutures on the head. Functionally, these structures are moulting sutures. Their primitive position, as seen in other Chelicerata, was at the margin of the head and their migration onto the dorsal surface is an apomorphy. Lauterbach connects the evolution of this apomorphy with stronger calcification of the dorsal cuticle in the Trilobita and resulting difficulties in moulting the primitively large compound eyes. However this may be, the roof of the head (cranidium) splits away from the lateral parts of the head (free cheeks) along the compound eyes at the dorsal margin of the cornea. The facial sutures and the remarkable method of moulting are seen as apomorphies evolved only once. And their identical expression is correspondingly stressed by Lauterbach as the substantive autapomorphy of the taxon Trilobita (in his sense).

The Trilobita are also welcome evidence for my argument that the stem-lineage concept can be extended, without the slightest logical difficulty, to purely fossil taxa.

Lauterbach has erected two units—the Emuellida and the Eutrilobita—as the highest ranking adelphotaxa within the Trilobita. In the Emuellida, which are relatively primitive (Fig. 68a), the trunk is divided into a prothorax of six segments and a long, distally tapering telosoma consisting of more than 50 segments. The fusion of segments 5 and 6 to form a diplosomite counts as an autapomorphy of the taxon. Constitutive features of the Eutrilobita, which are relatively more advanced than the Emuellida, are the marked reduction of the telosoma and the evolution of the pygidium by the fusion of trunk segments to form an intercalation in front of the telosoma. For a detailed justification of these results, I refer the reader to the original paper (Lauterbach 1980b). As concerns the stem-lineage concept the following situation is of interest. The basic change in the moulting mechanism of trilobites, for which the facial sutures provide fossilizable evidence, was evolved before the trilobites split into the Emuellida and the Eutrilobita. It thus occurred in a lineage which can incontrovertibly be called the stem lineage of the Trilobita (Fig. 67—the lineage branching off from the stem species w and carrying apomorphy 2). Logically, of course, unlike a systematization that combines fossil and recent taxa, the Emuellida cannot be seen as a plesion of the Eutrilobita—both of these fossil units must be treated, like recent adelphotaxa, as sister groups with identical rank.

This brings me to the next question: what are the arguments for placing the Trilobita in the stem lineage of the Chelicerata? From the synapomorphies listed by Lauterbach (1983b), I shall first cite the **apomorphous agreements** between the fossil **Trilobita** and the recent **Xiphosura** when both are compared with the Mandibulata.

(The Xiphosura are cited in this connection because they have the greatest number of plesiomorphous features among still surviving chelicerates.) The synapomorphies are as follows:
— The cuticle differs on dorsal and ventral surfaces—on the dorsal face it is greatly strengthened while on the ventral face it is a membrane of soft skin.
— The elevation of the axis of the body (rachis) above the lateral parts (pleura); the name of the taxon Trilobita refers to this well marked tripartition of the dorsal surface; among recent Chelicerata, it is pronounced only in the Xiphosura—in the prosoma and the anterior part of the opisthosoma.
— A well marked broadening of the anterior body compared with the posterior body—a feature in which the stem lineage of the Chelicerata differs basically from that of the Mandibulata.

These features must have arisen as evolutionary novelties in the early stem lineage of the Chelicerata (Fig. 67, apomorphy block 1 in the lineage between the stem species v and w).

Lauterbach has separated the fossil "**Olenellinae**" (Fig. 68c–e) from the taxon Trilobita and this is crucially important for a logically consistent phylogenetic systematization of the representatives of the stem lineage of the Chelicerata. The "Olenellinae" do not belong to the Trilobita, for they primitively lack the moulting structures which are the basic ground-pattern apomorphy of the Trilobita. There is another reason why the "Olenellinae" cannot belong to the stem lineage of the Trilobita. This is a complex of three apomorphous agreements with the recent Chelicerata which show that the "Olenellinae" definitely **belong above the Trilobita in the stem lineage of the Chelicerata**. This syndrome of features is set out by Lauterbach, as follows:

— The development of a postcephalic prothorax consisting of 15 broad segments freely articulated together. This portion clearly shows the evolution of the post-cephalic body of recent Chelicerata (two segments in the prosoma, 13 segments in the opisthosoma).
— Differentiation of the third prothoracic segment to form a macropleural structure. As a result of the evolution of strong pleurotergites in this segment, the first two trunk segments come to be, structurally and functionally, part of the head. Without doubt this represents the start of the evolution of the prosoma of the Chelicerata which involves the fusion of the first two trunk segments to the head.
— The evolution of a long spine extending from the roof of the 15th trunk segment; this becomes the terminal spine in the ground pattern of the Chelicerata (tail spine of king crabs, poison spine of scorpions).
These features must have arisen as evolutionary novelties in the stem lineage of the Chelicerata between the stem species w and x (Fig. 67, apomorphy block 3).

The fossils grouped together as "Olenellinae" are of general interest in yet another respect. Beneath the tail spine they primitively have a telosoma of many segments and this, naturally, corresponds to the telosoma of the Emuellida. Within the "Olenellinae" this telosoma is successively reduced (Fig. 68, c ⟶ d ⟶ e). But there

214

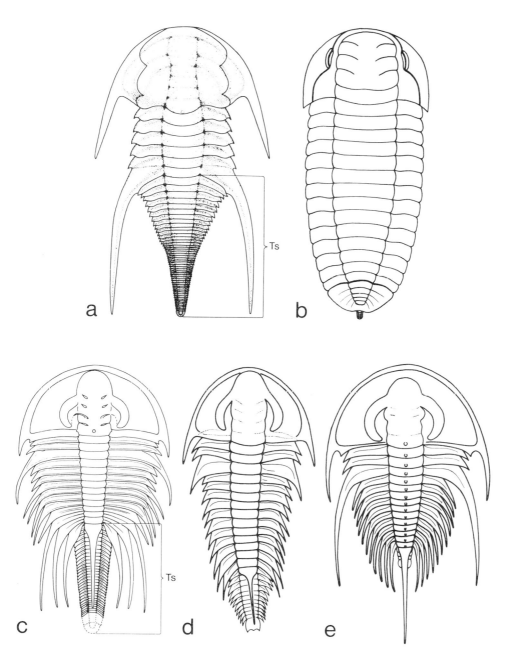

Fig. 68. Fossil Chelicerata. a + b) representatives of the monophyletic taxon Trilobita: a) *Balcoracania dailyi* (Emuellida); b) diagram showing the most widespread feature pattern of the Eutrilobita; c–e) representatives of the paraphyletic "Olenellinae" with increasing reduction of the telosoma: c) *Paedeumias robsonensis*, d) *Olenellus vermontanus*, e) *Olenellus thompsoni*. Ts = telosoma. From Lauterbach (1980b), after various authors.

is no single apomorphy common to the "Olenellinae". In other words, they are not a monophyletic taxon, but a paraphyletic grouping of fossil chelicerates. The three taxa figured obviously form a sequence in the stem lineage of the Chelicerata, as suggested schematically in Fig. 67 by three lines. They mark stages in the reduction of the telosoma which in the latest common stem species z of recent chelicerates had completely disappeared.

Above the "Olenellinae", in the part of the stem lineage leading to stem species z, the further development of the chelicerate ground pattern was then completed. This involved loss of the 1st antennae, the differentiation of the chelicerae and the division of the body into prosoma and opisthosoma (Fig. 67, apomorphy block 4).

Determining the correct relative position of the "Olenellinae" in the stem lineage of the Chelicerata also greatly strengthens the argument for regarding the Trilobita as representatives of the stem lineage of the Chelicerata. For between the "Olenellinae" and the Trilobita there are a whole series of agreements which can be seen as synapomorphies but which were lost in the higher parts of the stem lineage of the Chelicerata. These include the markedly crescentic shape of the compound eyes, the great widening of the doublure of the head in ontogeny, the presence of three pairs of spines in the larvae and the development of toothed structures in the labrum of the larvae (Lauterbach 1980a, 1983b).

I wish briefly to mention the adelphotaxon relationship proposed by Weygoldt & Paulus (1979b) between the Euchelicerata[1] and the Pantopoda. The only synapomorphy which can be quoted between the Xiphosura and the Arachnida is the reduction of the median eyes from four to two in the stem lineage of the Euchelicerata (Fig. 67, apomorphy 5). In this feature the Pantopoda, with four median eyes, retain the plesiomorphous condition. The great abundance of autapomorphies evolved in the stem lineage of the Pantopoda includes, to quote only a few examples, the proboscis on the prosoma, the division of the prosoma into an anterior part with three pairs of limbs and a posterior part with four pairs of limbs, the additional pair of limbs known as the ovigera, and the reduction of the opisthosoma to a peg-shaped vestige (Fig. 67, apomorphy block 6).

I mention the validation of the highest ranking adelphotaxa within the recent Chelicerata so as to reiterate a general point about the phylogenetic systematization of fossil taxa. I refer, namely, to the fact that the sister-group relationship postulated between the Euchelicerata and the Pantopoda is totally unaffected by whatever fossils are placed in the stem lineage of the Chelicerata, in whatever sequence. For example, the paraphyletic † "Aglaspida" may still belong in the stem lineage of the Chelicerata or they may have arisen only after the splitting of the stem species z of the chelicerate taxa that have recent representatives, in which case they should be systematized as representatives of the stem lineage of the Euchelicerata. Weygoldt & Paulus (1979b)

---

[1] Weygoldt & Paulus have proposed the new name Euchelicerata for a taxon comprising the Metastomata (fossil Eurypterida and Arachnida), which is likewise new, and the Xiphosura. The plesion concept, however, releases us from the straitjacket of coining names for the adelphotaxon relationships between fossil and recent taxa. Consequently, the name Metastomata is not needed and the name Euchelicerata becomes available for the combination of Xiphosura and Arachnida to form a monophylum.

Likewise, the name Arachnata is widely used to indicate the sister-group relationship between the Trilobita, the fossil taxa that follow them in the stem lineage, and the recent Chelicerata. It, too, is made logically superfluous by the plesion concept. It should be finally deleted.

discuss this question at length but the recent adelphotaxon relationships are totally untouched by the answer.

Mammalia

The mammals have already been used extensively as an example in this book and I do so again. There is an oft-repeated question, full of pathos: "What are mammals?" (Thenius 1979, p.32). I wish to eliminate the last suspicion in the reader's mind that this question can be answered by authority, by consensus, according to the demands of practice or somehow, arbitrarily, else. A subjective answer is no more acceptable here than for any other closed descent community. Nature gives the answer herself.

"The definition of a mammal" (Kemp 1982, p.293) on the basis of subjectively chosen features is simply a fallacy. For definitions of single descent communities in Nature (= monophyla of the system), like definitions of single evolutionary species or single individuals, are logically not possible (p.23). A single descent community, being ontologically an individual unity in Nature, cannot be defined. It can only be characterized on the basis of the novelties evolved in its stem lineage which were present as autapomorphies in the latest common stem species of the recent representatives. And in this characterization we are not allowed arbitrarily to choose evolutionary novelties so as to place the fossil taxa in position. We are not free, in the characterization of the monophylum Mammalia, to select the feature complex of the secondary jaw articulation (between the squamosal and dentary) or the three auditory ossicles in the middle ear, although this is widely advocated. Instead it is a question of the first evolutionary novelty (autapomorphy) which shows up in the fossil record and becomes a synapomorphy of the members of the descent community. Such an evolutionary novelty objectively establishes the extent of every single supraspecific taxon as being equivalent to a particular closed descent community in Nature.

In the example that I am discussing, all the fossil Amniota which show one, several or all the fossilizable synapomorphies of the recent Monotremata and Theria, but no autapomorphy of the Monotremata nor of the Theria, belong in the stem lineage of a particular closed descent community in Nature. And the equivalent of this closed descent community in the phylogenetic system of organisms is the supraspecific taxon Mammalia. The fossil units A—E in Fig. 69 are all, unconditionally, mammals! Furthermore, all fossil Mammalia which possess at least one synapomorphy of the recent Ornithorhynchidae and Tachyglossidae, but no autapomorphy of the platypus, nor of the echidnas, belong equally incontrovertibly to the stem lineage of the monophylum Monotremata (fossil units H—I). Fossils with features that we suppose to be synapomorphies of the recent Marsupialia and Placentalia, but which show no constitutive feature of the taxon Marsupialia nor of the taxon Placentalia, are representatives of the stem lineage of the taxon Theria (fossil units M–Q). And a corresponding argument applies in assigning the fossils T–U to the stem lineage of the monophylum Marsupialia or the fossils X–Y to the stem lineage of the monophylum Placentalia.

I shall now apply these general statements to the actual case of the amniote fossil record. Even the earliest "Pelycosauria" belong in the stem lineage of the Mammalia

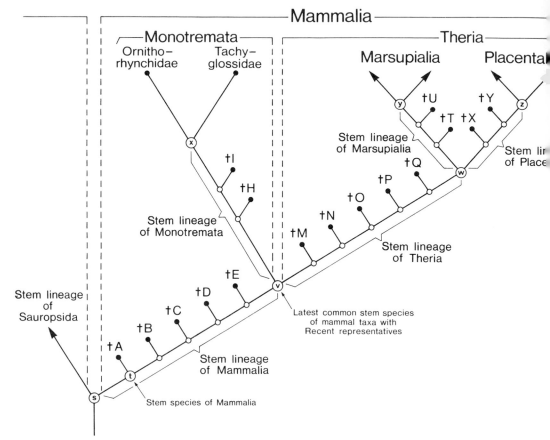

Fig. 69. Diagram of phylogenetic relationship for the mammals with the fossil taxa (A–Y) placed in the stem lineages of the Mammalia, Monotremata. Theria, Marsupialia and Placentalia. The alphabetical sequence symbolizes the temporal sequence of fossil units in the individual stem-lineages, according to the increasing number of evolutionary novelties which have come to be synapomorphies of the recent adelphotaxa.

because they have a synapsid skull with a lower temporal fossa (Fig. 70). They are, without if or but, representatives of the closed descent community Mammalia. And since the synapsid skull is the earliest recorded evolutionary novelty which constitutes an autapomorphy of the recent mammals, the pelycosaurs establish the latest possible date (the *terminus post quem non*) for the origin in geological time of this unity in Nature (p.229). The stem species of the Mammalia (t) arose when the latest common stem species of the recent Amniota (s) split into the stem lineages of the adelphotaxa Sauropsida and Mammalia. It must have lived in the Palaeozoic before the Upper Carboniferous.

Along with the Pelycosauria, it is obvious that all the other "mammal-like reptiles" of the late Palaeozoic and Mesozoic are not "reptiles", for these simply do not exist as a unity in Nature. These fossils are nothing but stem-lineage representatives of the Mammalia. Conventionally, they are usually included in a big paraphylum

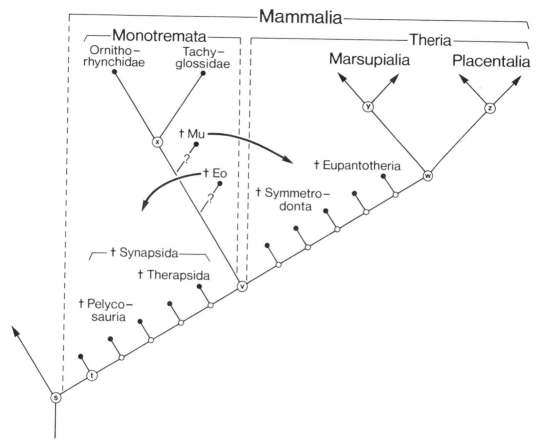

Fig. 70. Diagram of phylogenetic relationship for the mammals with the fossils inserted under the names of the conventional paraphyletic groupings. The arrows indicate: 1) the transfer of the Eotheria (Eo: Triconodonta + Docodonta) from the stem lineage of the Monotremata into that of the Mammalia; and 2) of the Multituberculata (Mu) from the stem lineage of the Monotremata to that of the Theria. The reasons for this are explained in the text.

"Therapsida" or divided into two paraphyletic "orders"—the "Therapsida" and the "Therosauria" (Kemp 1982). But they arose in the phylogenesis of the mammals after the "Pelycosauria" or, in other words, belong above these in the stem lineage of the taxon Mammalia. Within the "Therapsida", with its great number of species, the ground pattern of the recent Mammalia successively evolved with more than 60 features specific to them alone.

Near the Jurassic-Cretaceous boundary, therefore, there was a particular evolutionary species of mammal which possessed every one of the constitutive features of the Mammalia but which could not have had any autapomorphy of the Monotremata nor of the Theria. This species was not—I stress it again with emphasis—the stem species of the closed descent community Mammalia. Rather it stood at the end of the stem lineage of the Mammalia and was the latest common stem species (v) of the mammalian groups with recent representatives.

Thus I systematize all the species combined under the names "Pelycosauria" and "Therapsida" as representatives of the stem lineage of the taxon Mammalia. This contrasts with the conventional classification. For this takes all fossil species of these paraphyla that have a primary, articular-quadrate jaw articulation and a single auditory ossicle in the middle ear and unites them in a further paraphyletic assemblage under the name "Synapsida" (or "Theromorpha") (Fig. 70). These "Synapsida" are then arbitrarily cut away from the phylogenetic system of the Mammalia and seen as a systematic unit of the "Reptilia". In my view, nothing demonstrates more convincingly the collapse of the ordering attempts of traditional evolutionary classifications. For paraphyletic groupings have never existed as real unities in Nature but are the artificial products of Man. As such, they can claim no place in the phylogenetic system of organisms. There is no choice but to eliminate the names Synapsida (Theromorpha), Pelycosauria, Therapsida, Therosauria and all other paraphyla from the phylogenetic system of the Mammalia. This is not narrow-minded dogmatism, nor is there here an ambiguous situation in which different standpoints may be adopted after duly weighing the reasons. The insistence on paraphyletic groups results in much deeper errors in the basic interpretation of the course of phylogenetic development. For if the artificial grouping "Synapsida" is seen as a unity in Nature, is regarded as part of an equally artificial "class Reptilia", and if the "Reptilia" is then placed as a "class" of vertebrates alongside the "class" Mammalia, then grave misunderstandings may result. Thus Kemp (1982, p.1) has written: "Therefore this is the one example known where the evolution of one class of vertebrates from another class is well documented by the fossil record".

On the contrary, "... the evolutionary transition from one vertebrate class, the reptiles, to another, the mammals ..." (Kemp 1982, p.296) is plainly impossible. For in Nature there are neither supraspecific ancestors nor are there "classes" (as categories of rank). There are only evolutionary species as the potential ancestors of closed descent communities and closed descent communities as the historical products of phylogenesis with impassable limits. The categories of the Linnaean classification, such as genus, family, order and class are pure human inventions intended for the dubious indication of the different ranks of supraspecific taxa (p.234). The ostensible evolution of one class from another class, the postulated evolutionary transition from one genus or order into another genus or order, has never once happened in Nature. As purely human errors, formulations of this sort do great mischief in the phyogenetic literature.

This brings me back to the stem-lineage concept. The basic step is thus the allocation of the fossil units to the stem lineages of the Mammalia, Monotremata, Theria etc. The next step is then to sequence them, within the individual stem lineages, according to successive increase in the apomorphous features of the ground pattern of the recent representatives of the taxa in question. In Fig. 69 this is shown schematically by alphabetic order. I cite three attempts in this direction. Reisz (1980, Fig. 17) has arranged five fossil units of the paraphyletic "Pelycosauria" in a lineage which represents the basal part of the stem lineage of the Mammalia. Gaffney (1980, Fig. 3) has taken eleven fossil groups of the paraphyletic "Synapsida" and, by means of synapomorphies, has given evidence for their relative position in the stem lineage of the Mammalia. Finally in Kemp (1982, Fig. 113), 20 fossil mammalian taxa, belonging to the "Pelycosauria", "Therapsida", "Cynodonta" and Mammalia, are arranged in sequence. This diagram, however, does not show the branching-off of the adelphotaxa Monotremata and Theria and therefore fails to separate the stem lineages of the Mammalia and Theria.

In arranging fossil mammals sequentially in the stem lineages of the recent taxa, we must consistently reject the assertion, often thoughtlessly repeated, that various constitutive features of the Mammalia have arisen convergently (Olson 1959, Simpson 1959; see p.193). To the exact contrary, for each apomorphy we must begin, *a priori*, from the most parsimonious hypothesis that the feature has evolved once only. Only irresolvable conflicts, resulting from incompatibilities in the distribution of apomorphous features, can require, *a posteriori*, that individual apomorphous agreements be interpreted, point by point, as convergences.

I shall illustrate this by means of an example from the phylogenetic systematization of fossil mammals. In the traditional classification, the fossil Eotheria (Triconodonta + Docodonta) and the Multituberculata are combined with the recent Monotremata in a taxon Prototheria[1] (Non-Theria). The fundamental argument for this combination comes from particular agreements in the structure of the side wall of the skull. Thus the lamina obturans is a cartilage bone which ossifies independently in the skull wall; in the Monotremata it fuses with the periotic bone while in the Theria, on the other hand, it fuses with the ventral part of the alisphenoid (Kuhn 1971, Presley & Steel 1976, M. Griffiths 1978, Patterson 1981b, Kemp 1982). The skull wall of the Triconodonta and Multituberculata agrees with the monotreme alternative, and if these agreements are seen as synapomorphies, then they necessarily contradict the following empirical results: 1) In the Triconodonta a primary jaw articulation between the quadrate and articular has been shown to exist, which means that they can have had only one auditory ossicle in the middle ear; 2) the Multituberculata were probably viviparous as judged by the structure of their pelvis (Kielan-Jaworowska 1979). In other words, the hypothesis of a monophyletic taxon comprising Triconodonta + Docodonta + Multituberculata + Monotremata necessarily requires the evolution twice convergently, in the stem lineages of the Monotremata and of the Theria, of: 1) three auditory ossicles in the middle ear; and 2) vivipary.

As against this, the principle of parsimony requires a more economical and much more illuminating hypothesis. It implies, namely, that the structure of lateral skull wall of the Monotremata is the plesiomorphous condition for the Mammalia and that the structure seen in the recent representatives of the adelphotaxon Theria is the apomorphous alternative (Presley & Steel 1976, Patterson 1981b, Kemp 1982). But this interpretation eliminates the argument for systematizing the Eotheria and the Multituberculata as representatives of the stem lineage of the Monotremata. And if this argument fails, then speculations about the independent evolution of the middle-ear complex with three auditory ossicles and of vivipary in the Monotremata and Theria are null. Rather these two apomorphies remain what they were *a priori* supposed to be: 1) The three auditory ossicles in the middle ear form a synapomorphy of the Monotremata and Theria; the feature would have arisen, once only, as an evolutionary novelty in the stem lineage of the Mammalia. And 2) vivipary

---

[1] On the synonymy of the names Monotremata and Prototheria see p.223.

is a synapomorphy of the Placentalia and the Marsupialia; as such it evolved, once only, in the stem lineage of the Theria. Patterson (1981b) applies the results of these interpretations to the fossil taxa. He assigns the Triconodonta and the Docodonta to the stem lineage of the Mammalia (before the evolution of the auditory ossicles malleus and incus) and he assigns the Multituberculata to the stem lineage of the Theria (after the evolution of vivipary and before the evolution of the therian alternative for the side wall of the skull). These transfers of position have been expressed in Fig. 70.

Tabulation of mammalian relationships as in Fig. 69.
Mammalia
    Plesion † A
    Plesion † B
    Plesion † C
    Plesion † D
    Plesion † E
    ⋮
    Monotremata (= Prototheria)
        Plesion † H
        Plesion † I
        ⋮
        Ornithorhynchidae
        Tachyglossidae
    Theria
        Plesion † M
        Plesion † N
        Plesion † O
        Plesion † P
        Plesion † Q
        ⋮
        Marsupialia (= Metatheria)
            Plesion † T
            Plesion † U
            ⋮
            Recent adelphotaxa of the Marsupialia
        Placentalia (= Eutheria)
            Plesion † X
            Plesion † Y
            ⋮
            Recent adelphotaxa of the Placentalia

Thus hypotheses as to the first degree of phylogenetic relationship between organisms of the present day give the starting point for the phylogenetic systematization of fossils. And the postulated adelphotaxon relationships of recent species and species groups are not altered by assigning fossil units to the stem lineages of the adelphotaxa nor by transferring the fossils from one stem lineage to another. I have exemplified both these points in the case just discussed.

These results logically suggest that **redundant taxa should be deleted from parts of the system that contain both fossil and recent representatives.** I shall illustrate this in the case of the Monotremata and the Theria. The adelphotaxon of the Theria is a mammalian taxon containing the three recent species *Ornithorhynchus anatinus*, *Tachyglossus aculeatus* and *Zaglossus bruijni*. This taxon can have only one valid name, although, of course, it does not fundamentally matter what name is chosen. On the other hand, it is a gross factual error to advocate a taxon Prototheria to which a taxon Monotremata and also various fossil units are made subordinate. As concerns the recent representatives, the content of the Prototheria and the Monotremata is exactly the same and the names are synonyms. The situation is not altered in the slightest by placing fossils in the stem lineage of the monotremes. For these fossils are stem-lineage representatives of one and the same taxon which, as sister group of the Theria, takes a valid name. If we decide on Monotremata, then the younger name Prototheria has to be eliminated as a synonym.

These basic considerations obviously do not depend on whether the Triconodonta, Docodonta and Multituberculata belong in the stem lineage of the Monotremata or not. Moreover, the example of the Monotremata and Prototheria exactly corresponds to the mutual redundancy of the names Chelicerata and Arachnata within the arthropods. If the fossil Trilobita, the "Olenellinae" etc. are assigned to the stem lineage of the Chelicerata, a superordinate taxon Arachnata has no right to exist in the phylogenetic system of the Arthropoda (p.216).

I **summarize.** However many fossils are assigned to the stem lineage of the Mammalia, or to the stem lineages of the subordinate adelphotaxa within the Mammalia, there is only one correct way to tabulate the phylogenetic system of the taxon Mammalia, and this is shown on the opposite page. This tabulation exactly translates the diagram of phylogenetic relationship shown in Fig. 69.

I shall finish this discussion of the mammalian example by repeating a central thought. The "Mesozoic Mammals" do not represent "The first two-thirds of mammalian history", whatever the subtitle of the work by Lillegraven, Kielan-Jaworowska & Clemens (1979) may assert. The history of the mammals did not begin in the Mesozoic (Triassic) but about 100 million years earlier, in the Palaeozoic (Carboniferous). It began with an evolutionary species (t) which, after the splitting of the latest common stem species of the recent Amniota (s), stood at the beginning of the stem lineage of a closed descent community in Nature. In the phylogenetic system of organisms, this closed descent community forms the taxon Mammalia (Fig. 69).

V. Eliminating the "stem group" from the systematization of fossils

What we know of the process of phylogenesis shows that no supraspecific grouping of organisms, whether monophyletic or non-monophyletic, can as a group be the ancestor of any descent community in Nature. This statement holds for any group of extinct species, any group of mixed fossil and recent species and, obviously, also for any group of purely recent species. This statement is logically correct and incontrovertible. It implies the rejection of the widespread view that "supraspecific ancestors" exist, and of the word "stem group" whose meaning is inter-connected (p.32).

Despite the semantic burden of the traditional meaning of the word, Hennig attempted (1969, 1983) to make the term "stem group" useful in phylogenetic systematics by giving it a very different meaning. According to him (1983, p.15), a stem group should include all those fossil species "which can be shown probably to belong to a particular monophyletic group of the phylogenetic system but which are probably no closer related to one subgroup represented among recent animals than to another."

I shall now **compare** Hennig's definition of the **fossil stem group** with **the stem-lineage concept** consistently applied in this book. The fossils combined into a "stem group" correspond exactly to the sum of the "representatives of the stem lineage" of a monophyletic taxon. The stem lineage describes a particular segment in the history of a closed descent community in Nature. And all fossil units are to be arranged in it which lived from the origin of the stem species until the time when the latest common stem species split into the first sub-taxa that contain recent species.

Hennig wished it to be understood that the "stem group" was decidedly a compromise adapted to the limitations of palaeontology in acquiring data. Although unobjectionable in logic, it is unsatisfying in fact (cf. Wiley 1979c, 1981). My crucial objection is as follows: If the "stem group" comprises several fossil taxa—whether evolutionary species or monophyletic species groups—which share different numbers of apomorphous agreements with the recent representatives, then the "stem group" is necessarily a non-monophyletic grouping and, indeed, a paraphylum. Thus, in the schematic example of case 3 (Fig. 65), the species M and P and the monophylum γ (with R + S) would be combined together as the "stem group" of taxon II and perhaps given a name. Or the taxa *Australopithecus*, *Homo habilis* and *Homo erectus* in the lineage leading to *Homo sapiens* would be combined to form a "stem group". But these measures would produce classic examples of paraphyletic groups created for extinct species. Such artificial creations are unjustifiable, but unfortunately they have often happened. The †"Rhipidistia", †"Labyrinthodonta", †"Cotylosauria" and the above-mentioned †"Synapsida", †"Pelycosauria and †"Therapsida" are all infamous examples among vertebrates of the paraphyletic "stem groups" which penetrate the phylogenetic literature in large numbers.

Phylogenetic systematics, however, irrefusably requires that only monophyletic taxa have the right to exist. Thus it can in no way tolerate that fossil paraphyla and recent monophyla be mixed together in the structure of its system. It does not matter whether the "stem group" is claimed to signify a potentially non-monophyletic collection of fossil organisms or whether it results from erroneous notions of the existence of supraspecific ancestors. In either case, the word "stem group" must be eliminated from the terminology of phylogenetic systematics, even for the fossil record. Jefferies (1979, 1980) has proposed to set up a "crown group", contrasted with the "stem group". The "crown group" is supposed to comprise the latest common stem species of the subtaxa that have recent representatives plus all descendants of this stem species. But if stem groups are deleted, then crown groups are automatically redundant.

The evolutionary concept of the stem lineage replaces the stem group. The fossil representatives of the stem lineage of a particular taxon never form a "group" of their own and, as a totality, do not logically deserve a group name. The terms "representative of the stem lineage" or "stem-lineage representative" may sound somewhat long-winded. When compared with the word "stem group", however, they have the priceless advantage of standing for a theoretically unobjectionable concept which is logically adapted to the aims of phylogenetic systematics.

VI.   The identification and systematization of fossil stem species

I have interpreted the evolutionary species of the present day as the provisionally terminal taxa of phylogenesis (p.43). They may die out without descendants and then will be definitively terminal taxa. On the other hand they may, by the processes of splitting, become the stem species of new closed descent communities and, by means of their descendants, become part of the unbroken stream of phylogenetic development.

Going back into the geological past, we must reckon with both these conditions—with terminal species that died without issue, and also with the remains of real stem species. Any fossil organism may potentially be an individual of the stem species of a particular closed descent community in Nature.

In the preceding discussion of the phylogenetic systematization of fossil organisms, I have obviously not made appropriate allowance for this situation. For what did I do? I simply treated all fossil species as definitively terminal taxa and correspondingly systematized them as the terminal products, without descendants, of the stem lineages of closed descent communities.

Why did I do this and not otherwise? Why does phylogenetic systematics not place individual fossil species—such as *Archaeopteryx lithographica*—directly as stem species in the stem lineages of supraspecific taxa? After all, many biologists still see the discovery of stem species, and the documentation of direct ancestor-descendant relationships in the fossil record, as the culmination of phylogenetic research. I did not do so because it is extremely difficult to identify stem species incontrovertibly. As a rule, because the record is incomplete and given the methods now available, the difficulties can scarcely be overcome. I reject the widespread view that stem species can never be identified in principle (cf. Hull 1979). But nevertheless, testable hypotheses about stem species or about direct ancestor-descendant relationships will always be the result of rare lucky chances where an abundant record coincides with uninterrupted stratigraphical distribution and exhaustive biogeographical data (Prothero & Lazarus 1980, Wiley 1981).

What **conditions** must be realized in the feature pattern of fossil organisms if they are even to be considered as **possible candidates for the stem species** of closed descent communities? I shall focus on the actual, oldest stem species of the descent community and on the latest common stem species of the highest ranking adelphotaxa that have living representatives—these species stand, respectively, at the beginning and end of the stem lineage (Fig. 71).

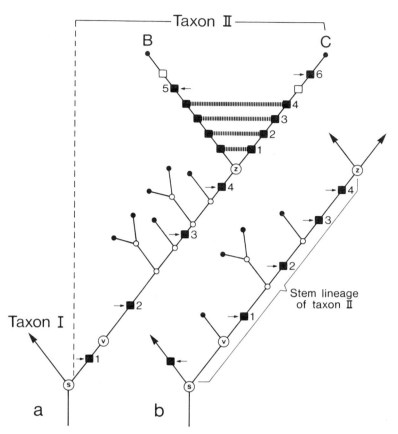

Fig. 71. Diagram of phylogenetic relationship to explain the methodological possibilities of identifying the stem species (v) of a monophyletic taxon and also the latest common stem species of its recent representatives (z). a) Stem species v with one autapomorphy of the taxon. b) Stem species v without any constitutive feature of the taxon.

The stem species of taxon II is the evolutionary species v. To be identified as such, it must possess at least one of the apomorphies which later become synapomorphies of the recent representatives of the taxon. This requires the early evolution of an autapomorphy 1 in the stem lineage of taxon II and assumes the existence of this autapomorphy in the stem species v (Fig. 71a). In addition, the stem species must be in every respect plesiomorphous compared with its descendants and must lack any evolutionary novelty of its own. But what happens if the stem species v still has none of the constitutive features of taxon II (Fig. 71b), since these were only evolved later in the stem lineage? In such case, it is methodologically impossible to recognize the evolutionary species v as being the stem species of taxon II.

The latest stem species z in the stem lineage of the taxon II must obviously have possessed the apomorphies 1–4 which count as synapomorphous agreements among the recent representatives B and C of the taxon. In addition, the latest common stem species z, like the oldest stem species v, would have had none but plesiomorphous features. It could not have had autapomorphy 5 of taxon B nor autapomorphy

6 of taxon C. Moreover, it could not have possessed any evolutionary novelties (autapomorphies) exclusive to itself.

As regards the potential identification of stem species on the basis of fossil feature patterns, the result of these considerations is very unsatisfactory. **It is, in principle, impossible positively to characterize stem species by way of features that they alone possess**.

Does this situation present special problems for phylogenetic systematics? Does the appropriate systematization of fossils in any way depend on whether past evolutionary species stood directly in the stem lineage or represented terminal taxa?

In searching for a binding answer, I refer back to *Archaeopteryx lithographica* as a representative of the stem lineage of the Aves (Fig. 72). On the one hand, it is extremely unlikely that this animal directly represents the stem species v of the closed descent community Aves. Why? Because the long series of constitutive bird features in *Archaeopteryx* (p. 230.) could scarcely have evolved within the limits of a single evolutionary species. On the other hand, it is impossible that *Archaeopteryx lithographica* represents the latest common stem species z of bird taxa that have recent representatives. This can be definitely excluded because a considerable number

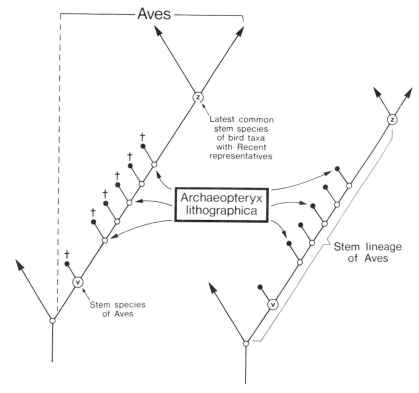

Fig. 72. The two possible positions of the evolutionary species *Archaeopteryx lithographica* with reference to the stem lineage of the Aves. Left: as a stem species in the middle part of the stem lineage. Right: as a terminal taxon.

of constitutive bird features were not yet present in *Archaeopteryx lithographica* but only evolved later in the stem lineage of the birds (p.230).

But these simple statements, of course, do not bring us to the kernel of the problem. For in the "middle part" of the stem lineage of the birds, *Archaeopteryx lithographica* could, in principle, be a stem species or a terminal taxon without descendants. It is very difficult to decide whether it is the stem species of all bird species that followed afterwards or the adelphotaxon of them. But the latter solution is likelier since *A. lithographica* shows an obvious autapomorphy in the structure of the ilium (Ostrom 1976; Rieppel 1984).

For the purposes of phylogenetic systematics, however, this question becomes a side issue if, fully conscious of the problems, it treats within the possibilities of its methodology all fossil organisms as terminal taxa in principle (Patterson & Rosen 1977). Thus, for example, it would consider *Archaeopteryx lithographica* as a terminal taxon and systematize the species in the stem lineage of birds as the adelphotaxon of all the units that follow it. For the phylogenetic systematization of the fossil record, it is of no importance whether *Homo habilis* forms the adelphotaxon of *Homo erectus* and *Homo sapiens*, or, as a stem species, was their immediate ancestor (p.210). However, there is a subtle, but essential, difference between these alternatives. The phylogenetic literature is overloaded with statements about stem species and direct ancestor-descendant relationships although, in general, such statements are speculations, untestable in principle. On the other hand, as concerns the validation of objective hypotheses about adelphotaxon relationships, a solid methodological basis now exists.

From the viewpoint of a logically consistent systematics, however, the stem-species problem results in another, seemingly serious difficulty. Every stem species belongs, without the slightest doubt, in the closed descent community of which it is the originator. But it is also certain that the stem species can belong to no subordinate sub-taxon of the closed descent community and cannot, itself, be a subordinate unit within the taxon.

Thus, if convincing arguments can be brought to identify a particular evolutionary species as the stem species of some descent community, how can this species be integrated into a phylogenetic system whose structure is solely based on the sister-group relationships of terminal taxa? Wiley (1979a, 1981) has made a simple and elegant suggestion. Like every evolutionary species, the stem species would receive a generic and specific name in accordance with the rules of nomenclature. The name of the stem species would then be placed after the name of the supraspecific taxon which arose from that stem species. Wiley's examples for the birds and mammals are as follows:

Taxon Aves (*Avus ancestorcus*)
Taxon Mammalia (*Mammalius primus*)

This procedure would show the equivalence between the supraspecific taxon and its stem species in an understandable and logically unobjectionable way. It is, however, more than doubtful whether we shall ever be in a position to make appreciable use of this proposal.

228

VII. The significance of the fossil record for phylogenetic research

In a general survey of the value of fossils in determining phylogenetic relation-ships, Patterson (1981a, p.218) concluded that: "It follows that the widespread belief that fossils are the only, or best, means of determining evolutionary relationships is a myth". His conclusion will no longer seem surprising after what I have just said about the conditions for integrating fossil organisms in stem lineages on the basis of a previously established phylogenetic system of recent organisms,.

However, I wish finally to emphasize three fields, closely connected with the anal-ysis and interpretation of features, in which fossil taxa, **after** they have been phylo-genetically systematized, give unique phylogenetic information.

The geological age of closed descent communities

The oldest known fossil of a taxon with the lowest number of apomorphies which count as synapomorphies with the recent representatives will mark what Hennig called the "*terminus post quem non*" of the taxon. This is the instant after which the unit would not have arisen as a closed descent community in Nature. Logically, this fossil establishes not only the latest possible geological age for the taxon to which it belongs, but also the latest possible age for the adelphotaxon.

Thus the oldest fossil which can be integrated in the stem lineage of the taxon Crocodilia is † *Protosuchus* from the Upper Trias (Hennig 1983). But this organism — and not *Archaeopteryx lithographica* from the Upper Jurassic — also establishes a corresponding latest possible date for the origin of the Aves, since these are the adelphotaxon of the Crocodilia and thus arose at one and the same time.

Again, the oldest well preserved taxon of the stem lineage of the Mammalia is the "pelycosaur" *Archaeothyris florensis* from the Pennsylvanian of Nova Scotia (Kemp 1982). In other words, the unity in Nature which forms the taxon Mammalia of our reference system, has its *terminus post quem non* in the Upper Carboniferous. Hennig's discussions (1982, p.181) in this connection do not refer to this instant but rather to the splitting of the latest common stem species of recent mammals into the adelphotaxa Monotremata and Theria (p.219). This split occurred in the Mesozoic near the Jurassic-Cretaceous boundary.

The sequence of evolution for the constitutive features of a taxon

The latest common stem species of the recent representatives of a closed de-scent community would have had the complete series of constitutive features of the taxon. Of course there will, as a rule, be a few features whose evolution would seem impossible unless other particular features were present, or whose origin is only conceivable in correlation with such features. Fundamentally, however, in the feature pattern inherited by recent representatives of a taxon, the constitutive fea-tures (autapomorphies) of the taxon — which exist side by side — say nothing about the temporal sequence in which they arose.

This is where the fossil record comes in. The more fossil stem-lineage representa-tives with different quantitative and qualitative expressions of the autapomorphies of the taxon, the more it is certain that a logically consistent systematization of the fossils within the stem lineage gives the temporal sequence for the evolution of the constitutive features.

As I shall show for *Archaeopteryx lithographica*, even a single extinct species, if well preserved and well dated, divides the fossilizable autapomorphies of a taxon into two sets which arose at different times. With reference to the list given above (p.93), a selection for the taxon Aves is as follows:

1st. set of constitutive ground-pattern features of the Aves

These were present in *Archaeopteryx lithographica* and therefore arose between the Upper Trias and the Upper Jurassic.
— Feathers and their arrangement on the fore limbs to form wings
— Elongate narrow scapula
— Union of the clavicles to form a furcula
— Transformation of the fore limbs to carry the wings, with loss of the 4th. and 5th. fingers; in the remaining three fingers the metacarpals are not yet fused
— Hind limb with differentiated intertarsal joint and with opposability of the hallux; as against this, the fibula is free and the metatarsals are also free and unfused.

2nd set of constitutive ground-pattern features of the Aves

These are absent in *Archaeopteryx lithographica* and therefore arose approximately in the interval between the Upper Jurassic and the Upper Cretaceous[1]
— Loss of the teeth
— Evolution of the saddle-shaped joints on the cervical vertebrae
— Formation of the synsacrum
— The fusion of the pelvic bones ilium, ischium and pubis to form the innominate bone
— Loss of the ventral ribs
— Flattening of the thoracic ribs and development of the uncinate process
— Further differentiation of the fore limb to form a flight organ with decrease in the number of phalangeal joints and fusions in the metacarpal region
— Further differentiation of the hind limb with the union of tibia and fibula and fusion of the four remaining metatarsals to form a single bone

In giving the periods of time for the evolution of the two sets of features, I have assumed as a simplification that the origin of the evolutionary species *Archaeopteryx lithographica* coincided with the existence of the individuals found in the Upper Jurassic. If *A. lithographica* already existed appreciably earlier, then the time for the evolution of the first set of constitutive features would be correspondingly shorter, and that for the second set would be longer.

---

[1] This is the *terminus post quem non* for the existence of the latest common stem species of the bird taxa that have recent representatives. Perhaps this instant was, in fact, considerably earlier (Cracraft 1982).

A contribution to deciding between plesiomorphy and apomorphy

If a particular feature exists in two expressions among recent representatives of a taxon, but is completely absent outside the taxon, then the difference is impossible to evaluate. In such case, fossils which can be integrated into the stem lineage of a taxon may show a condition of the feature which helps to decide between the alternatives of plesiomorphy and apomorphy (p.119).

I refer back to the example of the marsupial (epipubic) bones. This has been shown to exist in the †Tritylodontia which are stem-lineage representatives of the Mammalia. Because of this, the existence of the bones in the Monotremata and Marsupialia can be interpreted as a symplesiomorphy and its absence in the Placentalia as an autapomorphy.

K. Phylogenetic Systematization and the Formalities of Classification

"Systematic methodology, viewed as a whole, must proceed in one direction, from discovering the taxic hierarchy to classifying it. By classification, one simply means the assignment of the various taxa to the different categories (i.e. assignment of rank). By definition, then, taxa logically precede categories."

"A Linnean classification is nothing more than a system of names hierarchically arranged." (Eldredge & Cracraft 1980, p.168)

The aim of phylogenetic systematics is to discover the phylogenetic relationships between evolutionary species and closed descent communities and to portray them in a system whose structure corresponds to the order created by Nature. The logical activities of phylogenetic systematics come up against a series of problems of classification. Some of these problems are inevitable results of the International Rules of Nomenclature (Jeffrey 1973)[1]. Others take the form of conventions intended to meet various practical requirements. In any case, these are technical matters logically subsequent to the perception-processes of phylogenetic systematics. They have no scientific value of their own. I therefore emphasize what I said in the introduction to this book: putting the products of phylogenetic development into order should be called phylogenetic systematization; under this name, it must be clearly distinguished from the classificatory measures which may be needed, or may seem sensible, in the practice of systematics.

I. Rules and conventions according to Wiley

In his "annotated Linnean hierarchy" Wiley (1979c, 1981) has made a far-reaching attempt to combine, in the best practical way, the essential postulates of phylogenetic systematics with the various formal aspects of classification. Here I give Wiley's rules and nine of his conventions.

Rules

1) Taxa that are classified without any special indication are monophyletic groups in Hennig's sense (1966). Non-monophyletic groups may be inserted if they are unequivocally indicated as such.

2) The relationships of the taxa within the classification must be expressed exactly.

Conventions

1) In order to classify organisms, the Linnaean hierarchy will be used together with certain other conventions.

---

[1] International Code of Zoological Nomenclature, 3rd. edition (1985). — International Trust for Zoological Nomenclature, London.

International Code of Botanical Nomenclature (1975). — International Association for Plant Taxonomy, Vol.97, Bohn, Scheltema & Holkema, Utrecht 1978.

International Code of Nomenclature of Bacteriology (1966, 1969). — International Journal of Systematic Bacteriology 16, 459–490 (1966); 21, 111–118 (1971).

2) In erecting a classification, or modifying an existing classification, the number of taxonomic decisions should be kept to a minimum. This can be achieved in two ways. Firstly, no empty or redundant categories should be created, except where these categories are necessary taxonomic conventions, i.e. the five mandatory categories[1] are permitted to be redundant. Secondly, those natural taxa which are of essential importance for the classified group should, whenever possible, keep their traditional rank, in so far as this agrees with their phylogenetic relationships and the taxonomy of the group as a whole.

3) Taxa which form an asymmetrical part of a diagram of phylogenetic relationships ("phylogenetic tree") may be given identical rank (or else be inserted as plesia) and arranged in a sequence in accordance with their phylogenetic origin. If a list of taxa is introduced into a phylogenetic classification without further comment, then this implies that the list reflects a completely dichotomous sequence in which the first taxon is the sister group of all the others and the same is true for each included taxon down to the end of the list.

4) Monophyletic groups connected together at a trichotomy or polytomy take identical rank. Their uncertain mutual relationships are expressed by the term "sedis mutabilis" and they are placed at the level of hierarchy where their phylogenetic relationships to other taxa are known.

5) Fossil or recent monophyletic taxa of uncertain relationship are referred to as "incertae sedis". They are placed at the hierarchical level at which their phylogenetic relationships are best understood.

6) Paraphyletic and polyphyletic groupings, or groups of unknown nature, may be included in a phylogenetic classification if their names are put in quotation marks. These indicate that the included taxa are taxa *incertae sedis* at the hierarchical level to which the group is assigned. Such groups are given no rank—they neither take a formal rank nor do they have plesion status.

7) Fossil groups are classified differently to recent taxa. All monophyletic fossil taxa are given the status of a plesion. If a plesion is sequentially arranged in a classification which combines recent and fossil units, then it is the sister group of all other terminal taxa within its descent community that follow it in the classification. Plesia may also be *sedis mutabilis* or *incertae sedis* in relation to other plesia or recent taxa.

8) The stem species of a supraspecific taxon is classified in a monotypic genus and placed in brackets beside the taxon in which its descendants are included. The stem species of a genus is classified in this genus and placed in brackets beside the generic name.

9) A taxon of hybrid origin is placed with one or both of the parent taxa. Its hybrid nature is indicated by placing the names of the parents, in brackets, beside the name of the hybrid. The position where a hybrid taxon is inserted does not indicate where it split away in relation to the next following sequentially arranged taxa of non-hybrid origin.

The first rule and convention 6 connected with it have been repeatedly stressed in the course of this book as representing one of the central requirements of phylo-

---

[1] Compare p.239, 240.

genetic systematics. Non-monophyletic groupings, used provisionally until the phylogenetic relationships of their members are better understood, will as a rule be paraphyla. I emphasize the proposal that all groups which have not been validated as monophyla should be given quotation marks.

The second rule seems, at first sight, to be an unnecessary truism. Nevertheless, the arrangement of recent taxa in a sequence, as permitted by the rule and specified in convention 3, will become an important point in the discussion that follows here.

As a preliminary, I shall seek to state my basic position about the problematical assignment of the categories of the Linnaean hierarchy to the taxa of the phylogenetic system. This statement of position will require a discussion of conventions 1 and 2.

Conventions 4 and 5 concern how two totally different forms of unclarified phylogenetic relationships should be indicated.

Thus taxa sedis mutabilis are species or monophyletic species groups which derive from a stem species shared by them alone, but whose mutual relationships are unresolved. Logically at least three taxa must be involved. If, for example, it were thought that the relationships between *Gorilla*, *Pan* and *Homo*, or between Monotremata, Marsupialia and Placentalia, were insufficiently clarified by the hypotheses of relationship favoured above, then the corresponding parts of the phylogenetic system could be tabulated in the following provisional form:

Taxon Homininae	Taxon Mammalia
Taxon *Pan* sedis mutabilis	Taxon Monotremata sedis mutabilis
Taxon *Gorilla* sedis mutabilis	Taxon Marsupialia sedis mutabilis
Taxon *Homo* sedis mutabilis	Taxon Placentalia sedis mutabilis

In applying the term 'sedis mutabilis', it does not matter whether the taxa result from trichotomies or multiple speciations—and so in principle cannot be more closely analysed (p.61)—or whether they represent dichotomous adelphotaxon relationships which cannot be sorted out at the moment.

The term "incertae sedis", on the other hand, applies to species or monophyletic species groups which can be placed in a particular part of the phylogenetic system, but whose position within this part is unknown or, at least, uncertain.

In the chapter on the phylogenetic systematization of fossils I have dealt at length with the plesion concept (convention 7) and with the proposal for classifying stem species (convention 8).

II. Taxa and Categories

General survey

For the purpose of a comparative treatment of the logical and factual connections between group, taxon, rank and category, I shall state, once again, what these words mean.

1. Group

Biological systematics operates with groups of organisms. Using the word 'group' in the first instance as a theoretically neutral term, two fundamental groupings are distinguishable. **Groups of organisms** result:

a) from combining particular biological individuals together to form species—and that primarily independent of the respective definition of the word species.

b) from combining certain species together to form supraspecific groups—and again, primarily irrespective of the question whether such groups represent equivalents of real unities in Nature or not.

2. Taxon

Particular groups of organisms form taxa of the phylogenetic system.

The **species taxon** is a group of individuals which, being a closed reproductive community, represents an evolutionary species in Nature.

The **supraspecific taxon** (= monophylum) is a group of evolutionary species which, taken together with the stem species shared by them alone, form a closed descent community in Nature.

Like individuals, the taxa of the phylogenetic system are given **proper names** (*Nomina propria*). What is the logical legitimation for this measure? The taxa of the phylogenetic system are portrayals of real unities in Nature—and these in turn are, ontologically, unities with the characteristics of an individual.

Every species taxon (as the equivalent of a particular evolutionary species) takes a **binomen** which is a nomenclatorial combination of a generic and a trivial name. As against this, the names of supraspecific taxa (as the equivalents of closed descent communities) are **uninominal**, i.e. they consist of a single word.

3. Rank

Every taxon has a rank which places it at a particular hierarchical level of the phylogenetic system. The species taxon takes the lowest rank. (Possible subdivision into subspecies, i.e. geographical races, can be omitted here because of its non-hierarchical nature.)

Adelphotaxa (sister species, supraspecific sister groups) in principle take identical rank.

4. Category

Following Linnaeus, a category is assigned to every taxon in a classification. The category indicates the rank of the taxon.

In the traditional classification of animals there are more than 20 widely used categories (Simpson 1961, Eldredge & Cracraft 1980). To begin with, I shall content myself with a selection.

Phylum—phylum
Classis—class
Ordo—order
Familia—family
Genus—genus
Species—species

In this descending sequence, which is based on convention, one particular category is known as a class. Obviously, however, all categories of the Linnaean hierarchy have in logic the character of "classes".

The connection

Thoughtless confusion of the terms "taxon" and "category" is a well-spring of mistaken formulations which causes many misunderstandings. Blackwelder (1967) has compiled an abundance of impressive howlers of this sort. I now wish to discuss how the real unities in Nature above the level of the biological individual are connected with the meanings of the words group, taxon, rank and category (cf. Ax 1985a,c). As examples I tabulate, for the species *Apis mellifera* and *Homo sapiens*, a few supraspecific taxa at ascending levels of the systematic hierarchy together with their categories in the conventional classification just written down.

Real supra-individual unities in Nature	Taxa in the phylogenetic system		Categories in conventional classification	Interpretation of the categories
Closed descent communities as groups of species (supra-specific, mono-phyletic taxa of the system)	Arthropoda Insecta Hymenoptera Apidae *Apis*	Chordata Mammalia Primates Hominidae *Homo*	Phylum Class Order Family Genus	Arbitrary subjective indications of relative rank with no obligatory correspondence to unities in Nature or to taxa of the system.
Evolutionary species as groups of individuals (species taxa of the system)	*Apis mellifera*	*Homo sapiens*	Species	Mandatory category of the international codes of nomenclature. Equivalent to the evolutionary species and to the species taxon

1) The taxa of the phylogenetic system, as groups of organisms, are equivalent to unities in living Nature which really exist. Categories, on the other hand, are pure inventions of the human brain without any reality or objectivity. Categorial terms are nothing more than arbitrary labels which are subjectively attached to the proper names of taxa to indicate their rank in the traditional classificatory hierarchy (e.g. "order" Hymenoptera, "class" Mammalia).

In other words, there is in Nature, as a category, no "class" Mammalia, but only a particular descent community which is placed, as the taxon Mammalia, in the phylogenetic system of organisms. This taxon has one and the same rank as its sister group Sauropsida. Dubiously transferring the logical classes of the Linnaean hierarchy to a consistently phylogenetic system, the attempt has been made to apply a particular categorial term to this rank. The subjectivity of this procedure is shown by a short selection. Thus McKenna (1975) assigned the taxon Mammalia to the class category; Nelson (1969) and Gardiner (1982) chose the cohort category; and Wiley (1979c) chose the subdivision category. The only agreement lies in recognizing a particular descending sequence of categorial terms. The category allotted to the taxon Mammalia then depends solely on the categories that happen to be given to the taxa of the phylogenetic system above the Mammalia, i.e. the Amniota, Tetrapoda, Vertebrata etc.

The evolutionary species *Apis mellifera* does not belong to an "order" Hymenoptera, in so far as "order" implies category. Rather it belongs in a series of internested descent communities of increasing geological age. In the phylogenetic system, the taxon Hymenoptera is equivalent, at a particular level of hierarchy, to one of these descent communities. Once again, in point of fact, it does not matter in the slightest whether the rank of this taxon is signified by the order category or by any other category whatever.

2) As concerns the binomen, the international rules of nomenclature prescribe two **categorial terms** for each species taxon i.e. the **species category** and the **genus category**.

For the **species category**, there is an objectifiable connection to Nature and the system. This is the equivalence, already discussed (p.19), between the evolutionary species, the species taxon and the species category.

The "**generic name**" has become, in the historical development of systematics, a mandatory part of the binomen of species taxa. Nobody will ever wish to change this. But general agreement in this technical matter in no way indicates that there need be a correspondence between the genus category, any particular unity in Nature, and a supraspecific taxon of the obligatory extent. In fact, there is no such correspondence. All categorial terms applied to taxa above species taxa are nothing but arbitrary labels. The assignation of categories to supraspecific taxa of the phylogenetic system can, in principle, never be made objective.

The already discussed nomenclatorial requirements in the formal treatment of monotypic "genera", "families" etc. do not affect the truth of these statements in the slightest (p.41,76).

3) The accepted **sequence of categories** is supposed to show the **relative ranks** of taxa that are super- or subordinate to each other. So far as monophyletic taxa of the phylogenetic system are concerned, the relative rank logically implies several statements on the relative geological ages of the respective descent communities.

Thus, if categories are assigned subjectively, let the taxon *Homo* be seen as a genus, the taxon Hominidae as a family, and the taxon Primates as an order. If so, the rank sequence of genus, family and order implies that the stem species of the "genus" *Homo* (the latter comprising fossil species + *Homo sapiens*) lived later in geological time than the stem species of the "family" Hominidae; and the stem species of this family was younger, in its turn, than the stem species of the "order" Primates.

In the simple examples given in the table, the use of the Linnaean hierarchy to indicate the relative ranks of supraspecific taxa may seem perfectly plausible. But I shall show later that the attempt to apply categorial designations consistently to all levels of hierarchy in the phylogenetic system soon leads to intolerable inadequacies. Before doing so, I must point out some weaknesses of an entirely different sort, i.e. those revealed by comparing the genera, families, orders, etc. of a particular closed descent community with the same categories in some other descent community.

4) When **different descent communities** in Nature, equivalent to monophyletic taxa in the system, are **confronted with each other**, categories given the same categorial designation do not correspond at all in any real, objectifiable sense.

For example, if the cnidarian taxon *Hydra*, the plathelminth taxon *Taenia* or the vertebrate taxon *Homo* are each referred to as a "genus", then the result of this procedure is a purely semantic agreement without the slightest scientific value. Logically, the same is true for all categories of supraspecific taxa superordinate to the "genera"—thus, in the present example, no knowledge is conveyed by classifying the Hydroidea, the Cyclophyllidea and the Primates, with seeming conformity, as "orders". Osche (1961) has impressively demonstrated the inequality of the conventional "classes" Nematoda and Cestoda from the viewpoint of the evolution of parasitism.

The categories in different parts of the phylogenetic system simply cannot be compared to each other. This is because genera, families, orders, phyla, kingdoms and so forth do not, in fact, exist in Nature. They are nothing but human inventions.[1]

Logically, there can be **no binding definitions** of any of the **words** which have been applied **as categorial terms** to indicate the relative ranks of supraspecific taxa. Likewise, the "pragmatic definitions" (Mayr 1969, 1975) which emphasize the "gaps" between genera, families and still higher categories, are arbitrary formulations irrelevant to phylogenetic systematics.

Assigning categories

Because of this unsatisfactory situation, Hennig (1966) tried to use the classificatory categories to **characterize the absolute rank of monophyletic taxa**. The only objective parameter available is the absolute geological age of the closed descent communities in Nature. Hennig therefore proposed that the categorial terms of the conventionally accepted sequence should be given to descent communities according to the particular geological period when these arose. Thus descent communities (or the supraspecific taxa equivalent to them in the phylogenetic system) which arose in the time from Cambrian to Devonian would receive categorial ranks based on the term "class" (superclass, class, subclass, infraclass, etc.). If they originated in the following period, between the Mississippian and the Permian, they would take categorial ranks of the "ordinal stage" (superorder, order, suborder, etc.). And if they originated in the period from the Trias to the Lower Cretaceous, they would take categorial ranks of the "family stage". Thus the relative ranks of the mutually subordinate subtaxa of individual descent communities would be translated into absolute ranks based on a firm reference parameter. But also, and of special importance, the subtaxa of different descent communities—such as those of Mollusca, Arthropoda

---

[1] In this connection, I believe it will greatly help the argument to repeat something that I have said already (p.220): Widespread claims for the "evolutionary transition" of one class into the next class, of the "evolutionary transformation" of one classificatory unit into another of whatever category, or of the "intermediate phylogenetic position" of some particular class between other classes are all fictions. They do not, in any way, refer to anything real in living Nature.

or Vertebrata—having henceforth the absolute rank of families, orders, classes, etc. would acquire the character of objectively comparable unities of Nature.

The logic and legitimacy of this attempt cannot be faulted (Crowson 1970). In practice, however, the principle of using phylogenetic age to determine category is very difficult to apply (Kraus 1976). Nobody has adopted it in actual systematic work and Hennig himself (1969, 1983) did not use it in the phylogenetic systematization of insects and chordates.

Eldredge & Cracraft (1980) transfer the problem of establishing the rank of taxa by absolute age into an indefinite future when, perhaps, a complete phylogenetic system for all organisms will exist.

As things stand at present, therefore, we have to fall back on the "value" of categorial terms as described above. This, inevitably, raises the question: Can the **subjective assignment of the categories** of the traditional classification be **used**, in any way, to **signify the position of monophyletic taxa** in the hierarchy of the phylogenetic system? Wiley (1979c, 1981) defends the Linnaean hierarchy and advocates its use according to conventions 1 and 2 as already given. Above the genus, the labels regarded by Wiley as mandatory categories are the family, the order, the class and the phylum. I have respect for this attempt to retain a long-established principle of classification more than two hundred years old. But I feel compelled to point out, as clearly as I can, the problems of applying the Linnaean categories in a consistently phylogenetic system. I shall illustrate my argument with examples that show three different aspects of the matter.

For my first example, I refer back to the mammals *Ornithorhynchus anatinus* and *Orycteropus afer* (p.24,41). If, in the hierarchy of classification, they are assigned the categories that Wiley requires, then, for the platypus, the genus *Ornithorhynchus* and the family Ornithorhynchidae are redundant taxa. Likewise, in the case of the aard-vark, the genus *Orycteropus*, the family Orycteropidae and the order Tubulidentata are redundant. This seems unavoidable, and according to Wiley it is also permissible, but that is beside the point. Rather than discussing it, I shall consider the "manda-tory" stages of order, class and phylum—basing my discussion on the portion of the phylogenetic system of the Vertebrata as written down by Wiley himself, and the sequence of categories that he advocates.

Category	Platypus	Aardvark
Phylum	Chordata	Chordata
Class	Euosteichthyes	Euosteichthyes
Order	Monotremata	Tubulidentata
Family	Ornithorhynchidae	Orycteropodidae
Genus	*Ornithorhynchus*	*Orycteropus*
Species	*O. anatinus*	*O. afer*

According to Wiley, the platypus at the hierarchical level of the order is placed in the taxon Monotremata but the aardvark is not placed at this level in the equal-ranking adelphotaxon Theria (which simply does not appear among the mandatory categories). Rather it is put in the taxon Tubulidentata which, in the phylogenetic system of the Mammalia, forms a group subordinate to the taxon Placentalia. In both cases, in the ascending hierarchy the "mandatory categories" include neither

239

the taxon Mammalia (a subdivision according to Wiley) nor the taxon Vertebrata (a subphylum for Wiley). In Wiley's tabulation it is rather the taxon Euosteichthyes which takes the rank of class. And even the assignment of the phylum category to the taxon Chordata—as shown in the table—could be criticized because, between the phylum Chordata and the subphylum Vertebrata, the level of the adelphotaxa Acrania and Vertebrata is ignored. In my view, these results do not support the demand that some categories of the Linnaean classification should be seen as "mandatory" in the phylogenetic system.

Concerning the problem of the "desirable preservation of long-accepted categories" I shall discuss the arachnids as my next example. In the traditional classification, the monophylum **Arachnida** is divided into nine taxa, each of them assigned to the order category. Among these, the "Pedipalpi" have been recognized, meanwhile, as being paraphyletic. As an artificial combination of the Uropygi and the Amblypygi, they must be broken up and the name Pedipalpi must disappear. This gives the following ten "orders" for the Arachnida.

1. Scorpiones	6. Pseudoscorpiones
2. Uropygi	7. Solifugae
3. Amblypygi	8. Opiliones
4. Araneae	9. Ricinulei
5. Palpigradi	10. Acari

Group by group, each of these can be well validated as a monophylum. In traditional texts on systematic zoology, they are treated as of equal rank and discussed separately, one after the other. However, as equivalents of closed descent communities in Nature, these "orders" can in no way take identical rank. In a consistently phylogenetic system of the Arachida they must rather be placed at several different hierarchical levels. Taking account of the diagram of phylogenetic relationships suggested by Weygoldt & Paulus (1979b) and removing superfluous redundant taxa, I tabulate the system as follows, using the convention that the adelphotaxa are indented by one step for each successive level of subordination:

Arachnida
  Plesion Eurypterida [1]
      Scorpiones
    Lipoctena
      Megoperculata
        Uropygi
      Labellata
        Amblypygi
        Araneae
    Apulmonata
      Palpigrada
      Holotracheata
        Haplocnemata
          Pseudoscorpiones
          Solifugae
        Cryptoperculata
          Opiliones
        Acarimorpha
          Ricinulei
          Acari

The ten classical "orders" in this tabulation are printed bold for emphasis. They include three pairs of adelphotaxa in which each member of the pair necessarily has the same rank as the other member (Amblypygi—Araneae, Pseudoscorpiones—Solifugae, Ricinulei—Acari). These pairs, however, are placed at three different hierarchical levels. Each of the remaining four "orders" is sister-group to a complex of several classical arachnid orders. At one extreme are the Scorpiones which are the adelphotaxon of all other units of terrestial arachnids, these together forming a monophylum under the name Lipoctena. At the other extreme, and at the lower end of this part of the system, the Opiliones are sister-group to the taxon Acarimorpha which unites the Ricinulei and the Acari.

In the phylogenetic system of the Arachnida it is therefore impossible to retain the order category for all taxa called "orders" in the conventional classification. Rather, and in principle,the categorial rank of "order" can only be used for a particular pair of sister groups in the hierarchy of the system. Thus, if it were arbitrarily applied to the adelphotaxa Araneae and Amblypygi, then all categories above would become "higher categories" and all taxa below would be "lower categories".

These thoughts inevitably raise the question whether the classical concept of categorizing taxa has, in fact, enough categorial designations for all the levels of hierarchy of the phylogenetic system.

Farris (1976) proposed eight prefixes to be placed, in a fixed sequence, in front of the accepted designations for supraspecific categories. In the following table I give the prefixes, from +4 to −4, in the sequence in which they modify the primary categorial terms and I apply them to the categories of order and class.

giga-	+4	gigaorder	gigaclass
mega-	+3	megaorder	megaclass
hyper-	+2	hyperorder	hyperclass
super-	+1	superorder	superclass
—	—	order	class
sub-	−1	suborder	subclass
infra-	−2	infraorder	infraclass
micro-	−3	microorder	microclass
pico-	−4	picoorder	picoclass

Further categorial designations can be inserted in this sequence by putting two prefixes, or even several, to indicate the subdivisions. By means of compound words like gigainfraorder, supersuperorder or picosuborder it is then, in principle, possible to create any number of categorial labels above and below a single classical category.

---

[1] The widely accepted combination of the Eurypterida (Gigantostraca) with the Xiphosura to form a taxon Merostomata is mistaken. The fossil Eurypterida belong as a plesion in the stem lineage of the Arachnida. A synapomorphy with the Scorpiones (and thus a ground-pattern feature of recent Arachnida) is the fusion of the first pair of opisthosomatic limbs to form a structure called the metastoma. It may also be that the Eurypterida are paraphyletic. In any case, they have not yet been validated as a monophylum (Weygoldt & Paulus 1979b).

Is this approach practicable? I shall test it on a third selected example. Platnick (1977b) has developed a hypothesis on the phylogenetic relationships of high-ranking subtaxa of the Araneae and, in this connection, has applied Farris' terminological scheme to six hierarchical levels of the phylogenetic system. The result is as follows:

Superorder Labellata
  Order Amblypygi
  Order Araneae
    Suborder Mesothelae
    Suborder Opisthothelae
      Infraorder Mygalomorphae
      Infraorder Araneomorphae
        Microorder Palaeocribellatae
        Microorder Neocribellatae
          Gigapicoorder Hickmanithecae
          Gigapicoorder Bispermathecae
            Megapicoorder Gradungulospira
            Megapicoorder Tracheospira
              Hyperpicoorder Thaidoclada
              Hyperpicoorder Araneoclada

In **criticizing this procedure**, I stress three points:

1) As I have already said, placing the order category at the rank-level of the adelphotaxa Amblypygi and Araneae is arbitrary and cannot claim to be binding. Another arachnologist could, with equal right, demand that the order category be retained for the Scorpiones whereas an acarologist might believe that the subjects of his research were placed in the correct light if the taxon Acari were made an order. In applying prefixed categorial designations, the Araneae and Amblypygi, in the first case, would sink to the rank of a picoorder, but in the second case would be raised to a megaorder. Obviously the adelphotaxa at all other levels of hierarchy would likewise receive new and different designations.

2) But even if the order category could be made obligatory for the Araneae, the change from ordinal to class designations in going upwards would be subjective, as likewise the change from ordinal to family names in going downwards. Within one of these groups of categorial designations, it is similarly arbitrary whether one or two prefixes should be used. The six pairs of adelphotaxa subordinate to the "order" Araneae could equally well be given several other series of compound words such as gigasuborder, megasuborder, hypersuborder, supersuborder, infraorder or microorder.

3) With all such assignments it must not be forgotten that all parts of the phylogenetic system result from hypotheses which, with the advance of phylogenetic research, may at any time be replaced by better ones. Such changes in the phylogenetic system would require alterations in the assignment of categorial designations. With all these neologisms, this would almost certainly result in dreadful confusion.

The "classes" or categories of the Linnaean hierarchy are based on Aristotelean essentialism. These three examples show well enough that any attempt to apply them to designate the relative rank of supraspecific taxa in the phylogenetic system meets fundamental difficulties. Hennig, in his seminal discussion of the phylogeny of insects (1969) and in his posthumously printed account of chordate phylogeny (1983), refrained from giving categorial ranks to the higher-level groups. He did this "so as to avoid the fruitless discussions which, as experience shows, so often entangle the crucial and fundamental questions of phylogenetic systematics with the side-issue of the categorial rank of a group" (1969, p.10). Griffiths (1974, 1976) argued similarly and rejected the whole of Linnaean terminology. In his view, the practice of categorial classification should be entirely eradicated from the phylogenetic systematization of organisms as being empty formalism and because it continually led to confusion between the hierarchy of real units in Nature, on the one hand, and the sequence of classes in logic, on the other.

Are there other ways of designating the hierarchical levels of the phylogenetic system in a manner which is both logically correct and, at the same time, memorizable?

Numerical notation

Having given up categorial designations, Hennig (1969, 1983) indicated the position of taxa in the hierarchical system by means of series of numbers. In this book, I have given some examples of the use of such numerical notations for particular parts of the phylogenetic system. The last number always refers to the sister-group relationship of the group in question and, without doubt, this procedure clearly designates all the adelphotaxon relationships.

On the other hand, the weaknesses of this approach are obvious (Wiley 1979c, 1981; Eldredge & Cracraft 1980). Firstly, series of numbers are foreign elements in systematic work and ill-suited to good communication. Secondly, merely replacing categorial designations by groups of numbers does not eliminate the difficulties bound up with such designations. As in assigning a particular categorial rank, the hierarchical level at which the numeration starts is still a matter of purely subjective judgement. Moreover, exactly as with categorial designations, changes in the phylogenetic system on the basis of better knowledge of the phylogenetic relationships cause burdensome alterations in the series of numbers. Finally, in tabulating extensive portions of the hierarchical system, low-ranking taxa necessarily receive, instead of dubious categories like picopicoorder or micropicofamily, extremely long series of numbers which are highly impracticable to deal with. It would be quite impossible to assign a unique numerical designation to all taxa of plants or animals.

Indentation

This all leads to the question: Can we **free ourselves completely** from **the straitjacket of labelling the ranks of monophyletic taxa** by throwing out, as useless ballast, all forms of categorial designation or numerical notation?

This is totally possible. A very simple and practicable method, when tabulating the phylogenetic system, is to **indent adelphotaxa of equal rank consistently by the same**

amount. I have used this method above in describing the system of the Simiae (p.78) and to show the high-ranking sister-group relationships of the Arachnida (p.240). In the next chapter, I shall consistently apply the same method in tabulating the phylogenetic system of the Plathelminthes (p.283).

Objections have been raised against "pure indentation" also (Wiley 1979c, 1981). They are, however, purely technical scruples such as the necessity of using a ruler to discover two adelphotaxa far apart from each other, the question of how to continue a list of indented taxa on the next page or the waste of space and paper. But freedom from the troubles caused by labelling ranks with words or numbers certainly outweighs objections of this sort. Simply by indenting the highest-ranking adelphotaxa by the same amount, the tabulation of a segment of the phylogenetic system can begin at any hierarchical level whatever, without reference to categorial designations or groups of numbers. Alterations of the system required by favouring new hypotheses of relationship can be made easily, without changing any designation or learning it anew.

III. Subordination and sequencing

For each pair of adelphotaxa, phylogenetic systematics in principle demands a single name under which they are united to form a monophylum at the next higher hierarchical level of the system. This demand is neither new nor revolutionary and, for higher ranking taxa at least, it is usually taken as self-evident. I recall, for example, the combination of the Amphibia and Amniota to form the monophylum Tetrapoda, of the Sauropsida and Mammalia to form the Amniota, or of the adelphotaxa Marsupialia and Placentalia to constitute the Theria. Nevertheless, the principle of always giving a name, at all levels of the hierarchy down to the lowest-ranking taxa, has been bitterly disputed by the opponents of phylogenetic systematics. Their essential argument, in this connection, is the continual proliferation of new names for taxa.

Obviously this circumstance, even among phylogenetic systematists, is not a source of unclouded joy. If, however, a hierarchical order is to be adequately understood and recorded—an order which has been created not by us but by Nature—then the naming of taxa, which in logic is inevitably bound up with **subordination** of taxa within other taxa, must be approached unemotionally. In the first place, the following recommendations thoroughly deserve emphasis for work on any part of the phylogenetic system.

1) Every systematist should refrain from erecting and naming supraspecific taxa so long as the hypothesized phylogenetic relationships of the high-ranking subtaxa to be united under the new name have not been satisfactorily validated.

2) In systematizing taxa of low rank—those which are conventionally placed in the categories of family or genus—informal names may be used (Brundin 1972, Wiley 1981) by speaking of species groups within the larger taxa concerned. In my view, it is specially fitting to apply such "informal taxon names" (Wiley 1981, p.199) if a particular closed descent community can at present be systematized only in part. I recur to the example of the four evolutionary species of the plathelminth

taxon *Duplominona* which live today in the Galapagos Archipelago (p.65). As I have already shown, *D. galapagoensis, D. krameri, D. karlingi* and *D. sieversi* can be seen as a monophyletic group of species within the taxon *Duplominona*, which latter is categorized in traditional classification as a genus. In consequence of this hypothesis, I have united these four species in the "galapagoensis group" of the taxon *Duplominona*. This result stands whether other species of *Duplominona* can be united into monophyletic species groups or not.

This brings me to the alternative to subordination in the building up of the phylogenetic system. This **alternative is to arrange the taxa in a particular sequence** such that each individual taxon is the adelphotaxon of all the taxa that follow. A sequential arrangement of this sort can, in fact, precisely state the phylogenetic relationships of a whole series of taxa without needing any names for the superordinate units in the series of adelphotaxon relationships dealt with.

I have already discussed the principle of the sequential arrangement of taxa as concerns the insertion of fossils into the phylogenetic system (p.207). Here I repeat: In a systematization that concerns recent and fossil organisms, the extinct units in the stem lineage of the taxon are in principle organized in a sequence given by the step-wise increase in the number of synapomorphous agreements with the recent representatives.

In addition to this, the principle of sequential arrangement has also been considered for purely recent taxa (Nelson 1972, 1973; Cracraft 1974; Schuh 1976; Eldredge & Cracraft 1980; Wiley 1979c, 1981). I refer to the proposals in Wiley's convention 3 as given above.

I shall discuss the problems concerned by means of a schematic example in which six evolutionary species or monophyletic taxa A–F are together supposed to form the monophylum α. The relationships shown in the first diagram (Fig. 73) represent an

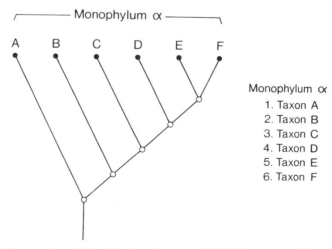

Monophylum α
1. Taxon A
2. Taxon B
3. Taxon C
4. Taxon D
5. Taxon E
6. Taxon F

Fig. 73. Diagram of phylogenetic relationship for the taxa A–F of the monophylum α and a complete tabulation of these relationships using the sequencing principle.

"asymmetrical part of the system" in Wiley's usage. On the right, I have translated the diagram of relationship into a list in which the six taxa are numbered and written down one after the other. In this case of a sequential arrangement, taxon A is the adelphotaxon of all the following units B–F; taxon B is the sister species or sister group of the taxa C–F that follow it; taxon C is, consequently, the adelphotaxon of D–F etc.

If, however, the phylogenetic relationship between the taxa A–F is judged to differ by only a single position, then arrangement in sequence becomes a dubious proposition. If we assume the relationships shown in the second diagram (Fig. 74), then in the equivalent tabulation the adelphotaxa I and II must be subordinated to monophylum α with equal rank at hierarchical level 2. Taxa C–D at hierarchical level 3 can then be numbered 1–4 and arranged sequentially.

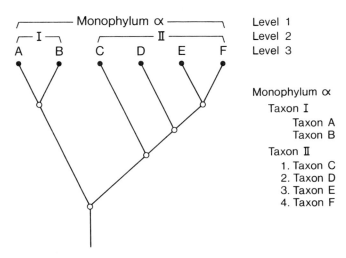

Fig. 74. Diagram of different relationships within the monophylum α and its tabulation by alternate subordination (at hierarchical level 2) and sequencing (at hierarchical level 3 for the taxa C–F).

Two further examples of possible phylogenetic relationships between the six taxa will close the argument. If A is the adelphotaxon of B, C that of D and E that of F, then it might be reasonable to insert a short sequence 1–3 at hierarchical level 2 for the taxa I, II and III (Fig. 75). On the other hand, if monophylum α is divided into subtaxa I and II, each with three units, then it would be possible, at hierarchical level 3, to give a sequence for taxa A–C and for D–F (Fig. 76).

The essential result of these thoughts is as follows: Even when setting out only a small portion of the system, the principle of sequential arrangement can be applied on its own only in the first example. For all other possible phylogenetic relationships between the six taxa, subordination must inevitably be used together with sequencing.

This brings me to criticize the use of sequential arrangement for recent taxa in the phylogenetic system. When systematizing a hierarchical order given by Nature, mixing two different principles of division simply asks for needless confusion. This is true, even if it is required that every change in a single part of the system from

246

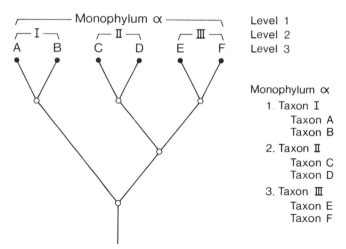

Fig. 75. The phylogenetic relationships in the monophylum α with three pairs of adelphotaxa (A + B, C + D, E + F). This portion of the system is tabulated by sequential arrangement of taxa I, II and III at hierarchical level 2.

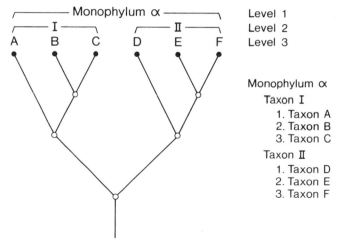

Fig. 76. Further possible relationships in the monophylum α with two groups, each containing three taxa. These are tabulated by the subordination of adelphotaxa I and II (at hierarchical level 2) and by sequencing for the taxa A + B + C and D + E + F at hierarchical level 3.

normal subordination to sequential arrangement be clearly indicated. Considerable parts of traditional classifications are long lists of taxa which, as genera, families, orders or classes are given an identical rank and are often numbered in sequence. Whether such arrangements perhaps show phylogenetic relationships, or are purely arbitrary procedures, can usually not be decided. Given this situation, it seems to me that confusion with a visually identical sequential arrangement based on the evaluation of adelphotaxon relationships is unavoidable.

247

A further important aspect must be mentioned. Sequencing seeks to avoid the naming of various closed descent communities in Nature but, in doing so, it loses information (Platnick & Gertsch 1976, Platnick 1977b). I shall illustrate this statement using the vertebrate examples in the annexed table. In the phylogenetic system of the Tetrapoda, the Sauropsida and the Mammalia can be placed sequentially after the Amphibia; and in the system of the Mammalia, in a corresponding way, the taxa Marsupialia and Placentalia can be placed after the Monotremata. I shall now compare the tabulations given by sequencing and subordination.

Sequencing	Subordination
Tetrapoda	Tetrapoda
1. Amphibia	Amphibia
2. Sauropsida	Amniota
3. Mammalia	Sauropsida
	Mammalia
Mammalia	Mammalia
1. Monotremata	Monotremata
2. Marsupialia	Theria
3. Placentalia	Marsupialia
	Placentalia

In the sequential arrangement, taxon 1 is correctly indicated, no doubt, as the sister group of the two following taxa and taxon 2, with equal clarity, is shown as the adelphotaxon of taxon 3. But with the suppression of the taxon Amniota (Sauropsida + Mammalia) and the taxon Theria (Marsupialia + Placentalia) important elements for information and communication are lost. For example, we are no longer in a position to emphasize the constitutive features of a closed descent community in Nature under the name of Amniota. Instead it is necessary to speak, long-windedly, of the synapomorphies of Sauropsida and Mammalia which show these two taxa to be sister groups of an unnamed taxon in the phylogenetic system of the vertebrates. The same follows logically for the taxon Mammalia, where sequential arrangement would eliminate the unit Theria from the system of the Mammalia.

There are therefore good reasons why the principles of subordination and sequencing should not thoughtlessly be mixed together in the phylogenetic system. Moreover, there are conclusive reasons why information should not be abandoned voluntarily, if it can be stored only by consistently subordinating taxa within taxa. Consequently, in striving towards the best possible phylogenetic systematization, we must be consistent and **must reject sequential arrangement as not being adequate to show the phylogenetic relationships of recent taxa.**

Finally, I should like to stress the real reasons why fossil taxa should, to some extent, be treated differently from recent taxa in phylogenetic systematization.

In a part of the system containing fossil and recent taxa, the proposal to arrange fossil units sequentially in the stem lineage of closed descent communities is logically justified because of the additive evolution of apomorphous features in time. The seriation of fossil taxa in a stem lineage exactly follows the successive addition of evolutionary novelties in the closed descent community in Nature to which the fossils belong.

The evolutionary species of the present day, and the monophyletic units made up of these species, exist at one and the same time and have a mosaic of plesiomorphous and apomorphous features which varies from taxon to taxon. In systematizing them, the appropriate procedure is to place at the same level of hierarchy those adelphotaxa which arose simultaneously. The logical result is increasing subordination of sets of equal-ranking adelphotaxa as these sets become younger in geological time.

In accordance with these considerations, the principle of subordination can likewise be used when systematizing extinct monophyletic taxa that consist of fossil organisms alone. The Trilobita belong, in the first place, in the stem lineage of the Chelicerata as a plesion without rank (p.212). But, in systematizing the monophylum Trilobita itself, the Emuellida and Eutrilobita form subordinate subtaxa. As sister groups which arose simultaneously they take equal rank within the Trilobita.

L. The Plathelminthomorpha

In this chapter I shall apply the principles and methods of phylogenetic systematics to the taxon Plathelminthomorpha. Under this name I unite the taxa Gnathostomulida (Ax 1956b) and the Plathelminthes, considering them as adelphotaxa of equal rank (Ax 1985b).

My argument takes account of the results of many years of intensive collaboration with Dr. U. Ehlers of Göttingen. The essential data for establishing the phylogenetic system of the Plathelminthes are taken from his works (Ehlers 1984, 1985a,b.).

Some of my earlier views (Ax 1961, 1963a) disagree with the interpretation of features given here and with the hypotheses of relationship which can now be formulated. To this extent, I expressly declare that these earlier views are outdated.

On the basis of still unpublished results from our ultrastructural studies in progress, *Xenoturbella bocki* Westblad 1949 (taxon Xenoturbellida) does not belong to the Plathelminthmorpha. *Xenoturbella* will therefore not be discussed in the following account.

I. The rejection of empirically untestable speculations

Before beginning a rational discussion of the phylogenetic relationships of the taxon Plathelminthomorpha within the Bilateria, and of the relationships of its monophyletic sub-groups to one another, a confused tangle of outdated speculations must be thrown out as ballast since they are incompatible with modern insights into the process of phylogenesis.

1. The Plathelminthes as "stem group" of the Bilateria

I shall start my discussion with a whole series of controversial speculations which seek the "phylogenetic derivation" of the Plathelminthes from certain non-bilaterian taxa or the "derivation" of all other Bilateria from the Plathelminthes.

Already in the nineteenth century, the Polycladida, as a single subtaxon within the Plathelminthes, were emphasized in attempts to derive all plathelminths from creeping Ctenophora (taxon Platyctenidia) (Lang 1881, 1884; Wilhelmi 1913).

These speculations are now of nothing but historical interest. On the other hand, a totally different subtaxon of the Plathelminthes still plays a central role in such discussions. These are the Acoela. They are usually interpreted as being, in all their features, the most primitive of free-living Plathelminthes and have thus been regarded, even very recently, as the basal group of the Bilateria (Salvini-Plawen 1978). Some workers have derived the Acoela from a hypothetical "Phagocytella" (Ivanov 1968, 1970; Ivanov & Mamkaev 1973) but this view is *a priori* untestable and so I shall ignore it. Apart from this, essentially two different "phylogenetic connections" of the Acoela with actually existing groups of animals have been proposed. On the one hand, they have been derived from ciliates or ciliate-like protozoans (Hadži 1944,

1963; Steinböck 1958, 1963; Hanson 1963, 1977). On the other hand, the Bilateria have been derived, by way of the Acoela, from a planula of the Cnidaria which had become sexually mature (Hyman 1951, 1958; Beklemishev 1958, 1960, 1963; Reisinger 1970; Salvini-Plawen 1978 with an extensive survey of the "problem").

This complex of mutually contradictory speculations can be criticized on two grounds:

1) Neither between the ciliates and the plathelminths, nor between the Cnidaria (or their planula larva) and the plathelminths, are there any agreements which can be established as synapomorphies. I shall illustrate this here with one often-mentioned group of features only. The trichocysts of the Ciliata, the nematocysts of the Cnidaria and the rhabdites of the Plathelminthes (p.287) are fundamentally different in ultrastructure. There is no shadow of a reason to suppose a homology between any two of these three secretion products, let alone to interpret them as a synapomorphy of the Ciliata and the Plathelminthes (in the case of the trichocysts and the rhabdites) or between the Cnidaria and the Plathelminthes (as concerns the nematocysts and rhabdites).

2) More important in the present connection, however, is another objection. The competing speculations in no way represent objective hypotheses of relationship, between which a well founded probability decision can be made by applying an objective empirical test. To be more precise, all these speculations look for "phylogenetic connections" between particular supraspecific taxa (Platyctenidia—Polycladida, Ciliata—Acoela, Cnidaria—Acoela) and thus they all operate with the artificial concept of supra-specific ancestors. And this is true, not only for the "derivation" of Plathelminthes from ciliates, cnidarians or ctenophores, but also for the subsequent claim, in all these speculations, that the Plathelminthes are the supraspecific stem group of the other Bilateria.

I fully understand why none of these views, incompatible as they are with the theory of evolution, has ever prevailed over the others—indeed they have all run a dead heat. They are subjective, untestable speculations and thus can be entirely forgotten.

2.   The Plathelminthes as "reduced Coelomata"

The gaudy palette of speculations on the "phylogenetic origin" of the Plathelminthes has another aspect. The venerable "enterocoel theory" of the previous century, as more recently defended by Remane (1963) and opposed by Hartman (1963), still continues as the "archicoelomate concept" (Siewing 1972, 1976, 1980, 1982). It postulates the existence of coelomic sacs in the stem species of the Bilateria. Accordingly, it must interpret the Plathelminthes as "reduced coelomates" which, in evolution, have lost a hydroskeleton enclosed in epithelium and some other organs such as nephridia, blood-vascular system and anus.

Phylogenetic systematics rejects this interpretaion, also, as one of the whole bundle of untestable speculations. It makes no difference whether, in deriving certain trimerous bilaterian taxa (Tentaculata, Hemichordata, Echinodermata) from the level of

251

the Cnidaria, the Scyphozoa with their four gastral pouches are preferred, or the Anthozoa with their octomerous ground pattern. Whatever the case, the artificial concept of the supraspecific ancestor is employed.

Obviously, this particular obstacle could be surmounted by postulating a particular gut morphology for the ground pattern of the Cnidaria and then supposing that, farther back, it was a ground-pattern feature of the latest common stem species of the adelphotaxa Coelenterata and Bilateria[1]. But, even then, the enterocoel theory fails in one of its basic assumptions. For the hypothesis of a homology between the gastral pouches of the Coelenterata and the secondary body cavity of particular taxa in the Bilateria is an empirically untestable speculation. There is total divergence between these organs in structure and function and, consequently, not one single discussable starting point for hypothesizing the gastral pouches of particular Coelenterata as the plesiomorphous condition and the coelomic compartments of oligomerous Bilateria as the apomorphous condition of one and the same ground-pattern feature of the Eumetazoa.

If, despite all this, such a homology were asserted to exist, and assumed as a "proven fact" preceding all other considerations, then, of necessity, the Plathelminthes could only be reduced coelomates.

But I do not accept such an assertion. Moreover, I see no way of establishing the homology between gastric pouches and coelomic sacks *a posteriori*, on the basis of a phylogenetic system of the Bilateria set up by using other features. We cannot, in other words, use the argument shown to be legitimate when homologizing the endostyle of the Acrania with the thyroid gland of the Vertebrata (p.161).

The enterocoel theory and the archicoelomate concept are not compatible with an analysis of the ground patterns of the Plathelminthomorpha and Bilateria that uses the methodology of phylogenetic systematics.

II. The ground pattern of the Plathelminthomorpha

I shall therefore consign to the archives all speculations deriving real unities in Nature from "supraspecific antecedents" and adopt the rational approach to the study of phylogenetic relationships which alone is methodologically viable. For the present problem this consists in: 1) establishing the Plathelminthomorpha as a monophyletic taxon of the Bilateria; and 2) searching for the adelphotaxon of the Plathelminthomorpha, i.e. the group which, with the Plathelminthomorpha, can be derived from a stem species common to the two groups alone.

This approach begins by working out, without prejudice, the ground pattern of the Plathelminthomorpha. This means putting together, without any logical contradictions, the mosaic of features of that evolutionary species which was the latest common stem species of the Gnathostomulida and the Plathelminthes.

The feature-by-feature analysis and the reconstruction of the ground patterns must obviously be preceded by a satisfactory validation of the Gnathostomulida and Plathelminthes as adelphotaxa within the phylogenetic system.

---

[1] Eumetazoa = Coelenterata + Bilateria. (Hennig 1980) established the Coelenterata as being a monophyletic taxon. The Ctenophora, in his opinion, are possibly only a subordinate subtaxon of the Hydrozoa.

1. Establishing individual features of the ground pattern

Shape and size of body

Definite correlations between the shape and size of the body, on the one hand, and the structure of the mouth region, on the other, are the basis for justifiable hypotheses on the body-habit of the latest stem species common to the Gnathostomulida and the Plathelminthes.

In plathelminth taxa that have a plain mouth pore in the epidermis or a pharynx simplex the body length always lies between fractions of a millimetre and a few millimetres (Nemertodermatida, Acoela, Catenulida, Macrostomida).

Body lengths measured in centimetres or decimetres occur only in combination with a complex pharynx compositus and are correlated, in particular, with the pharynx plicatus (Polycladida, Seriata) and the pharynx bulbosus (Rhabdocoela). Compared with the mouth pore or the pharynx simplex these certainly represent apomorphous conditions within the Plathelminthes.

These definite correlations necessarily imply the following conclusion: the body sizes of the big Polycladida and Tricladida (Seriata) are apomorphous states for the taxon Plathelminthes. The evolution of protrusible pharynges as sucking or swallowing apparatuses for improved predatory feeding were preconditions for the secondary increase in body size and volume.

The microscopic Gnathostomulida can readily be fitted into this picture. So far as we know at present, they feed on bacteria and perhaps also on diatoms and fungal hyphae. They graze the organic film on sand grains, using a hard basal plate and cuticular jaws (Sterrer 1971).

A further important correlation must be considered. Gnathostomulida and freeliving Plathelminthes of millimetre-size usually have bodies of circular transverse section. The name "flatworms" is generally most unsuitable for the majority of freeliving Plathelminthomorpha. A flat, dorso-ventrally compressed body occurs only in connection with a considerable increase in the size and volume.

Consequently we can postulate for the ground pattern of the Plathelminthomorpha that the body was about a millimetre in length and circular in transverse section.

Pharynx and gut

In the Plathelminthes, a ciliated pharynx simplex, being an invagination of the epidermis, exists in the Catenulida and the Macrostomida (Fig. 84b). They also possess a rod-shaped cellular gut.

In the taxon Acoelomorpha, on the other hand, pharynges of this sort are found only occasionally. The Nemertodermatida and the great majority of the Acoela merely have a mouth pore in the epidermis. The digestive tissue, more or less without a lumen, is joined directly to this aperture; for the taking up of food the digestive tissue can be thrust forwards out of the mouth pore (Ehlers 1985b).

Which is the plesiomorphous and which the apomorphous alternative as concerns the mouth region? The probability judgement is difficult because the consequences are hard to explain in either case.

Postulating the more obvious alternative of a plain mouth pore for the stem species of the Plathelminthes, then, on the basis of the phylogenetic system of the

253

Plathelminthes established on other features (p.281), the pharynx simplex must have been evolved from the mouth pore several times convergently. This would need to have happened in the Catenulida, the Macrostomida and in various taxa of the Acoela.

If, conversely, an invaginated pharynx simplex is seen as a ground-pattern feature of the Plathelminthes, then the plain mouth opening of the Nemertodermatida and the majority of the Acoela would be the apomorphous result of repeated reductive evolution. There is no convincing argument for this supposition.

Comparison with the adelphotaxon Gnathostomulida is of no help in making this probability judgement. For the muscular pharyngeal apparatus of the Gnathosto-mulida, with a pair of cuticular jaws and a cuticular basal plate (Fig. 77), is a highly apomorphous structure with respect to both forms of the mouth region mentioned in the Plathelminthes.

New ultrastructural data support the view of Karling (1974) that a simple mouth pore represents the original condition in the ground pattern of the Plathelminthes (Doe 1981; Rieger 1981; Smith 1981; Ehlers 1985b). I now favour this hypothesis because, as concerns the consequences just touched upon, it probably represents the more parsimonious interpretation of the facts.

Evaluating the alternatives in the structure of the gut presents fewer problems. The Gnathostomulida, Catenulida and Macrostomida agree in having an epithelial gut formed of cells. In outgroup comparison with the Coelenterata, as well as in the ground pattern of the other Bilateria, we find a cellular endoderm that forms the gut lining. An epithelial gut can therefore be postulated for the ground pattern of the Plathelminthomorpha while the condition, found in the Acoelomorpha, of a gut tissue without a lumen or even developed as a "syncytium" can be seen as an apomorphy (Karling 1974; Tyler & Rieger 1977; Smith 1981).

Gut configuration and dorso-ventral musculature

Extensive lateral gut diverticula occur only in the plathelminth taxon Trepaxo-nemata. Compared with the unbranched gut of the adelphotaxon Macrostomida, it is reasonably certain that the gut with diverticula represents the apomorphous condition. The evolution of a branched gastro-vascular system was closely correlated with the evolution of a large, flattened body. A precondition for the latter was the existence of a pharynx compositus (p.253). A gastro-vascular system with lateral diverticula was evolved several times convergently within the Trepaxenomata—in combination with a pharynx plicatus in the Polycladida and Tricladida and with a pharynx bulbosus in the parasitic Neodermata.

Well developed dorso-ventral musculature is lacking in the Gnathostomulida, Catenulida, Acoelomorpha and Macrostomida. Serially arranged bundles of dorso-ventrally oriented muscles occur only in combination with the flat body structure of the taxa previously mentioned. The bundles are inserted between the lateral gut diverticula and serve to brace the flattened body.

Thus lateral gut diverticula and the interpolated dorso-ventral musculature can be excluded from the ground pattern of the Plathelminthomorpha and also from that of the Bilateria. The attempt to derive the dorso-ventral muscula-ture of the Nemertini and of the Mollusca from apomorphous conditions within

254

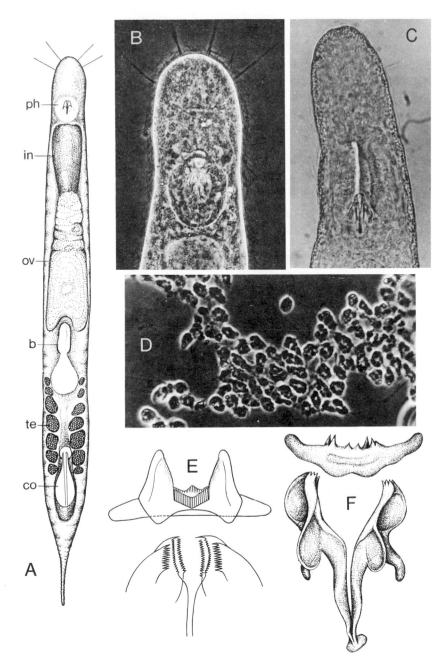

Fig. 77. Gnathostomulida. A. Organization of *Gnathostomula paradoxa* Ax. B. Anterior end of *Gnathostomula paradoxa* with a pair of cuticular jaws in the muscular pharynx. C. *Gnathostomaria lutheri* Ax. Long oral fissure; the jaws are in a terminally located muscle bulb. D. Monociliary epidermis of *Gnathostomaria lutheri*, separated from the body; in the upper middle part of the picture the connections between the cells are loosened and an isolated epidermal cell shows a single cilium (phase contrast). E. Basal plate and jaws of *Gnathostomula paradoxa*. F. Basal plate and jaws of *Pterognathia swedmarki* Sterrer. Abbreviations: b = bursa; co = copulatory organ; in = gut; ov = ovary; ph = pharynx; te = testis. (From Ax 1964, 1965; Müller & Ax 1971).

255

the Plathelminthes (Salvini-Plawen 1972; Hennig 1980) is unworkable. It must be rejected.

Anus

The Plathelminthomorpha are the only extensive monophylum of the Bilateria in which all the subtaxa always lack an anus.

In working out the ground pattern, we are confronted with the following alternatives: compared with the remaining "anus-Bilateria" (Hennig 1980), is the lack of an anus in the Plathelminthomorpha an apomorphy, or is the agreement with the Coelenterata, which likewise lack an anus, a shared primitive feature?

The only argument in favour of the first hypothesis is the scattered absence of an anus in particular taxa of the Bilateria. I wish to make two points in this connection:

a) Quoting examples at random does not contribute in the slightest to a justifiable probability judgement in the actual case before us.
b) The usually cited examples of forms lacking an anus are the Testicardines (Brachiopoda) or the Ophiuroidea (Echinodermata). These are both subordinate subtaxa of well validated monophyla that have an end-gut anal region as a ground-pattern feature. Since this is not true for the Plathelminthomorpha, the situations are not comparable.

A scientific argument based on the principle of parsimony, if not weighted down with pre-ordained hypotheses, must *prima facie* take the condition without an anus, as seen in the Coelenterata and Plathelminthomorpha, as being a shared primitive feature. The next requirement is carefully to examine whether this interpretation is mutually compatible with the other features assigned to the ground pattern.

At this point I therefore postulate a blind-ending gut for the ground pattern of the Plathelminthomorpha. As a plesiomorphy it would derive from a comparable structure in the stem species of the Bilateria and in that of the Eumetazoa.

As concerns the problem of the evolution of the "one-way gut" with an anus, various isolated structures in the Plathelminthes must be discussed. These can be referred to neutrally as gut adhesions or gut openings. In the gnathostomulid taxon *Haplognathia* (Knaus 1979) and in a few free-living Plathelminthes (*Haplopharynx*, *Tabaota*, *Archimonocelis*; see Marcus 1950; Karling 1965, 1966) local fusions between the gut tissue and the epidermis have been described, as well as cases where the gut opens through the epidermis. It may seem tempting to derive the anus of the Bilateria from structures of this sort in particular Plathelminthomorpha, but such thoughts must definitely be rejected. It does not matter which particular case in the Plathelminthomorpha is cited, since such speculations necessarily involve a grave error. For no species can pass the limits of the closed descent community to which it belongs. In other words, no single species of the monophylum Plathelminthomorpha can be the stem species of any other taxon whatsoever of the Bilateria. Here only one explanation is logical and does not contradict the facts. The isolated occurrences of "gut openings" in the Plathelminthomorpha are independently evolved novelties (autapomorphics) in a few subordinate subtaxa. They have nothing to do with the anus of other Bilateria.

256

Consequently, I favour the following hypothesis. At the end of the stem lineage of the Bilateria there was an evolutionary species which had a blindly ending gut with no anus. As a plesiomorphy, this condition goes back to the ground pattern of the Eumetazoa (Coelenterata + Bilateria). A gut with no anus thus passed, as a primitive feature, into the stem lineage of the Plathelminthomorpha and was transferred unchanged to the stem species of the Gnathostomulida and of the Plathelminthes. In another stem lineage, shared by all other Bilateria, the evolution of a "one-way gut" with an anus took place (Hennig 1980). Movement of the gut contents only in one direction was the pre-condition for new mechanisms in the catabolism of food. Without doubt it represents an important evolutionary step forward. On the most parsimonious assumption that the anus evolved only once, in this stem lineage, the "one-way gut" is the crucial autapomorphy in validating and setting up a large monophyletic taxon within the Bilateria. I name this taxon the Eubilateria (Fig. 82).

The blastopore

After discussing the gut, it is logical to consider the primitive relationship of the embryonic blastopore to the structure of the adult.

In ontogeny the definitive mouth opening of the Plathelminthomorpha, as in the Coelenterata, arises from the blastopore. That is self-evident, for with only one gut opening there is no alternative.

Accordingly, one rationally justifiable result must be taken as self-evident. The origin of a single mouth opening from the blastopore region is a plesiomorphous ground-pattern feature of the Plathelminthomorpha and of the Bilateria. It must have been derived from the stem species of the Eumetazoa. All deviations from this condition which arose within the Eubilateria after the evolution of the one-way gut are, without doubt, apomorphies.

This statement results in the deletion of the traditional group "Protostomia" from the phylogenetic system of the Bilateria. This is not, in the slightest, because the blastopore varies in its fate in the bilaterian taxa conventionally placed in the "protostomian line"—though the group Protostomia was recently questioned for this reason by Fioroni (1980) and Siewing (1980b). For the features of the stem species of every closed descent community can vary in the descendent species to any extent whatever, with an unlimited amount of convergence, without at all affecting the status of the species groups which form monophyletic taxa of the phylogenetic system. The elimination of a group "Protostomia" from the phylogenetic system of the Bilateria is imperative solely and entirely because the feature of the blastopore region becoming the mouth is a plesiomorphy. As such, it is simply not relevant to constituting a monophyletic taxon.

It would certainly be throwing out the baby with the bath water, however, if the rejection of the traditional split into Protostomia and Deuterostomia led to the deletion of the Deuterostomia as a taxon of the phylogenetic system of the Bilateria. Without reference to the quashing of a taxon "Protostomia", the formation of the anus from the blastopore region can still count as an apomorphy which evolved, once only, in the stem lineage of the Hemichordata + Echinodermata + Chordata. And, along with other constitutive features, it helps to validate a monophylum Deuterostomia within the Eubilateria.

257

Epidermis

A cellular, ciliated ectoderm undoubtedly belongs to the ground pattern of the Plathelminthomorpha and to the ground pattern of all Bilateria. These general statements can now be made more precise on the basis of extensive ultrastructural data.

The Gnathostomulida are the only taxon of the Bilateria in which the adult is completely covered by an epidermis of monociliary cells with a locomotory function (Fig. 77D, 78A; Ax 1956b, 1964a). In them the individual cilium of each epidermal cell is accompanied by a diplosomal basal apparatus. This consists of the basal body of the cilium and of an accessory centriole perpendicular to the basal body (Rieger & Mainitz 1977). As against this, the epidermal cells of the free-living Plathelminthes always have several or many cilia. Furthermore, they constantly lack the accessory centriole—the cilia of the Plathelminthes are equipped with a basal body only (Fig. 78B, C).

Once again, it is a question of evaluating an obvious pair of alternatives. Is it the monociliary epidermis of the Gnathostomulida or the multiciliary epidermis of the Plathelminthes and other bilaterian taxa which belongs in the ground patterns of the Plathelminthomorpha and Bilateria?

A probability judgement follows from an out-group comparison. On the one hand, monociliary cell layers with a diplosomal basal apparatus are specific structural elements in the Porifera, the Placozoa (*Trichoplax adhaerens*) and the Coelenterata. On the other hand, among the Bilateria, the ventral creeping sole of part of the Gastrotricha consists of monociliary cells exactly as in the Gnathostomulida. But about half of the gastrotrich species which have been studied by electron microscopy have a multiciliary creeping sole with one or two basal bodies per cilium (Rieger 1976).

This distribution of features gives an unequivocal answer. The monociliary epidermal structure is the plesiomorphous alternative. As an extremely ancient feature, cells with one cilium and a diplosomal basal apparatus were taken over from the stem species of the Metazoa, and from the stem species of the Eumetazoa, into the ground pattern of the Bilateria. Within the Bilateria, the plesiomorphous condition of a body completely covered with monociliary cells, is preserved unchanged only in the taxon Gnathostomulida. In the stem lineage of the Plathelminthes, a change occurred to the multiciliary epidermis with the loss of the accessory centriole. Comparable changes have occurred, independent of the Plathelminthes, within the Eubilateria, i.e. several times convergently within the taxon Gastrotricha and also in other stem lineages.

Dermal muscle layer

The locomotion of the primarily microscopic Gnathostomulida, Catenulida, Acoela and Nemertodermatida depends mainly on the ciliary beat of the epidermal cells. In clear correlation with their small body size and their mechanism of locomotion, the representatives of these taxa all have a weakly developed subepithelial dermal muscle layer consisting of loosely packed outer, transverse muscles and inner longitudinal muscles. Oblique muscles may also be present.

Accordingly, this condition of the dermal muscle layer can be fitted without difficulty into the ground pattern of the Plathelminthomorpha. Furthermore I

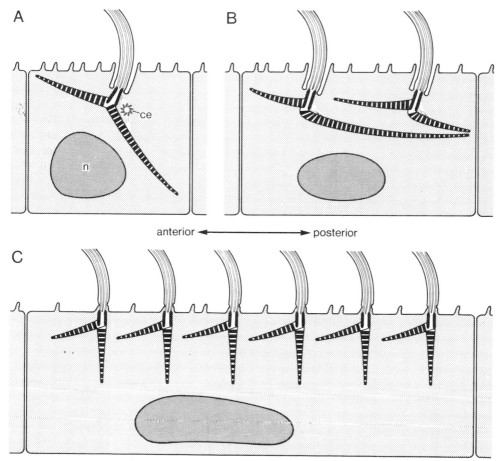

A

ce

n

anterior ←——————→ posterior

C

Fig. 78. The epidermis in the Plathelminthomorpha (diagrams of longitudinal sections). A) Monociliary polygonal epidermal cells of the Gnathostomulida with a nucleus (n) and an accessory centriole (ce) next to the basal body of the cilium. B) Sparsely multiciliary epidermal cells of the Catenulida (*Retronectes*). C) More strongly multiciliary epidermal cells of the Rhabditophora. Density of the ciliation: Gnathostomulida 0.15–0.2 cilia/μm^2 of the body surface; Catenulida 0.2–1.8 cilia/μm^2 .of body surface; Euplathelminthes (Acoelomorpha + Rhabditophora) 3–6 cilia/μm^2 of body surface. (A after Rieger & Mainitz 1974; B and C original by U.Ehlers).

postulate smooth muscle cells for the dermal muscle layer of the stem species of the Plathelminthomorpha, on the basis of the wide distribution of these cells within the Plathelminthes. The striated musculature of the Gnathostomulida (Rieger & Mainitz 1977), as well as isolated cases of striated muscles in the dermal muscle layer of the Plathelminthes (taxon Otoplanidae), must have evolved independently, convergent to similar muscle patterns within the Eubilateria (Gastrotricha etc., Ehlers 1985b).

By comparison with the characteristic epithelial muscle cells of the Coelenterata (Cnidaria), the subepithelial dermal muscle layer of the Plathelminthomorpha is definitely an apomorphy. Reports of the existence of epithelial muscle cells in the plathelminth taxon Catenulida are false (Rieger 1981; Ehlers 1985b). The muscle cells of the Plathelminthomorpha are, without exception, individualized "true" muscle cells.

On the basis of this situation, the evolution of a subepithelial dermal muscle layer, comprising transverse and longitudinal muscles, must be postulated to have happened in the stem lineage of the Bilateria. This feature existed as an autapomorphy in the latest common stem species that was shared by Plathelminthomorpha and Eubilateria.

Nervous system and sensory apparatus

At the present stage of analysis of the Plathelminthes by optical and electron microscopy, the following statements can be made (Ehlers 1985b). In the ground pattern of the Plathelminthes there was a brain-like nerve centre, consisting of a mass of neurons at the anterior end of the body, and a peripheral nerve plexus with an intra-epithelial to subepidermal position. On the other hand, a particular orthogonal structure of the nervous system (Reisinger 1972) which occurs in various "orthogons" of particular plathelminth taxa, cannot be assigned to the stem species of the Plathelminthes.

As to the adelphotaxon Gnathostomulida, there is only one ultrastructural study (Kristensen & Nøvrrevang 1977). This states that in *Rastrognathia* the organ corresponding to the "brain" of Plathelminthes consists of a mass of ganglion cells and nerve fibres in the basal part of the epidermis at the anterior end of the body. From this centre, six "longitudinal nerves", showing in different parts of the transverse section of the body, run rearwards. A further pair of ventral longitudinal nerves enter the buccal cavity and unite with each other at the posterior end of the pharyngeal bulb to form a buccal ganglion. The "body nerves" as well as the "brain" are differentiated entirely within the epidermis. They lie between normal epidermal cells and are always completely external to the basal lamina. Further studies are needed to discover whether the longitudinal nerves described represent separate neuronal strands, or whether, perhaps, they form part of a superficial intraepidermal nerve plexus.

In any case, the minimal components of the nervous system which can be postulated for the ground pattern of the Plathelminthomorpha are a brain-like nerve centre at the anterior end, and a nerve plexus extending over the body in the epidermis. The same features can also be assigned, farther back in time, to the latest stem species that was shared by Plathelminthomorpha and Eubilateria.

This peripheral, plexus-like arrangement corresponds to the structure of the nervous system in the Coelenterata. From the ground pattern of the Eumetazoa, it was taken over into the stem lineage of the Bilateria as a plesiomorphy. On the other hand, the concentration of neurons at the anterior end of the body is an autapomorphy of the Bilateria. This "brain" arose in the stem lineage of the Bilateria in connection with the evolution of a rostral sensory pole to control the activities of a bilaterally symmetrical organism with locomotion in one direction.

Monociliary sense cells are the only sensory organs which can definitely be assigned to the ground pattern of the Plathelminthomorpha. They exist scattered in the epidermis in the Gnathostomulida—several monociliary receptors combine in groups to form the taste buds of the rostral sensorium (Fig. 77B; Kristensen & Nørrevang 1977). Similar monociliary sense cells are widely distributed in the Plathelminthes (Ehlers 1977).

Monociliary receptors are also found in the Coelenterata and within the Eubilateria (Nørrevang & Wingstrand 1970; Kristensen & Nørrevang 1977). Accordingly, they must also have existed in the latest common stem species of the Eumetazoa. From it they passed as plesiomorphies into the ground pattern of the Bilateria, of the Plathelminthomorpha and of the Eubilateria.

Statocysts and rhabdomeric ocelli with pigment cups are absent in the taxon Gnathostomulida. There is no reason to think they have been lost secondarily. Likewise, a statocyst can no longer be postulated for the ground pattern of the Plathelminthes. According to Ehlers' ultrastructural studies (Ehlers 1985b), statocysts must have arisen independently in the stem lineages of several individual plathelminth taxa (Catenulida, Acoelomorpha, Proseriata).

Body cavity

Several fine-sounding names are available to characterize the condition of the body cavity in the Plathelminthes. We could choose the term "schizocoel" or describe the Plathelminthes as acoelomate organisms, speak of the tissue in the body cavity as parenchyma or mesenchyme, and regard the structure of this tissue as simple or complicated.

The only things that count, however, are the following bare facts. In the Plathelminthes, various cells fill the spaces of the body cavity between the epidermis, gut and genital organs. On the basis of electron-microscopical studies, there are, in addition to cell bodies that have sunk in from the ectoderm and endoderm, two types of cell that are characteristic for the body cavity of the Plathelminthes—stem cells (neoblasts, replacement cells) and parenchyma cells in different subtaxa (Rieger 1980, 1981).

In the Gnathostomulida, no exhaustive study of the body cavity has yet been made. Over wide areas, endoderm and ectoderm are in immediate contact with each other. Parenchymatous tissue has been described from the head of *Gnathostomulida* and from the region of the copulatory organ in *Gnathostomula* and *Semaeognathia* (Riedl 1971; Mainitz 1979). In *Rastrognathia*, in addition to muscle cells and protonephridia, there are interstitial cells wedged, in places, between the epidermis and the gut (Kristensen & Nørrevang 1977).

These observations lead to the following central question: Are the extensive absence of a body cavity in the Plathelminthomorpha and the filling-up of the small interstitial spaces with cells to be seen as a plesiomorphy or an apomorphy when compared with the unlined body cavities (pseudocoel) or the epithelium-lined cavities (coelom) of other bilaterian taxa?

As concerns the attempt to interpret the body-cavity condition of the Plathelminthomorpha as being derived from an epithelial hydroskeleton, only one discussable aspect exists. It results from the hypothesis that the Gnathostomulida and the majority of the Plathelminthes are small organisms secondarily. With decrease in body-size and volume, an epithelium-lined hydroskeleton would have ceased to be useful in the service of locomotion. The coelom would consequently be lost for

261

reasons of space. Epidermis and gut pushed up against each other and the remaining interstitial spaces came to be filled with packing tissue.

These suggestions can justifiably be rejected because there is a strict correlation, in the Plathelminthes, between small body size and plesiomorphous mechanisms for ingesting food (p.253). With a simple mouth pore or a pharynx simplex, the plathelminths can reach a body length of only about a millimetre. The small dimensions of the Catenulida, Acoela, Nemertodermatida and Macrostomida show the order of size required as the plesiomorphous condition for the ground pattern of the Plathelminthes. Since the Gnathostomulida, which bear jaws, are also of roughly this size, the same statement applies, farther back in time, to the stem species of the Plathelminthomorpha (p.253). Speculations that the Plathelminthomorpha derive from "coelomate ancestors" of larger body size are not based on empirical evidence and must be abandoned.

Nevertheless, it is true that, for reasons of space, an organism of millimetre size can only have tiny interstices between ectoderm and endoderm.

In other words, the conditions of the body cavity which exist in the primarily microscopic Plathelminthes must be regarded as a plesiomorphy and assigned to the ground pattern of the Plathelminthomorpha. The primitive condition is to have a weakly developed primary body cavity with a few cells bracing occasional local spaces. Massive parenchymatous connective tissue was evolved within the taxon Plathelminthes only in connection with secondary increase in size and volume in the Polycladida, Tricladida and Rhabdocoela.

This plesiomorphous condition of the plathelminthomorph body cavity was taken over, farther back in time, from the ground pattern of the Bilateria. It can be postulated, without contradictions, as existing in the feature mosaic of the latest common stem species of the Plathelminthomorpha *plus* Eubilateria. Within the Eubilateria, a parenchymatous body cavity is retained as a plesiomorphy in the Nemertini. Unlined and epithelium-lined body cavities were first evolved, within the Eubilateria, as apomorphies.

This analysis leads me to criticize various speculations which, on the basis of ultrastructural observations, see the body cavity of the Gastrotricha as reduced "coelom" (Teuchert 1977; Teuchert & Lappe 1980). Local spaces in the gastrotrich body, which are lined with muscle cells but not with epithelium, may, of course, be dubbed with any purely descriptive name that seems appropriate. However, any attempt to homologize the gastrotrich body cavity with an epithelial hydrostatic skeleton demands empirical evidence. Such evidence is absent, for Teuchert & Lappe themselves point out that: "When compared with a diagram showing the classical coelomate type, the differences are obvious. Essential contrasts are the lack of mesenteries, the absence of true cavities and, in the latter connection, also the lack of nephridia" (Teuchert & Lappe 1980, p.436).

For the purposes of phylogenetic systematics, however, more is needed. Indeed, a testable hypothesis of phylogenetic relationship is required, asserting the precise coelomate adelphotaxon of the Nemathelminthes. As compared with the body-cavity morphology of such a taxon, the situation in the gastrotrichs would have to be justified as an apomorphy. Understandably enough, such a hypothesis of relationship has never been formulated.

Moreover, an appeal to the small size of the Gastrotricha, in explaining their body-cavity condition as caused by regressive evolution, is just as weak as for the Plathelminthomorpha. There is no shadow of a reason why all organisms living in the interstitial system of marine sands should be lumped together as secondarily miniaturised forms. Indeed, I know of no reason for supposing that the Gastrotricha were ever anything but microscopic in size.

Blood vessels

Except in the Nemertini, channels of transport for the directed circulation of body fluid are only found in association with epithelium-lined body cavities.

The Nemertini, however, are of interest not only because of the existence of a blood-vascular system in the parenchymatous primary body cavity but also because, in striking contrast to all other invertebrates, their blood vessels have their own endothelium (Turbeville & Ruppert 1981). A blood-vascular system of this sort must be interpreted as an autapomorphy of the nemertines. It would have evolved in the stem lineage of the Nemertini, independent of the blood vessels of the other Eubilateria.

The blood vessels of all invertebrate "coelomates", on the other hand, including the Acrania, are primary cavities without any endothelial lining. The lumina of the vessels are limited by the basal lamina of the surrounding coelomic epithelium or by connective tissue (Rähr 1981; Ruppert & Carle 1983).

The absence of a blood-vascular system in the Gnathostomulida and the Plathelminthes cannot be derived from the condition in the Nemertini, nor from that of the "coelomates". There is not the slightest trace of secondary loss caused by regressive evolution.

In other words, the lack of a transport system for the circulation of body fluid must be assigned to the ground pattern of the Plathelminthomorpha as a plesiomorphy.

Those Plathelminthes characterized by a secondary increase in body size and volume (Polycladida, Tricladida, parasitic Neodermata) have solved the problem of distributing feeding products by evolving a gastrovascular system with extensive gut diverticula.

A blood-vascular system is likewise completely lacking in the taxon Nemathelminthes. Accordingly, the primary absence of blood vessels can be validly postulated, not for the Plathelminthomorpha only, but also for the ground pattern of the taxon Bilateria.

Nephridial organs

Nephridial organs for osmoregulation and for the removal of dissolved excreta occur only in the taxon Bilateria. Compared with their absence in the adelphotaxon Coelenterata, these structures represent, with reasonable certainty, an evolutionary novelty of the Bilateria. But the structural and functional condition of the nephridia evolved in the stem lineage of the Bilateria is a matter of sharp controversy.

I shall analyse this problem on the basis of the following clear correlations. The above-mentioned body-cavity conditions in the Plathelminthomorpha, the parenchymatous body cavities of some Eubilateria (Nemertini, Entoprocta) and the unlined body cavities of other Eubilateria (Gastrotricha, Rotifera, Kinorhynchia, Acanthocephala) are always and exclusively combined with blind-ending protonephridia. They are never associated with nephridia having open ciliated funnels. In addition, protonephridia sometimes occur in adult polychaetes, are developed in the trochophore larva of the polychaetes and are known in the ontogeny of a few molluscs. A particular form of solenocyte organ is represented in the excretory system of the Acrania (Kümmel 1962; Wilson & Webster 1974; Brandenburg 1975).

Nephridia with open ciliated funnels, on the other hand, are strictly correlated with the existence of epithelium-lined body cavities.

Although the body-cavity condition of the Plathelminthomorpha can without doubt be interpreted as a plesiomorphy (p.261), I shall once again raise the alternative of an epithelium-lined hydroskeleton as the primitive condition, discussing it, this time, as concerns the nephridia.

If coelomic spaces are assigned to the ground pattern of the Bilateria, then the nephridia with open ciliated funnels, which are associated with these spaces, must represent the primitive form of nephridium for the Bilateria. A few consequences follow at once:

a) Blind-ending protonephridia would be an apomorphous condition within the Bilateria.
b) The Plathelminthomorpha, Nemertini, Nemathelminthes and Entoprocta would, together, have to be validated as a monophyletic taxon of the Bilateria, i.e. they would have to be established as descendants of a stem species common to these taxa alone which possessed protonephridia as an apomorphy. Why? Because extensive agreements in the fine-structural details allow no serious alternative to the evolution, once only, of the protonephridium represented in these taxa.

None of these consequences can be defended by any argument worth discussing. I know of only one verbal attempt to extract the enterocoel theory from the problems implied by the distribution of nephridia. This was made by Remane (1967, p.605) in the following words: "Whether protonephridia existed first, to which ciliated funnels gained connection as gonoductal openings, or whether nephridia first existed as gonoducts, and ciliated side-branches developed from their ectodermal, canal-like portion and then became independent protonephridia after the loss of the ciliated funnel, is hard to decide and perhaps not very important".

Deciding between plesiomorphy and apomorphy, however, *is* of absolutely basic importance for the logically consistent phylogenetic systematization of organisms. And, in the present case, a decision can be satisfactorily made. I start from the following considerations.

a) In order to empty a body cavity that has no considerable interstices, the only nephridial organ which could function would need to enclose a space in which the cilia, as motors for the distally directed current of liquid, have room to act. Such a space exists in the blind-ending terminal cell of protonephridia. It makes no difference, in this connection, whether the cell is directly inserted between ectoderm and endoderm as in the Gnathostomulida, or whether it acts in a parenchymatous body cavity.

264

b) On the other hand, the combination of protonephridia with more voluminous body cavities is not obligatory from the functional point of view—it is in no way obvious why protonephridia should evolve in such conditions. The existence of protonephridia in the Nemathelminthes, where the body cavity is unlined, and within the polychaetes which have an epithelial hydrostatic skeleton, can only be rationally interpreted under the following suppositions. Protonephridia were evolved from the ectoderm of an organism having a primary body cavity and inserted between an endoderm and ectoderm which were in immediate contact with each other or into crevices between the cells of parenchyma. As plesiomorphous nephridial organs, they were then inherited by taxa with a unlined body cavity and partly, also, by taxa with an epithelium-lined coelom.

This brings us to a definite conclusion. Protonephridia ending blind, with closed terminal cells, represent the original state for nephridial organs in the Bilateria. They evolved in the stem lineage of the Bilateria, in a primary body cavity which was weakly developed in the form of minute interstices between stem cells.

This rather general statement can be made much more precise on the basis of ultrastructural studies on the protonephridia of the Gnathostomulida, Plathelminthes and Gastrotricha. In this way we can arrive at a detailed picture of the protonephridia of the ground pattern of the Bilateria.

The simplest morphological condition for the nephridia in the Bilateria occurs in the taxon Gnathostomulida. In these animals several unconnected protonephridia are found in two lateral longitudinal rows in the body (Graebner 1968; Kristensen & Nørrevang 1977). Ultrastructural studies on *Gnathostomula paradoxa* and *Haplognathia rosea* (Lammert 1985) give essentially the following results (Fig. 79A, B). The individual protonephridium consists of three cells—the terminal cell, a canal cell and a nephropore cell—which latter is situated among the monociliary epidermal cells. In the terminal cell, a cilium[1] is inserted which, exactly as in the epidermal cells, has a diplosomal basal apparatus with a basal body and an accessory centriole. In the periphery of the terminal cell, a distally directed cytoplasmic margin with clefts is developed, and through these clefts fluid from the body cavity can enter the protonephridium. An extracellular matrix fills the clefts and so forms the filtration area of the protonephridium. Inside the terminal cell, eight microvilli grow up around the cilium as cytoplasmic protrusions to form a supporting apparatus of rods. The result is a stable hollow cylinder in which the cilium can undulate freely.

From the condition in the Gnathostomulida, the primary pattern of the protonephridia of the adelphotaxon Plathelminthes can easily be derived. On the basis of the structure of the terminal cells of the Catenulida, the following evolutionary steps in the stem lineage of the Plathelminthes may be postulated: doubling of the number of cilia, loss of the eight-rod supporting apparatus, and loss of the accessory centriole. The result is a terminal cell which, in all Catenulida (*Retronectes, Catenula, Stenostomum*), always carries two cilia (Kümmel 1962; Moraczewski 1981; Rieger 1981; Ehlers 1985b). As in the Gnathostomulida, the filter area is differentiated from a peripheral cytoplasmic border of the terminal cell (Fig. 79C, D).

---

[1] In principle there is no difference between cilia and flagella, and I therefore use the single term cilium when discussing both protonephridia and sperms.

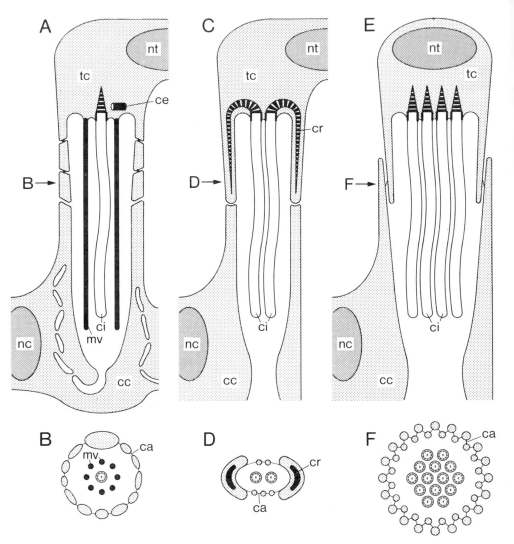

Fig. 79. The ultrastructure of protonephridia in the Plathelminthomorpha—the terminal cell and the adjacent canal cell. A, C and E are diagrammatic longitudinal sections. B, D and F are transverse sections. Arrows show the levels of the transverse sections. A and B, Gnathostomulida (*Gnathostomula*); C and D, Catenulida (*Retronectes*); E and F, the pattern of the Rhabditophora. Abbreviations: ca = filtration area; cc = canal cell; ce = accessory centriole; ci = cilium; cr = ciliary rootlet; mv = microvilli; nc = nucleus of canal cell; nt = nucleus of terminal cell; tc = terminal cell. (Original by U. Ehlers)

A striking autapomorphy of the Catenulida is the rod-like elongation of the two ciliary rootlets. These bend through 180° and run distalwards. They probably represent new supporting elements replacing the lost primary supporting apparatus.

Further evolutionary changes led to the protonephridial system of the plathelminth taxon Rhabditophora (Fig. 79E, F). Firstly, compared with the Catenulida, the number of cilia was doubled again and then increased many times (from four cilia to over 100). And secondly, in certain subtaxa of the Rhabditophora additional elements were added to the canal cell by forming a new compound double weir or grill. A second set of plasmatic rods was developed by the canal cell. These grew towards the cytoplasmic border of the terminal cell—a border which itself had differentiated into longitudinal rods. The processes of the two cells overlapped each other and together formed a regular girdle of alternating rods with the filter area stretched between them. And, at the same time, the new filter area took on a supporting function in that it formed a constant space for the action of the multiciliary propeller.

As concerns the stem lineage of the Eubilateria, the protonephridial system of the Gastrotricha is of special interest. The ultrastructure is known for one representative of the marine Macrodasyoidea (*Turbanella*) and of the predominantly limnic Chaetonotoidea (*Chaetonotus*) (Brandenburg 1962, 1966, 1975; Teuchert 1973). The structure of the terminal cells shows detailed agreements with the protonephridia of the Gnathostomulida. In both cases there is a single cilium in an identical supporting apparatus formed of eight microvilli. As in the Gnathostomulida, the filter area is differentiated from a peripheral cytoplasmic border of the terminal cell—in the Gastrotricha, as in the Gnathostomulida, it is provided with pores and clefts. However, a definite apomorphy compared with the condition in the Gnathostomulida is that several terminal cells (3–4 in *Turbanella*, 2 in *Chaetonotus*) open into a shared collecting cell.

These observations allow us to specify, without any contradictions, the structure of the protonephridia in the latest common stem species shared by the Plathelminthomorpha and Eubilateria. In the stem lineage of the Bilateria, a protonephridium was evolved consisting of a terminal cell, a canal cell and a nephropore cell. The terminal cell carried a cilium with a diplosomal basal apparatus, an internal supporting apparatus of eight microvilli (rods) and a peripheral filter area.

In the phylogenesis of the Bilateria, this pattern was transmitted to the stem lineages of the Plathelminthomorpha and of the Eubilateria. Only in the taxon Gnathostomulida is it retained in the unaltered plesiomorphous condition. But very primitive morphological conditions likewise persisted in the stem lineage of the Eubilateria. For the protonephridial system of the Gastrotricha has undergone only very slight changes from the bilaterian ground pattern (compare new results of Neuhaus in *Microfauna Marina* 3, 1987).

This brings me to a last important point. The structure of the organ in the Gnathostomulida gives a convincing model for the evolutionary origin of the protonephridia from ectoderm. The terminal cell of the gnathostomulid protonephridium is nothing but a slightly modified monociliary epidermal cell. In sinking into the primary body cavity, it took with it the cilium and the diplosomal basal apparatus. The eight rods which support the action-space of the cilium are merely extreme elongations of eight microvilli which always surround the cilium of every normal epidermal cell, forming there a circlet of short cell processes (Fig. 78A; Rieger & Mainitz 1977).

Mode of fertilization
The free release of gametes into the surrounding water, and the fusion of ova and

sperm outside the body, both belong in the ground pattern of the Bilateria. These conditions also hold for the ground pattern of the Eubilateria. External fertilization is an extremely ancient phenomenon in multicellular animals which was taken over from the latest common stem species of the Metazoa, as a plesiomorphy, into the stem lineage of the Bilateria and afterwards passed into the stem lineage of the Eubilateria.

By contrast, the Gnathostomulida and the Plathelminthes agree in having direct sperm transfer and an identical method of internal fertilization. I postulate that these advanced mechanisms evolved, once only, in the stem lineage of the Plathelminthomorpha and consequently assign both features to the ground pattern of this taxon. When compared with the ground pattern of the Eubilateria, they represent a first complex of derived features which can be claimed as an autapomorphy in constituting a monophyletic taxon Plathelminthomorpha (Gnathostomulida + Plathelminthes).

The spermatozoan

The plesiomorphous mode of external fertilization is correlated with a particular basic structure for the spermatozoa. The primitive sperm of the Metazoa consists of a rounded oval head containing the nucleus, a short middle piece with a few mitochondria (basic number four) and a tail section with a cilium having the primitive 9 + 2 number of microtubuli (Franzen 1956, 1977).

On this basis, two questions can be put when determining the ground pattern of the Plathelminthomorpha:

Firstly, are there apomorphous sperm structures connected with internal fertilization which can be interpreted as synapomorphies of the Gnathostomulida and the Plathelminthes?

Secondly, have structural elements of the primitive metazoan sperm been retained within the Plathelminthomorpha?

Among the Gnathostomulida, representatives of the taxa *Pterognathia* and *Haplognathia* have elongated thread-like sperms with a long spiral head piece, a rod-like middle piece and a cilium with a 9 + 2 pattern. The dwarf sperms without cilia, and also the large conuli which likewise lack cilia, of various other taxa must be interpreted as the evolutionary results of changes within the Gnathostomulida (Graebner 1969a; Sterrer 1968, 1974; Sterrer, Mainitz & Rieger 1985).

Among the Plathelminthes, observations of particular importance have recently been made on the ultrastructure of the Nemertodermatida. *Nemertoderma* and *Meara* agree in having thread-like sperms with an elongated head, a long cylindrical middle piece and a cilium of 9 + 2 pattern (Tyler & Rieger 1975, 1977; Hendelberg 1977).

Among the Plathelminthes, all departures from this sort of sperm can validly be seen as apomorphies. I shall confine myself to a few points in this connection. Within the Plathelminthes, aciliary sperm have been evolved several times convergently. Thus they occur: in the Catenulida (Borkott 1970; Sterrer & Rieger 1974), though vestiges of ciliary structures have been detected in the spermatogenesis of *Retronectes* (Rieger 1978, Ehlers 1985b); in the Macrostomida; and in the Prolecithophora (Ehlers 1985b). On the other hand, sperm with two cilia arose independent of each other in the stem lineage of the Acoela (cilia with 9 + 2, 9 + 1 and

9 + 0 patterns; Hendelberg 1977) and in that of the Trepaxonemata (Polycladida + Neoophora—cilia with nine peripheral tubuli and a complicated axial rod, fig. 89).

The observations just mentioned suggest the following answers to the questions posed above:

1) A thread-like sperm with a long head, a cylindrical middle piece and a cilium (with a 9 + 2 pattern of the microtubuli) must be assigned to the ground pattern of the Plathelminthomorpha.

2) The elongate thread-like sperm arose as an evolutionary novelty in the stem lineage of the Plathelminthomorpha by alteration of the primitive metazoan sperm. This ground-pattern feature was transmitted to the stem lineages of the sister-groups Gnathostomulida and Plathelminthes but in the Gnathostomulida it is retained only in a few subtaxa (*Pterognathia, Haplognathia*). In other words, the identical condition of a thread-shaped sperm can be interpreted, without any contradictions, as a synapomorphy of the Gnathostomulida and the Plathelminthes. This gives a further constitutive feature to justify the monophyly of the Plathelminthomorpha. I stress that this hypothesis is logically and factually independent of the fact that comparable "modified" sperm have arisen convergently in various taxa of the Eubilateria which have changed to internal fertilization.

3) The existence of sperm having a single cilium with the primitive 9 + 2 pattern is a plesiomorphy of the Plathelminthomorpha. This morphology was taken over, unchanged, from the primitive metazoan sperm and transmitted to the stem species of the Plathelminthomorpha, of the Gnathostomulida and of the Plathelminthes. The above-mentioned evolutionary changes in the sperm tail (loss of cilia, doubling of cilia, changes in the number and structure of the central microtubuli) arose separate from each other within the Gnathostomulida and in various stem lineages of the Plathelminthes. These apomorphies will be used below in validating various monophyletic subgroups of the Plathelminthes.

Egg cell

In the Gnathostomulida and in several subtaxa of the Plathelminthes (Catenulida, Acoelomorpha, Macrostomida, Polycladida) simple, endolecithal eggs are produced. Without doubt this is a primitive feature of the Metazoa which was taken over, as a plesiomorphy, into the stem lineage of the Plathelminthomorpha. Endolecithal eggs certainly belong to the ground pattern of the taxon.

The characteristic separation of the ovary into germarium and vitelline gland and the resulting formation of ectolecithal compound "eggs" made up of germ cell and yolk cells (Fig. 80) is an apomorphous complex of features which first evolved, within the Euplathelminthes, in the stem lineage of the taxon Neoophora.

Sexual condition

The production of sperm and ova in different individuals of a species is the primitive condition in Metazoa. The phemonenon of separate sexes was transmitted from the stem species of the Metazoa to the ground pattern of the Bilateria.

269

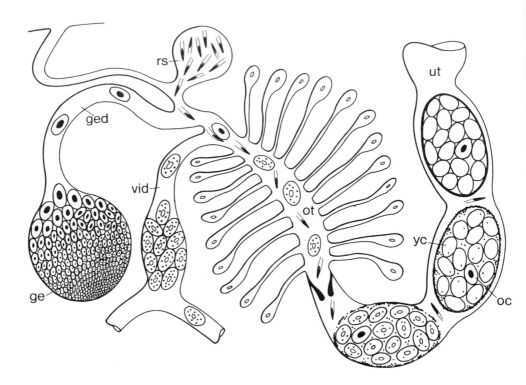

Fig. 80. Neoophora. Diagram of the production of ectolecithal eggs as exemplified by the Digenea. In the uterus, various stages of the formation of the egg capsule are shown with the secretion of shell substance by the yolk cells. Abbreviations: ge = germarium; ged = germarial duct; oc = germocyte (oocyte); ot = ootype; rs = receptaculum seminalis; ut = uterus; yc = yolk cell. (Modified after Smyth & Clegg 1959; from U. Ehlers 1985b.)

At the splitting of the latest stem species shared by Plathelminthomorpha and Eubilateria, this plesiomorphous gonochorism at first passed unchanged into the stem lineage of the Eubilateria and is conserved in many of its sub-taxa. In the stem lineage of the Plathelminthomorpha, on the other hand, a change to hermaphroditism occurred. Since all Gnathostomulida, and almost all Plathelminthes, are hermaphroditic, the evolution, once only, of hermaphroditism in the stem lineage of the Plathelminthomorpha can readily be postulated. As an autapomorphy, hermaphroditism is thus a third feature that can be used to validate the monophyly of the Plathelminthomorpha.

Mode of cleavage

So far as known, all Plathelminthomorpha which produce endolecithal eggs have the spiral mode of cleavage. The basic expression of this pattern of cleavage shows the following group-specific alternatives:

a) In the Gnathostomulida (Riedl 1969), as well as in the three plathelminth taxa Catenulida, Macrostomida and Polycladida, micromere quartets are formed in identical manner (surveyed in Ax 1961; Ax & Borkott 1969).

270

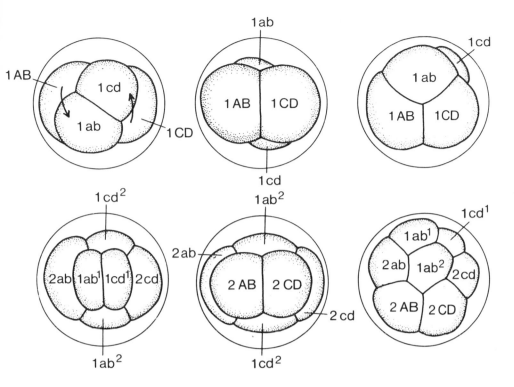

Fig. 81. Duet spiral cleavage in the Acoela, as shown by *Oligochoerus limnophilus* Ax & Dörjes. Upper row: four-cell stage; left—view from animal pole, middle—view from vegetative pole, right—view from side. Lower row: eight-cell stage with orientations as above. From Ax & Dörjes (1966).

b) In the Acoela, on the other hand, the cleavage is characterized by the formation of micromere duets at the animal pole (Fig. 81; Bogomolov 1960; Ax & Dörjes 1966; Apelt 1969). No studies have yet been made for the adelphotaxon Nemertodermatida.

Accordingly the first question is as follows: does the widespread quartet spiral cleavage belong in the ground pattern of the Plathelminthomorpha, or the duet spiral cleavage as found only in Acoela (and possibly in the Nemertodermatida)? In other words, which alternative is the plesiomorphous condition for the taxon Plathelminthomorpha?

The arguments for a justifiable probability decision come from three different sources.

1) In the phylogenetic system of the Plathelminthes, which is based on other features, the Catenulida and the Euplathelminthes (including Acoelomorpha) are placed alongside each other as adelphotaxa (p.283).
2) As to out-group comparison, the quartet mode of spiral cleavage is widespread in the Eubilateria, e.g. in the Nemertini, Entoprocta, Sipunculida, Echiurida, in the ground pattern of the Mollusca and in the ground pattern of the Articulata.

271

3) There are extensive agreements between the Polycladida (Kato 1940) and the above-named taxa of Eubilateria in the cell lineage up to the origin of micromeres 4a–4d and in the prospective significance of the individual blastomeres.

In this situation, an attempt to interpret the duet mode of spiral cleavage as the plesiomorphous alternative would inevitably have the following consequences: Firstly, the quartet mode of spiral cleavage would have arisen convergently three times within the taxon Plathelminthomorpha—in the stem lineages of the Gnathostomulida, Catenulida and Rhabditophora. Moreover, quartet spiral cleavage would also have evolved from duet spiral cleavage within the Eubilateria, independent of the situation within the Plathelminthomorpha.

By contrast with this extremely unparsimonious interpretation, the view that duet spiral cleavage is the apomorphous alternative is much more economical. It involves nothing more than the transition, once only, from quartet to duet spiral cleavage in the stem lineage of the Acoela (or somewhat farther back, in the stem lineage of the Acoelomorpha if duet cleavage should prove to be characteristic in the Nemertodermatida). There are two conceivable ways in which this change may have happened. Either the first micromere duet of the Acoela corresponds to stem blastomeres B and D of quartet cleavage, or else the 2nd. division of the quartet cleavage was suppressed in Acoela. In the latter case, two blastomeres of the quartet cleavage would correspond to one blastomere of the duet cleavage. Following Bogomolov (1960), I show this interpretation for the four- and eight-cell stages of cleavage in Acoela by labelling the cells with the corresponding cell notations (Fig. 81; AB, CD, ab, cd)

I therefore favour the following hypotheses:

1) The quartet pattern of spiral cleavage evolved once only in the phylogeny of Metazoa.
2) Quartet spiral cleavage is the primitive mode of cleavage for the Plathelminthomorpha. The duet spiral cleavage of the Acoela is a secondary modification.
3) Beyond that, quartet spiral cleavage belongs in the ground pattern of the taxon Bilateria. It must already have existed in the latest stem species shared between the Plathelminthomorpha and Eubilateria. In this stem species it would have had a cell lineage like that of Polycladida and various taxa of Eubilateria.

I shall now list, without prejudice, the logical consequences of these hypotheses:

— There is no monophyletic subtaxon Spiralia within the Bilateria. As a synonym of Bilateria, the name Spiralia must be deleted from the phylogenetic system of the Metazoa.
— Within the Bilateria, all the other modes of cleavage represent secondary modifications of quartet spiral cleavage. This statement holds without exception. It covers the bilateral cleavage of the Nemathelminthes as well as the radial cleavage seen in the Tentaculata and in the ground pattern of the Deuterostomia.

This conclusion is based partly on theoretical considerations of various sorts (Schmidt 1966; Costello & Henley 1976; Hennig 1980). But it is also based on clear

272

cases where quartet spiral cleavage has completely given place to other modes of cleavage in well based monophyla of the Eubilateria. Thus, in the taxon Mollusca, quartet spiral cleavage was replaced in the stem lineage of the Cephalopoda by incomplete, discoidal cleavage. And in the taxon Arthropoda, apomorphous changes extend to the equal holoblastic cleavage of particular Decapoda. In *Penaeus trisulcatus*, for example, the arrangement of the eight blastomeres after the 3rd. division exactly corresponds to a "classical" radial cleavage (Zilch 1979).

This leads me to a last important point. If quartet spiral cleavage is validly assigned to the latest stem species shared between Plathelminthomorpha and Eubilateria, then it must have evolved, at latest, in the stem lineage of the Bilateria. But could it have arisen even earlier in the phylogeny of the Metazoa? This question can be decided by a comparison with the taxa Porifera and Coelenterata. In neither of these is there a pattern of cleavage agreeing in detail with quartet spiral cleavage. This outgroup comparison therefore suggests, on good grounds, that quartet spiral cleavage evolved in the stem lineage of the Bilateria. Seen as an evolutionary novelty (autapomorphy) of the taxon Bilateria, it becomes a synapomorphy of the adelphotaxa Plathelminthomorpha and Eubilateria.

This consideration of quartet spiral cleavage allows me to illustrate a mistake often made when discussing the evolutionary transformation of features. Ivanova-Kasas (1959, 1982) tried to derive the spiral cleavage of particular Bilateria from the cleavage patterns of certain coelenterate species which show some possible indications of spiral cleavage. She referred to these indications as "pseudo-spiral cleavage". I have already explained the artificiality of deriving whole monophyletic taxa from the middle of a single monophylum: But, equally, the derivation of a particular apomorphous structure (spiral cleavage of the Bilateria) from particular feature states (pseudo-spiral cleavage) of particular single species in some other monophyletic taxon is not compatible with what we know of the process of phylogenetic development. The attempt to derive the spiral cleavage of the Bilateria from so-called pseudo-spiral cleavage would only be permissible, in logic and fact, if a definite pseudo-spiral sequence of cleavage could be established for the ground pattern of the whole of the Coelenterata and for the mosaic of features which existed in the latest stem species shared between Coelenterata and Bilateria. Then, and only then, would the hypothesis be legitimate that spiral cleavage evolved from pseudo-spiral cleavage. But, of course, the widespread distribution of radial cleavage in the Porifera and Coelenterata argues against such a view. Indeed, it gives good grounds for supposing that the latest stem species of the Eumetazoa (Coelenterata + Bilateria) had a radial pattern of cleavage. In other words, in thinking about the evolution of quartet spiral cleavage in the stem lineage of the Bilateria, we have to start from radial cleavage.

Life cycle

According to Jägersten (1972), a "pelago-benthic life cycle", with two phases separated by a metamorphosis, is supposed to belong to the ground pattern of the Metazoa. The pelagic phase would be represented by a primarily planktotrophic larva, whereas the benthic phase was the adult. If, in accordance with this widespread view, the two-phase mode of development of the Porifera, Coelenterata and many Bilateria is seen as a plesiomorphy of the Metazoa, then the lack of larvae in all metazoan taxa that have direct development would be a secondary apomorphous state.

I shall now confront this hypothesis of a primary pelago-benthic life cycle with the situation in the taxon Plathelminthomorpha.

The Gnathostomulida have a direct development without larvae. Among Plathelminthes, marine pelagic larvae occur only in one subgroup of the free-living Trepaxonemata. These larvae are the juvenile stages of Polycladida referred to as Müller's and Goette's larvae. They do not occur, however, in all Polycladida but only in certain species. Because of their striking lobe-shaped processes, ranging in number from four to ten, Salvini-Plawen (1980b) groups these larvae together as lobophora larvae.

I shall not discuss the subordinate question, still completely open, whether Polycladida with larvae derive from a common stem species with a "lobophora" larva or not. Any attempt to see the life cycle of certain Polycladida with larvae as being the primary mode of development in Plathelminthomorpha implies that loss of larvae and the change to direct development must have happened several times independently. On the basis of the phylogenetic system of the Plathelminthomorpha, which can be established by means of other combinations of features, this logical requirement can be stated very precisely. A primarily two-phase life cycle must have been lost convergently at least five times, i.e. once in the stem lineage of the Gnathostomulida and, within the adelphotaxon Plathelminthes, in the stem lineages of the Catenulida, Acoelomorpha, Macrostomida and Neoophora. There is no shadow of justification for such a conclusion, neither in the ontogeny nor in the biology of the Plathelminthomorpha.

What is the alternative? The following clear arguments can be given for interpreting Müller's and Goette's larvae as evolutionary novelties (apomorphies) which first arose within the taxon Polycladida.

a) The Polycladida are the only plathelminth taxon with endolecithal eggs in which the body has appreciable enlarged above the primary size of about one millimetre. The correlation between the two-phase life cycle and large benthic plathelminths producing endolecithal eggs definitely suggests that larvae evolved in connection with a secondary increase in size and volume.

b) There is not a single agreement between the larvae of polyclads, on the one hand, and any larval form in the Eubilateria, on the other, which can be interpreted as a synapomorphy. The only feature which might be discussed in this connection is the existence of an "apical organ" as in the trochophore larva (Salvini-Plawen 1980b). This, however, can be rejected as a purely semantic agreement. Indeed, in the polyclad larva (Ruppert 1978) the apical complex of sensory and gland cells is not even a larval feature. For it is nothing but the frontal organ, widespread in the Euplathelminthes, which passed from the adult into the juvenile stage when a pelagic phase was evolved within the Polycladida. This interpretation even makes it doubtful whether the term "larva" can validly be used within the Polycladida. Apart from the lobe-like processes of locomotory function, the pelagic developmental stage of the polyclads has no genuine larval features whatever.

274

There are strong arguments, therefore, for assigning direct development without a pelagic larva to the ground pattern of the Plathelminthomorpha.

I shall go a step further and concentrate, within the Eubilateria, on the Nemathelminthes. Here the gastrotrichs, rotifers and nematodes, which are primarily of millimetre size, all have direct development without exception. Once again, I see no reason to hypothesize the absence of larvae as being a derived, apomorphous condition. As in the Plathelminthomorpha, therefore, a direct mode of development seems to be required for the stem species of the Eubilateria and also, logically, for the stem species of all Bilateria.

A last question comes within the scope of the present discussion: Is the lack of a larva in the ground pattern of the Bilateria to be seen as a plesiomorphy or an apomorphy in comparison with the adelphotaxon Coelenterata?

This question can be answered by way of the planula larva of the Cnidaria. The existence of a pelagic planula is strictly correlated with the development of a sessile polyp phase. I interpret the planula larva as an autapomorphy of the descent community Cnidaria, for its evolution as a dispersal phase can only be understood in connection with the evolution of a sessile polyp.

If, however, the planula larva first evolved in the stem lineage of the Cnidaria, then all speculations must fail which derive the Bilateria, as the "further development" of a planula, out of the Coelenterata or Cnidaria. For such attempts involve, once again, the artificial concept of "deriving" one taxon (Bilateria) out of the middle of another monophylum (Coelenterata) (p.32).

Here I shall argue according to the principles of phylogenetic systematics. I postulate that the latest common stem species shared between Coelenterata and Bilateria was a vagile, microscopic, benthic organism with direct development. This species could have had a diploblastic level of organisation like *Trichoplax adhaerens* (of Grell's taxon Placozoa). For this, so far as known, has no pelagic larva and may represent the adelphotaxon of Coelenterata + Bilateria. However this may be, a direct mode of development passed as a plesiomorphy into the stem lineage of the Bilateria whereas a two-phase life cycle, with a planula larva as an evolutionary novelty (apomorphy), arose in the stem lineage of the Cnidaria.

By thus rejecting the hypothesis of a primary, pelago-benthic life cycle for the Metazoa, there is no longer any need to derive very different forms of larva, such as the trochophore or the tornaria, from a common origin. If pelagic larvae have arisen more than once within the Eubilateria, then they can be used to validate the monophyly of the larger subgroups of the taxon Eubilateria.

2. The composition of the ground pattern

In accordance with what I have said on the composition of the ground pattern of closed descent communities in Nature (p.147), I shall now arrange the ground-pattern features of the Plathelminthomorpha into two groups which differ in time of origin. In the first group there are primitive features which already existed in the stem lineage of the Bilateria and were taken over from this lineage as plesiomorphies. While in the second group are those evolutionary novelties which first evolved in the stem lineage of the Plathelminthomorpha. Interpreted as autapomorphies, these latter can be used to validate the taxon Plathelminthomorpha as equivalent to a closed descent community in Nature (Fig. 82, apomorphy block 2).

In arranging the ground-pattern features in this manner, the latest common stem species shared between the Gnathostomulida and the Plathelminthes can be seen, without any contradictions, as made up of the following elements.

a) Plesiomorphies

— The organism was vagile, benthic and about one millimetre in size. The body was bilaterally symmetrical and circular in transverse section.
— There was a simple mouth pore on the ventral face which arose in ontogeny from the blastopore region.
— The gut was epithelial, rod-shaped and lacked an anus.
— The epidermis consisted of monociliary cells. Each cilium had a diplosomal basal apparatus with a basal body and accessory centriole.
— The dermal muscle layer was weak and sub-epithelial and consisted of outer transverse muscles and inner longitudinal muscles. The muscles were smooth and the muscle cells were individualized i.e. there were no epithelial muscle cells.
— The nervous system had a brain-like nerve centre and a peripheral intra-epidermal nerve plexus.
— The body cavity was primary. Over wide areas the ectoderm was in immediate contact with the endoderm. In addition to the dermal muscle layer, the protonephridia and the genital organs, there was perhaps only a weak development of stem cells (and parenchyma cells) in the narrow interstices of the primary body cavity.
— A blood-vascular system was absent.
— Paired protonephridia were present and consisted of three cells—a terminal cell, a canal cell and a nephropore cell. The terminal cell had a cilium and a diplosomal basal apparatus, contained an outer cytoplasmic filtration area and an inner supporting apparatus of eight rods (elongated microvilli).
— Egg formation was endolecithal.
— Cleavage was spiral with the formation of quartets.
— Development was direct. There was no pelagic larva.

b) Autapomorphies = synapomorphies of the Gnathostomulida and Plathelminthes

— Sperm transfer was direct, with internal fertilization of the egg cells.
— The sperms were thread-shaped with a long head piece, a cylindrical middle piece and a long tail section (with a primitive cilium of 9 + 2 pattern).
— Hermaphroditism was present.The animal was hermaphroditic with male and female gonads.

276

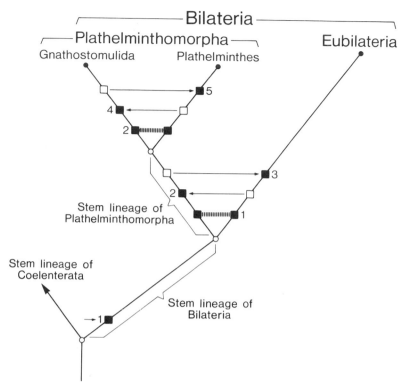

Fig. 82. Diagram of the phylogenetic relaionships of the Bilateria. The black squares 1–5 summarize the features which arose as evolutionary novelties (autapomorphies) in the stem lineage of the Bilateria (1), of the Plathelminthomorpha (2), of the Eubilateria (3), of the Gnathostomulida (4), and of the Plathelminthes (5).

III.   Position of the Plathelminthomorpha in the taxon Bilateria

1.   The monophyly of the Bilateria

In determining the position of the Plathelminthomorpha within the Bilateria, it is logical that the taxon Bilateria should itself first be established as equivalent to a closed descent community in Nature.

In this connection, I refer back to the ground pattern of the Plathelminthomorpha as already discussed. The plesiomorphies of the latest common stem species that was shared between Gnathostomulida and Plathelminthes (features of group a) include some **features which can justifiably be regarded as having evolved in the stem lineage of the Bilateria**.

These features, apomorphous for the Bilateria, are as follows:

— The body had bilateral symmetry in the antero-posterior axis.
— There was a weak, sub-epithelial dermal muscle layer consisting of individualized

277

muscle cells. The layer was made up of outer transverse and inner longitudinal muscles.

— There was a brain-like nerve centre, being a concentration of neurons at the anterior pole of the body.
— Protonephridia were present and consisted of a terminal cell, a canal cell and a nephropore cell in the epidermis. The first two cells evolved from monociliary ectodermal cells which had migrated into the primary body cavity.
— Spiral cleavage occurred with the formation of quartets.

In accordance with the principle of parsimony, I postulate that these features evolved, once only, in the stem lineage of the Bilateria. From this they passed into two new stem lineages and thus the autapomorphies of the Bilateria became synapomorphies of the Plathelminthomorpha and Eubilateria (Fig. 82, apomorphy block 1).

The remaining ground-pattern plesiomorphies of the Plathelminthomorpha (features of group a) are likewise plesiomorphies for the taxon Bilateria. They were taken over, farther back in time, from the stem lineage of the Eumetazoa.

2. The adelphotaxon relationship between Plathelminthomorpha and Eubilateria

In the course of the argument I now need to justify the phylogenetic systematization of the Bilateria into the adelphotaxa Plathelminthomorpha and Eubilateria. This is possible because the following four plesiomorphous ground-pattern features of the Bilateria evolved divergently.

Bilateria	Plathelminthomorpha	Eubilateria
1. External fertilization	Internal fertilization	External fertilization
2. Primitive sperms	Modified sperms	Primitive sperms
3. Gonochorism	Hermaphroditism	Gonochorism
4. Blind-ending gut without anus	Blind-ending gut without anus	One-way gut with anus

In the stem lineage of the Plathelminthomorpha there occurred the changes already discussed, i.e. from external fertilization to internal fertilization with direct transmission of sperm, from the primitive metazoan sperm to a thread-shaped sperm, and from separate sexes to hermaphroditism (Fig. 82, apomorphy block 2). In the stem lineage of the Eubilateria, on the other hand, these three features were at first taken over unchanged.

As against this, in the stem lineage of the Eubilateria the one-way gut with an anus evolved as a novelty (Fig. 82, apomorphy 3) whereas an alimentary canal with no anus was retained as a plesiomorphy in the stem lineage of the Plathelminthomorpha.

The Plathelminthomorpha and the Eubilateria can thus both be validated as monophyla by means of autapomorphies which each group alone possesses. In both

278

taxa the alternative features are present as unchanged plesiomorphies. Thus the Plathelminthomorpha and Eubilateria fulfil the conditions required by phylogenetic systematics when establishing monophyletic taxa as sister groups. The correct fulfilment of these conditions is absolutely essential for work on the phylogenetic system, so I turn to a different instance of exactly the same question.

IV. Gnathostomulida and Plathelminthes as adelphotaxa within the Plathelminthomorpha

Two closed descent communities in Nature constitute adelphotaxa of the phylogenetic system if: 1) all species of both taxa can be traced back to a stem species common to these two taxa alone; 2) if both taxa can be validated as monophyla, and distinguished from each other, by their own autapomorphies; and 3) if the autapomorphies of the one group contrast with the plesiomorphous alternatives in the other.

With these requirements in mind, I shall now examine the postulated adelphotaxon relationship between the Gnathostomulida and the Plathelminthes.

The derivation of all species of Plathelminthomorpha from a stem species common to them alone has just been validated on the basis of three autapomorphies.

In fulfilment of the second and third requirements, I cite five plesiomorphous ground-pattern features of the Plathelminthomorpha which all passed unchanged from the stem lineage of the Bilateria into that of the Plathelminthomorpha.

Plathelminthomorpha	Gnathostomulida	Plathelminthes
1. Monociliary epidermis	Monociliary epidermis	Multiciliary epidermis
2. Cilium with diplosomal basal apparatus	Cilium with diplosomal basal apparatus	Cilium with one basal body, without accessory centriole
3. Protonephridia with one cilium	Protonephridia with one cilium	Protonephridia with at least two cilia
4. Terminal cell of protonephridium with supporting apparatus of eight rods	Terminal cell of protonephridium with supporting apparatus of eight rods	Supporting apparatus of terminal cell lost
5. Simple stomodaeum	Pharynx with a pair of jaws and a basal plate	Simple mouth pore

Features 1 to 4, taken from the ultrastructure of the epidermis and the protonephridia, all underwent obvious evolutionary changes in the stem lineage of the Plathelminthes. As concerns each of these features, on the other hand, the Gnathostomulida retained the plesiomorphous alternative. For multiciliary epidermal cells with one basal body per cilium and protonephridia with a doubling of the cilia and loss of the internal supporting apparatus are unequivocal apomorphies which firmly establish the Plathelminthes as a monophyletic taxon (Fig. 82, apomorphy block 5).

On the other hand, the stem species of the Plathelminthes inherited a simple mouth region from the ground pattern of the Plathelminthomorpha, and of the Bilateria. The apomorphous alternative for this feature is represented by the complex pharyngeal apparatus of the Gnathostomulida with a pair of jaws and a basal plate in the buccal cavity (Fig. 77). Both structures arise as peripheral secretions of particular formative ectodermal cells. In *Haplognathia rosea*, the cuticular basal plate is formed by five cells of the ventral epithelium of the buccal cavity, whereas 14 cells produce the paired jaw apparatus (Lammert, unpublished). The mode of formation for the hard structures of the buccal cavity and pharynx thus differs in principle from the basal or intracellular production of convergent grasping hooks in the proboscis organ of the plathelminth taxon Kalyptorhynchia (Rieger & Doe 1975, Doe 1976) and also from the intracellular differentiation of penile hard structures in the Plathelminthes (Mainitz 1977, Ehlers & Ehlers 1980, Doe 1982). I postulate that the pharyngeal hard structures were evolved once only in the stem lineage of the Gnathostomulida and I validate the monophyly of the taxon by way of this remarkable autapomorphy (Fig. 82, autapomorphy 4).

Thus the requirements of phylogenetic systematics are satisfied. Having already recognized the sister-group relationship between the Plathelminthomorpha and Eubilateria, we can now establish the Gnathostomulida and Plathelminthes as equal-ranking adelphotaxa within the Plathelminthomorpha.

In this systematization, however, another phenomenon is of high interest—namely the **absence of any trace of mitosis** in the somatic cells of the Plathelminthes. In this group, the worn-out body cells are always replaced by stem cells which lie, ready and waiting, in the interstices of the body cavity (Ehlers 1985b). The fact that the differentiated somatic cells of Plathelminthes cannot mitose can certainly be seen as a well marked apomorphy, as shown by an out-group comparison with the Coelenterata as well as the Eubilateria. Unfortunately the behaviour of the Gnathostomulida in this respect has not yet been analysed. For the moment it is uncertain whether the inability of somatic cells to divide developed in the stem lineage of the Plathelminthes or had already evolved in that of the Plathelminthomorpha. In the present state of knowledge, the feature can best be taken as an autapomorphy of the Plathelminthes and positively helps, along with the arguments already mentioned, to validate them as a monophyletic taxon.

As a result of this analysis, the basic diagram of basal phylogenetic relationships of the Bilateria can be translated into the following hierarchical tabulation:

> Bilateria
> > Plathelminthomorpha
> > > Gnathostomulida
> > > Plathelminthes
> > Eubilateria

Without assigning any categorial rank, this phylogenetic systematization can be carried out with complete clarity and without any objections.

V. The phylogenetic system of the Plathelminthes

The traditional classification of Plathelminthes into the three equal-ranking "classes" of Turbellaria, Trematoda and Cestoda is not compatible with the known phylogenetic relationships within the taxon. I stress four essential points.

1) The **"Turbellaria"** are a paraphyletic collection of primitively free-living species and must, once and for all, be eliminated from the phylogenetic system of the Plathelminthes (p.156). The name "Turbellaria" has to be deleted definitively.

2) The **Trematoda**, if understood in the traditional fashion as comprising Monogenea and Digenea, are likewise an artificial grouping. However, the name Trematoda can, in fact, still be used in a consistent phylogenetic systematization of the parasitic flatworms, to cover Aspidobothrii and Digenea.

3) The **Cestoda**, as traditionally delimited, are fully accepted since they can be validated unobjectionably as a monophylum. However, they constitute only a low-ranking, subordinate subtaxon of the Plathelminthes.

4) The old controversy as to whether the **parasitism** of the traditional Trematoda and Cestoda evolved once only, in a stem lineage shared by both, or whether trematodes and cestodes became parasites independently of each other, can now be settled. The discovery of a remarkable apomorphous body covering (the neodermis) shows that Trematoda and Cestoda can both be validly traced back to a stem species common to the two groups alone. Both can therefore be united in a monophyletic taxon Neodermata.

A first attempt to produce a system of the Plathelminthes based on the principles of phylogenetic systematics was that of Karling (1974). He limited himself, however, to the free-living "orders" of Plathelminthes.

Ehlers (1984, 1985) has aimed at a **complete phylogenetic systematization of the Plathelminthes**, taking account of ultrastructural observations and considering, as is indeed necessary, the parasites as well as the taxa with free-living representatives. The result is the system written as a hierarchical tabulation on p.283[1] and presented as a diagram of phylogenetic relationships in Fig. 83.

Once again, the categorial terms of the Linnaean hierarchy have been thrown out as useless ballast which bring nothing but confusion.

In the present state of knowledge, this phylogenetic system of the Plathelminthes is inadequate in the following points:

---

[1] In this system for the Plathelminthes, the authors of the subtaxa are cited. Taxa assigned to Ehlers were introduced in Ehlers (1984).

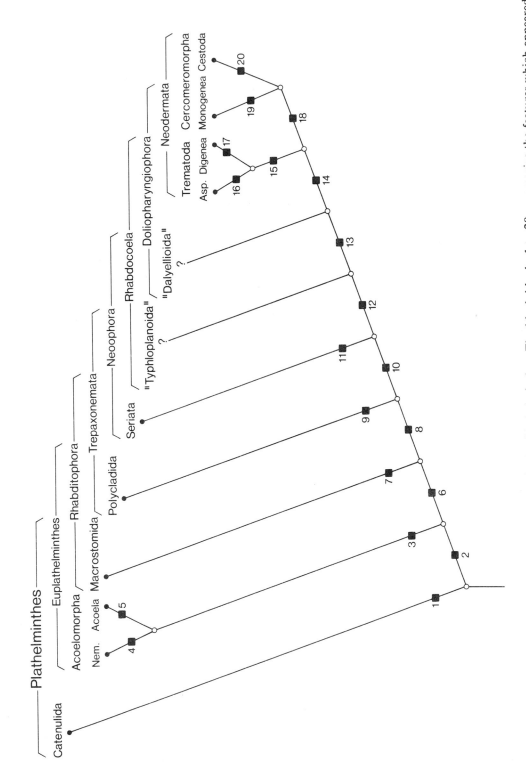

Fig. 83. Diagram of the phylogenetic relationships in the Plathelminthes. The black blocks 1 to 20 summarize the features which appeared as evolutionary novelties (autapomorphies) in the stem lineages of the respective monophyletic subtaxa of the Plathelminthes.

1) The **Prolecithophora** and the **Lecithoepitheliata** are two monophyletic taxa of uncertain position within the Plathelminthes.

The taxon **Prolecithophora** has aciliary sperms of complicated structure as an autapomorphy (Ehlers 1981). Because of their ectolecithal egg formation they belong, with reasonable certainty, in the monophylum Neoophora. However, their adelphotaxon cannot yet be determined.

The **Lecithoepitheliata** are very difficult to evaluate. As autapomorphies they have the germocytes (oocytes) and vitellocytes arising in a common layer and the developing germocytes enclosed by vitellocytes (Karling 1974). However, the yolk cells of the Lecithoepitheliata lack the droplets of shell substance which are characteristic for the Neoophora. At present it seems possible that the Lecithoepitheliata should be assigned to the Neoophora. On the other hand, it may equally be that the separation of germ cells from yolk cells in the stem lineage of the Neoophora happened independent of evolution of the germovitellarium of the Lecithoepitheliata, which contains both germocytes and vitellocytes. For the time being there is no way of deciding between these alternatives.

Because of these unsolved problems, the Prolecithophora and the Lecithoepitheliata can, for the moment, only be placed in the system as *taxa incertae sedis*.

2) No autapomorphies can yet be assigned to the traditional groups of the "**Typhoplanoida**" (including the monophyletic subtaxon Kalyptorhynchia) and the "**Dalyellioida**". These traditional groups can therefore not be validated as monophyletic units of the Plathelminthes. Within the monophyla Rhabdocoela and Doliopharyngiophora they represent mere **provisional arrangements**.

```
Plathelminthes
    Catenulida      von Graff
    Euplathelminthes      Bresslau & Reisinger
        Acoelomorpha      Ehlers
            Nemertodermatida      Steinböck
            Acoela      Uljanin
        Rhabditophora      Ehlers
            Macrostomida      von Graff
            Trepaxonemata      Ehlers
                Polycladida      Lang
                Neoophora      Westblad
                    Seriata      Bresslau
                    Rhabdocoela      Ehrenberg
                        "Typhloplanoida"      von Graff
                        Doliopharyngiophora      Ehlers
                            "Dalyellioidea"      Meixner
                            Neodermata      Ehlers
                                Trematoda      Rudolfi
                                    Aspidobothrii      Burmeister
                                    Digenea      von Beneden
                                Cercomeromorpha      Bychowsky
                                    Monogenea      von Beneden
                                    Cestoda      Gegenbaur
```

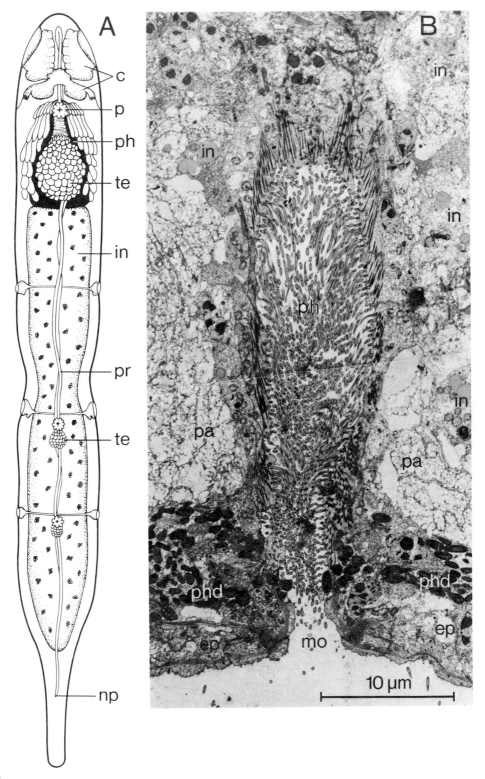

284

New studies are needed aimed at resolving these open questions of relationship.

Otherwise, the monophyly and the adelphotaxon relationships of all the systematic units presented can be established with reasonable certainty from the highest ranking level of the adelphotaxa Catenulida and Euplathelminthes to the lowest ranking pair of sister-groups, i.e. the Monogenea and Cestoda. Group by group, I here set out a series of striking features each of which must have arisen in the stem lineages of the respective taxa as evolutionary novelties. In the appended diagram of phylogenetic relationship these have been inserted as autapomorphy blocks 1 to 20 (Fig. 83).

After stating the adelphotaxon relationship in question, I shall now discuss the chosen autapomorphies for each of the blocks, taxon by taxon.

Catenulida–Euplathelminthes

A1: Catenulida

— Unpaired protonephridium

The Catenulida are the only taxon of Plathelminthes which possess an unpaired protonephridium. It lies in the mid-line, in the longitudinal axis of the body (Fig. 84a). By out-group comparison with the Gnathostomulida, which have paired protonephridia, this condition can be seen as an apomorphy.

— Ciliary supporting elements in the terminal cells.

The rootlets of the two cilia in the terminal cells of the protonephridia are elongated in the form of rods. They extend distalwards as supporting pillars on the narrower sides of the filter area (Fig. 79c, d). Comparable structures are absent from the protonephridia of all other taxa of Plathelminthes.

A2: Euplathelminthes

— Epidermis strongly multiciliary

There is an obvious increase in the number of cilia per unit area of epidermis compared with the epidermis of the Catenulida which, by comparison, is weakly multiciliary (Fig. 78b, c).

— Frontal organ

This is a specific complex of sensory and gland cells at the anterior end of the body (Fig. 85a). It is absent in all Catenulida, whether marine or fresh-water. We postulate that the frontal organ was evolved, once only, in the stem lineage of the Euplathelminthes.

Fig. 84. Catenulida. A) Organisation of *Stenostomum sthenum* Borkott. Dorsal aspect of a chain of zooids. Note the unpaired protonephridium in the long axis of the body and the nephropore at the posterior end. B) A transverse section through the pharynx simplex of *Retronectes* cf. *sterreri* Faubel. Abbreviations: c = brain, ep = weakly multiciliary epidermis, in = intestine, mo = mouth opening, np = nephropore, p = male genital pore, pa = parenchyma, ph = pharynx with strongly multiciliary epithelium, phd = pharyngeal glands, pr = protonephridium, te = testis. A) after Borkott 1970, B) original by U. Ehlers.

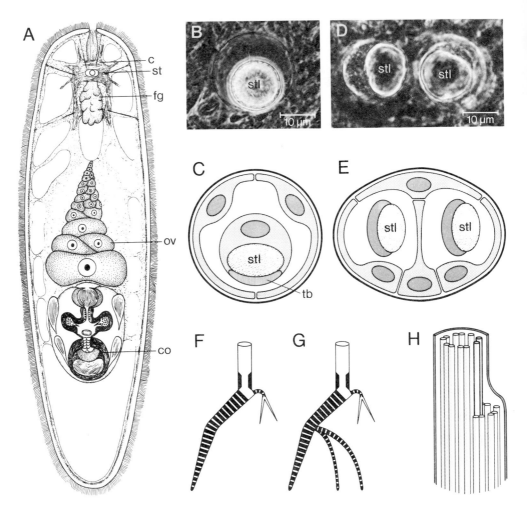

Fig. 85. Acoelomorpha (Nemertodermatida and Acoela). A) Organisation of *Haplogonaria syltensis* Dörjes (Acoela). B) and C) statocyst of the Acoela showing the single statolith and the "tubular body". B) *Anaperus tvaerminnensis* (Luther); dorsal aspect in phase contrast. C) Diagram of transverse section on the basis of ultrastructural studies. D) and E) Statocyst of the Nemertodermatida showing the two statoliths. D) *Nemertoderma* cf. *bathycola* Steinböck; dorsal aspect in phase contrast. E) Diagrammatic transverse section. F) The structure of the rootlets in the epidermal cilia of *Nemertoderma* (Nemertodermatida)—the posterior rootlet splits below the basal body. G) The structure of the rootlets in epidermal cilia of the Acoela—there are two additional lateral rootlets on the posterior rootlet. H) The distal end of the epidermal cilia of the Acoelomorpha—the cilium terminates in a shaft. Abbreviations: c = brain, co = copulatory organ, fg = frontal gland, ov = ovary, st = statocyst, stl = statolith, tb = tubular structure. A) From Ax 1966; B) to E) from Ehlers 1985b; F) to H) originals by Ehlers.

Acoelomorpha–Rhabditophora

A3: Acoelomorpha

— Interconnection of ciliary rootlets in the epidermis
  In the ground patterns of the Gnathostomulida and the Plathelminthes, the two
  ciliary rootlets had the following orientation. One rootlet ran horizontally forwards from the basal body while the second rootlet ran obliquely rearwards or
  vertically downwards (Fig. 78).
  In the stem lineage of the Acoelomorpha, this plesiomorphous condition of
  the ciliary rootlets was considerably altered. Both in the Nemertodermatida and
  Acoela, the middle of the anterior ciliary rootlet is bent at an angle. The posterior rootlet divides into two bundles of fibres immediately below the basal body
  (Fig. 85F) and these bundles both make contact with the knee-like angle of the
  rostral rootlet situated obliquely behind them. As a result, adjacent cilia of the
  epidermis make contact with each other in a remarkable network. This has no
  counterpart in other Metazoa (Hendelberg & Hedlund 1974; Tyler & Rieger
  1977, Tyler 1979).

— Shafts at the distal ends of the cilia
  As a general rule, the 9 + 2 pattern of the axonemal microtubuli extends to the
  very end of the cilium.
  In Nemertodermatida and Acoela, on the other hand, the epidermal cilia are
  characterized by a special terminal shaft region. For four of the nine peripheral
  double tubuli end suddenly, on the posterior side, a short distance below the
  distal end of the cilium. Only five of the peripheral tubuli and the two central
  tubuli pass upwards into the narrow shaft (Fig. 85H; Tyler 1979).
  This unique structure of the ends of the cilia provides a further well marked
  apomorphy for constituting a monophyletic taxon Acoelomorpha within the Euplathelminthes.

— Absence of protonephridia
  The protonephridial system is absent in all Nemertodermatida and Acoela. This
  fact has been confirmed by electron-microscope studies. In view of the existence
  of protonephridia in the Gnathostomulida, the Catenulida and all other taxa
  of the Euplathelminthes, the lack of protonephridia in the Nemertodermatida
  and Acoela must be seen as an apomorphous condition. The most parsimonious
  explanation postulates the loss, once only, of protonephridia in the stem lineage
  of the Acoelomorpha.

A4: Rhabditophora

— Lamellar rhabdites
  Free-living Plathelminthes produce rod-like secretions (rhabdoids, rhabdites,
  rhamnites) in the epidermis or in sub-epidermal gland cells. Specific details
  of their ultrastructure can now be used to establish phylogenetic relationships
  (Smith, Tyler, Thomas & Rieger 1982; Ehlers 1985a,b).
  The rod-like secretions of the Macrostomida, Polycladida and Neoophora can
  be traced back to a common ground pattern. In these taxa the cortex of the
  rhabdites consists of concentrically arranged lamellae (Fig. 86). Comparable fib-
  rillar cortices are completely absent in the structures secreted by Catenulida and

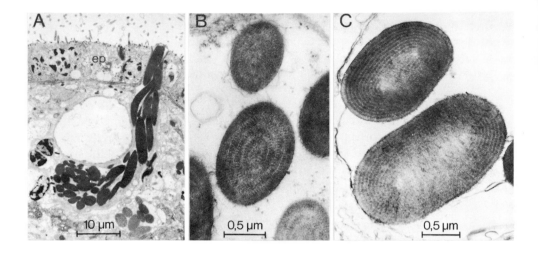

Fig. 86. Lamellated rhabdites of the Rhabditophora. A. *Cirrifera aculeata* (Ax) (Seriata). Production of the rod-shaped secretions in glandular cells beneath the epidermis (ep). B. *Paromalostomum fusculum* Ax (Macrostomida). Cross-section. C. *Notoplana* cf. *atomata* O. F. Müller) (Polycladida). Cross-section. (A. Original by Sopott-Ehlers: B–C from Ehlers 1985b.)

Acoelomorpha. On the basis of this characteristic distribution among free-living Plathelminthes, lamellar rhabdites can be interpreted, without any contradictions, as a synapomorphy of the Macrostomida + Trepaxonemata (= Polycladida + Neoophora). We postulate that they evolved, once only, in the stem lineage of a closed descent community Rhabditophora.

— Duo-gland adhesive organ

The Macrostomida, the Polycladida and most taxa of free-living Neoophora possess adhesive organs of identical structure whose ground structure comprises three specifically differentiated cells (Tyler 1976):

a) A gland cell with large, electron-dense secretions which have the function of adhering to the substrate.

b) A gland cell with smaller granules, supposedly with a lytic function and serving to destroy the adhesion.

c) An anchor cell which is a modified epidermal cell. The anchor cell is penetrated by the outlet ducts of the two gland cells. It forms circlets of microvilli around the outlet ducts of the adhesive gland cells (Fig. 87, 88).

Fig. 87. The duo-gland adhesive organ of the Rhabditophora. A) *Xenotoplana tridentis* Ax & Sopott-Ehlers. B) *Pesudominona dactylifera* Karling (Seriata). In both these taxa of Seriata, the posterior end of the animal has been differentiated to form a three-lobed adhesive apparatus densely provided with duo-gland adhesive organs. *Pseudominona dactylifera* is illustrated in two rapidly alternating phases—anchored to the substrate with spread-out adhesive lobes, and in locomotion with the lobes abutting. (Original by Ax.)

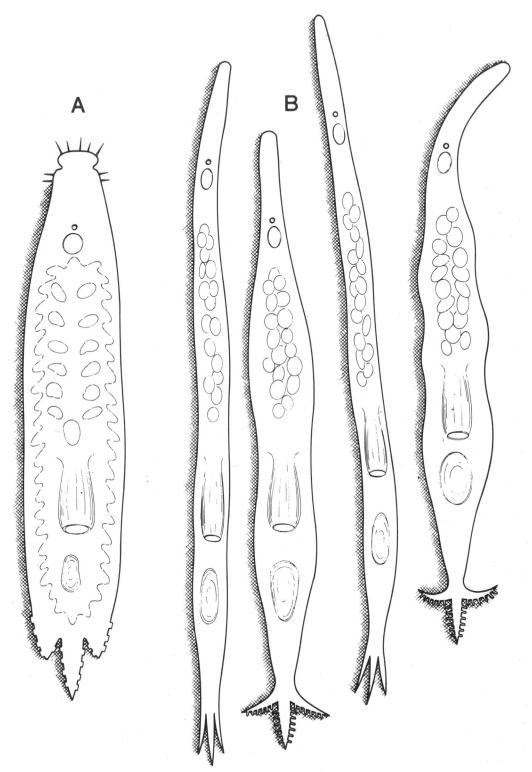

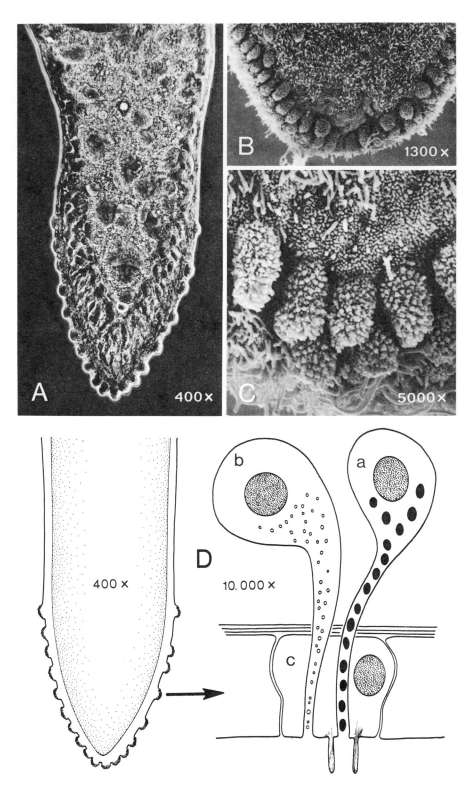

The absence of such duo-gland adhesive organs in the Catenulida and the Acoelomorpha must be seen as a plesiomorphy. On the other hand, its absence among the Doliopharyngiophora (including all the parasites) can only be interpreted as a secondary condition.

In agreement with the distribution of lamellar rhabdites, we postulate that the complicated duo-gland adhesive organ is a further autapomorphy of the Rhabditophora which evolved, once only, in the stem lineage of the taxon.

— Increase in number of cilia in the protonephridia
   As already mentioned, the cilia in the terminal cells of the Catenulida are always two in number. As against this, an increase in the number of cilia has occurred in the Rhabditophora. The "ciliary flames" in the terminal cells here consist of a minimum of four cilia (as found in a few Macrostomida and Polycladida). In certain parasitic Plathelminthes, they may contain more than a hundred cilia.

Nemertodermatida–Acoela

A5: Nemertodermatida

— Statocyst with two statoliths
   By contrast with the sister-group Acoela, which have only one statolith, the statocysts of the Nemertodermatida have two statoliths next to each other (Fig. 85d, e). The evolution of the two statoliths must have happened in the stem lineage of the Nemertodermatida. It certainly represents an autapomorphy of the taxon.

A6: Acoela

— Statolith cells with tubular structure
   In principle the Acoela possess only a single statolith in each individual. In all Acoela this is completely enclosed by the statolith cell which secretes it. Ventrally in this cell a tubular structure is differentiated on which the statolith rests (Fig. 85c). This "tubular body" is a remarkable structure unknown elsewhere and is an autapomorphy of the taxon Acoela.
— Additional lateral rootlets of the cilia
   The pattern of ciliary rootlet described for the Acoelomorpha, under A3 above, underwent a further apomorphous change in the stem lineage of the Acoela. For in this taxon the anterior rootlet developed two lateral rootlets as an evolutionary novelty (Fig. 85G). The lateral rootlets arise in the "knee" region and run

Fig. 88. Duo-gland adhesive organs of the Rhabditophora. A) Posterior end of *Bothriomolus balticus* Meixner (Seriata) with attachment organs projecting as papillae (phase contrast). B) and C) Posterior end of *Coelogynopora axi* Sopott (Seriata) with anchor cells; the enlarged microvilli are visible (SEM pictures). D) Diagram of an attachment organ comprising a gland cell (= a) with large electron-dense secretions, a second gland cell (= b) with small granules, and an epidermal anchor cell (= c) with a circlet of microvilli. A) Original of Ax; B) to D) originals of Sopott-Ehlers.

rearwards to contact the terminal points of the adjacent anterior rootlets (Hendelberg & Hedlund 1974). Because of these new, additional points of contact, the intra-epidermal network of ciliary rootlets in the Acoela achieves a unique autapomorphous structural complexity.

— Biciliary sperms

The Nemertodermatida is the only taxon of Plathelminthes which retains the ground-pattern sperm with one cilium only (p.268). By contrast, all Acoela which have been studied in detail have two cilia on each sperm. By comparison with the adelphotaxon Nemertodermatida, biciliary sperm are a clear autapomorphy of the Acoela.

Macrostomida–Trepaxonemata

A7: Macrostomida

— Aciliary sperm

Compared with the sister-group Trepaxonemata, the Macrostomida represent a taxon with numerous primitive features. An autapomorphy which can be given for them is the absence of cilia from the sperm. The Macrostomida certainly belong in the subtaxon Rhabditophora, so this situation in the Macrostomida can only be interpreted as an apomorphy.

A8: Trepaxonemata

— Biciliary sperms

As in the Acoela, biciliary sperms are widely distributed in the subtaxa of Plathelminthes grouped here under the name Trepaxonemata. At first sight it seems plausible to postulate that sperm with two cilia existed in the latest common stem species of all Plathelminthes. This hypothesis fails, however, with the discovery of plesiomorphous monociliary sperm in the Nemertodermatida, for this taxon is well established as the sister group of the Acoela. It is now necessary to assume that biciliary sperm evolved twice independently of each other, once in the stem lineage of the Acoela and once in that of the Trepaxonemata.

— Spiral axial rod in the cilia of the sperm

An important autapomorphy of the taxon Trepaxonemata is the replacement of the primitive central double tubuli of the sperm cilia (9 + 2 pattern) by an unpaired axial rod (9 + "1" pattern). According to Thomas (1975), the axial rod in both cilia of the sperm comprises three components: a) a central electron-dense element; b) an intermediate bright zone; and c) a peripheral electron-dense shell with spirally running longitudinal bands (Fig. 89).

— Pharynx compositus

The pharynx simplex, as represented in the adelphotaxon Macrostomida, evolved in the stem lineage of the Trepaxonemata into the protrusible pharynx compositus. In the Polycladida and the Seriata this is found in the plesiomorphous condition as a pharynx plicatus. In the stem lineage of the Rhabdocoela the latter evolved to the apomorphous condition known as the pharynx bulbosus.

292

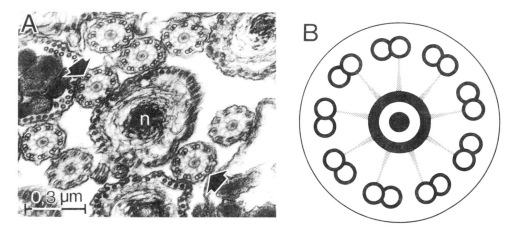

Fig. 89. Ultrastructure of sperm in the Trepaxonemata. A) *Notocaryoplanella glandulosa* (Ax) (Seriata). Sperm in transverse (n = nucleus). The arrows point to the two cilia of the sperm. B) Diagrammatic transverse section of a sperm cilium to show the central unpaired axial rod and the nine peripheral double microtubuli. (From Ehlers 1985b.)

Polycladida–Neoophora

A9: Polycladida

— Body shape
  The body is dorso-ventrally flattened. This shape is connected with an increase in body size above the primitive length of about a millimetre.
— Structure of gut
  The primitive rod-shaped gut has evolved into a gastro-vascular system with lateral diverticula (the polyclad gut).
— Follicular ovaries
  The paired ovaries have divided into numerous follicles.

A10: Neoophora

— Production of ectolecithal eggs
  In the stem lineage of the Neoophora, the undivided primitive ovary became separated into the germarium and the vitellarium. The germarium produces ova that are capable of developing. The vitellarium produces yolk cells with droplets of yolk and of shell substance. Ova and yolk cells come together to form ectolecithal eggs (Fig. 80).

Seriata–Rhabdocoela

A11: Seriata

— Serial arrangement of follicular gonads
  The testis and vitellaria are subdivided into numerous follicles. The gonads form longitudinal rows with a regular serial sequence of follicles.

293

— Pharynx tubiformis
The pharynx plicatus, primitively directed to the ventral surface, came to be stretched out into a tube and was directed rearwards in the long axis of the body. In this new condition it is known as the pharynx tubiformis.

A12: Rhabdocoela

— Pharynx bulbosus
The plesiomorphous pharynx plicatus has evolved into a barrel-shaped pharynx bulbosus. The pharynx bulbosus is completely shut off from the parenchyma of the body by the formation of a new septum.

"Typhloplanoida"–Doliopharyngiophora

?: "Typhloplanoida"

— This is a paraphyletic grouping of primitive Rhabdocoela which has no autapomorphies. The pharynx rosulatus, directed perpendicular to the ventral surface, represents the plesiomorphous ground pattern of the pharynx bulbosus.

A13: Doliopharyngiophora

— Pharynx doliiformis
The pharynx has moved to the anterior end of the body and has acquired a subterminal mouth opening. Other apomorphous details are the loss of cilia and the migration of the nucleated parts of the pharyngeal epithelium to form a crop (Ax 1961).

"Dalyellioida"–Neodermata

?: "Dalyelloida"

— A paraphyletic grouping of those Doliopharyngiophora which have a pharynx doliiformis. They are primitively free-living but sometimes parasitic. No autapomorphies have been shown to exist.

Fig. 90. Neodermata. Degeneration of the ciliated epidermis and the formation of the neodermis in the ontogenesis of the Monogenea.
A) Young embryo: the primary body cover formed of completely ciliated epidermal cells and areas of syncytium without cilia; the nuclei are intra-epithelial.
B) Older embryo: in the ciliated epidermal cells the nuclei are degenerating while the nuclei of the syncytia are being expelled; beneath the basal lamina, which is in process of differentiation, two stem cells are shown.
C) Free-swimming larva (oncomiracidium): the epidermis is now without nuclei while flat cytoplasmic processes of the stem cells are spreading out beneath it.
D) The juvenile worm: the primary epidermis has now been sloughed off and the body is covered with unciliated syncytial neodermis originating from the stem cells; the nuclei of the neodermis remain subepithelial in position.
Abbreviations: bl = basal lamina, ci = cilia. epc = epithelial cells. n = nuclei, neod = neodermis, stc = stem cells, sy = syncytium. (After Lyons 1973, from Ehlers 1985b.)

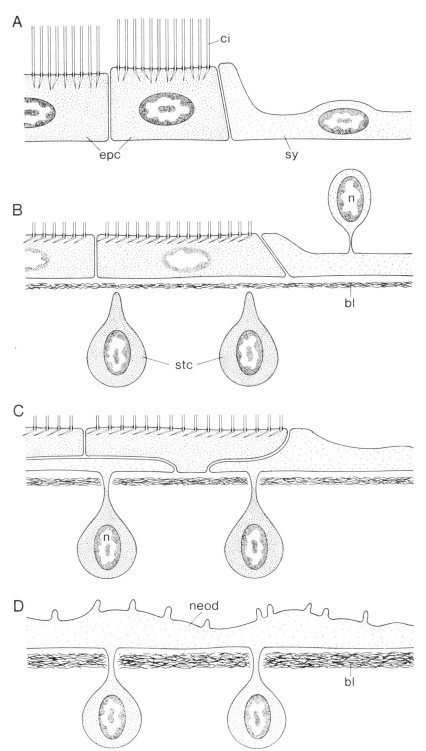

295

A14: Neodermata

— Loss of the ciliated ectodermal epidermis
The plesiomorphous multiciliate superficial epithelium of free-living Plathelminthes is, in the taxon Neodermata, in principle confined to the larval stages. Thus among Trematoda it is found in the cotylocidium and miracidium larvae; among Monogenea in the oncomiracidium; and among Cestoda in the lycophora and the coracidium. By the end of the larval stage, at latest, the ciliated ectoderm has been lost completely.

— Evolution of the neodermis (secondary body covering)
The stem cells (neoblasts) inside the body send out cytoplasmic extensions which break through the basal lamina. After the larval epithelium has been sloughed off, these extensions spread over the "naked" body and fuse with each other to form a syncytial peripheral layer named as the neodermis by Ehlers. The nucleated cell bodies remain in a subepithelial position (Fig. 90).
This remarkable mode of formation is, in principle, the same in the Trematoda, Monogenea and Cestoda. It can therefore validly be postulated that the secondary body cover evolved once only, and was present, at latest, in the last common stem species of these parasitic taxa (Ehlers 1985a,b). The neodermis is thus the crucial autapomorphy uniting the Trematoda, Monogenea and Cestoda in a monophylum Neodermata.

Trematoda — Cercomeromorpha

A15: Trematoda

— Partial ciliation of the larvae
The larvae of the Aspidobothrii and the Digenea are not completely covered with cilia, for the multiciliate ectodermal cells are always separated from each other by syncytial cytoplasmic areas of neodermis.
By contrast, the larvae of the Cestoda are at first completely covered by ciliated ectodermal epidermis. Comparison with the free-living Plathelminthes shows, without doubt, that this is the plesiomorphous condition for the larval body covering. The partial ciliation of the larvae of trematodes can be seen, without any contradictions, as a synapomorphy of the Aspidobothrii and the Digenea.

A16: Cercomeromorpha

— Sickle-shaped hooks at the posterior end of the larva
A comparison of the oncomiracidium larve of the Monogenea with the lycophora larva of monozoic tape worms (*Gyrocotyle, Amphilina*) and with the oncosphaera larva of merozoic Eucestoda reveals a high degree of structural agreement in the larval hooks; they are produced with identical form intracellularly in oncoblasts. The number of hooks (16) realized in the Monogenea belongs in the ground pattern of the Cercomeromorpha.

296

Aspidobothrii–Digenea

A17: Aspidobothrii

— Larva (cotylocidium) with a posterior sucker
— Sucker of adult divided into alveoli
  The adult Aspidobothrii are characterized by a huge sucker which may take up the greater part of the ventral face of the body. This sucker is divided into a large number of little sucking organs.
— Neodermis with microtubercles
  In Aspidobothrii, the microvilli of the neodermis consist of characteristic microtubercles of a structure unknown in any other taxon of Plathelminthes with secondary body covering.

A18: Digenea

— The larval ectoderm cells arranged in tiers
  The ciliated ectodermal cells of the miracidium are separated from each other by narrow strips of cytoplasm so as to form regular transverse bands.
— Life cycle with several stages of development (miracidium, sporocyst, redia, cercaria, hermaphroditic adult).
— Miracidium and Sporocyst without a digestive system.
— Obligatory change of host.
— Cercaria and adult with a central sucker.

Monogenea–Cestoda

A19: Monogenea

— Ciliation of the oncomiracidium in three zones
  In the ground pattern of the Monogenea the multiciliate epidermal cells of the oncomiracidium, about 60 in number, are aggregated to form three complexes, i.e. an apical, a middle and a caudal complex (Lambert 1980). The remaining, unciliated areas are formed of a syncytium with intraepithelial nuclei.
  This differentiation is fundamentally different from the partial ciliation of the miracidium and cotylocidium of the trematodes (A15). For the aciliate areas of the oncomiracidium of the Monogenea obviously do not represent neodermis. Rather they correspond to the original ectoderm in which the cilia have been lost and the cell walls have disappeared.
— Existence of four rhabdomerous photoreceptors

A20: Cestoda

— Larvae and adult with no gut
  The original alimentary canal has been completely lost. Food is absorbed through the neodermis.
— Larvae with ten sickle-shaped hooks
  Compared with the Monogenea, the number of sickle hooks was reduced to ten

in the stem lineage of the cestodes. This ground pattern is found in the lycophora larva of the Gyrocotylidea and Amphilinidea. Within the Cestoda a further reduction has occurred to give the six hooks of the oncosphaera (Caryophyllidea + Eucestoda).

———————

The goal of phylogenetic systematics is two-fold: first, to discover the order in living Nature which has arisen in the process of phylogenesis; and second, to erect a system exactly corresponding in structure to the phylogenetic relationships which Nature has created between evolutionary species and closed descent communities — and to do this in a logically consistent manner.

In zoology we shall take a great step towards this double goal if an analysis like the preceding is applied to all the "phyla" of the traditional classification and hypotheses are formulated, step by step, on adelphotaxon relationships. For these hypotheses, being based on the methodology of phylogenetic systematics, are open to objective scientific test.

M. The Potentialities of the Phylogenetic System

I. The claims of phylogenetic systematics

There are millions of recent and fossil organisms. The task, therefore, of discovering the first degree of phylogenetic relationship of all evolutionary species and all closed descent communities in Nature must seem immense, only approachable by small steps and perhaps never fully achievable (Hennig 1984). Nevertheless, the laying bare of every adelphotaxon relationship in living Nature is and remains the mission of phylogenetic systematics—whatever the limits to reaching it in individual cases and regardless of questions about the usefulness of the insights that are sought.

It cannot be denied that phylogenetic systematics is qualified to gain insights into phylogenetic relationships. And it is therefore illogical to question the legitimacy of presenting the favoured hypotheses in a form adequate to our scientific knowledge. Our efforts are strictly orientated towards the order created in Nature by the process of phylogenesis. It is this order only, therefore, which determines the structure of a logically consistent phylogenetic system. Presenting our knowledge must be primarily divorced from all considerations of applying the system for any other purpose whatsoever.

Phylogenetic systematics, however, claims more than this. As I have already argued, the phylogenetic system of organisms is the sole objective system of reference for biology which is based on an epistemologically unobjectionable subdivision of the organic diversity in Nature. I repeat, once more, the simple reason for this. The connection by descent of all evolutionary species in Nature, together with the step-wise sequence of shared origin in time, gives the only objectifiable measure for the systematization of organisms and does so by providing objectively testable hypotheses.

The objections still raised to this claim have nowadays almost ceased to cavil at the theory or methodology of phylogenetic systematics. Rather they are directed, above all, at the supposedly intolerable results of applying the phylogenetic system as the universally obligatory reference system of biology. Hennig (1984) answered in detail a variety of objections. Here I wish to concentrate my discussion on two important aspects—the explanatory power of the phylogenetic system and its practicability compared with traditional evolutionary classification.

II. Explanatory power

The explanatory power of a theory or hypothesis is measured by its logical consequences (Vollmer 1975). The phylogenetic system is nothing else than an infinite series of mutually compatible hypotheses and so we need precisely to examine what the consequences are which this system necessarily produces or which can be derived from it. In other words, what capacity does it have as a store of information and to what extent can it allow the formulation of correct predictions? For these two characteristics are widely believed wholly to determine the explanatory power of a system or a classification.

1. Storage of information

From any and every order-structure for organisms which has been created by Man taking account of the International Rules of Nomenclature, the only things that can be directly extracted are symbols. These take the form of proper names for species and for supraspecific species groups. The next question, therefore, is as follows: What information, bound to these symbols, is exactly stored in the phylogenetic system and retrievable from it?

Phylogenetic relationship

The system can fulfil the function of storing hypotheses on the first degree of phylogenetic relationship between real unities in Nature only if: 1) the supraspecific unities are consistently and without exception validated as monophyletic taxa; and 2) they are in principle each given their own name, in their character as replicas of closed descent communities in Nature. If, in addition, a particular arrangement of symbols is chosen, then the phylogenetic system stores, in either of its two forms of presentation, all the adelphotaxon relationships in living Nature. Moreover, it stores them in an absolutely unmistakeable way so that they are always retrievable. Thus, in the diagram of phylogenetic relationships, each taxon is placed alongside its adelphotaxon at the corresponding level of hierarchy. And in the hierarchical tabulation, each taxon together with its adelphotaxon is indented by a fixed amount according to the sequence of subordination — and this is true whether or not rank is further specified by numeration or categorial designation. I have already explained in detail that the relative ranks of monophyletic taxa in the systematic hierarchy cannot be indicated, in any logically consistent way, by using the conventional Linnaean categories (p.239). In academic teaching, I have had no difficulty in expounding the logical structure of the phylogenetic system without using categorial designations.

On the other hand, all attempts at ordering are useless as bearers of information on phylogenetic relationship if: 1) they accept non-monophyletic groupings as units of classification; or 2) they are not logically consistent in joining every pair of adelphotaxa into higher ranking systematic units with proper names. I have already justified these statements, in detail, by reference to the paraphyletic "Apterygota" and the equally paraphyletic "Reptilia" and the conventional subdivisions of both these groupings (p.185,189). I shall now illustrate points 1) and 2) by reference to the taxon Plathelminthes.

Anyone who wishes to retain the traditional classification of the Plathelminthes into the three "classes" of Turbellaria, Trematoda and Cestoda, i.e. as three units of the same rank, must be clear in his mind that he is approving an order-structure that stores no trace of phylogenetic information. If the paraphyletic class of the "Turbellaria" is to be tolerated, with its traditional subdivision into numerous equal-ranking "orders" such as Catenulida, Acoela, Macrostomida, Polycladida, Seriata, Rhabdocoela etc., then it will not hold any information whatever on how the taxa of free-living flatworms are phylogenetically related to each other. And, on the other hand, tolerating a classification with two disconnected "classes" of parasitic flatworms

300

arranged alongside each other covers over the important question of whether the Trematoda and Cestoda go back to a parasitic stem species shared by them alone, or have become parasitic independently in separated stem lineages.

On the other hand, in the system of the Plathelminthes expounded in the previous chapter (Fig. 83), a taxon Trematoda (in its new delimitation) is placed alongside and at the same level as all the other parasites and these are together given the name Cercomeromorpha (= Monogenea + Cestoda). And the Trematoda + Cercomeromorpha are then, logically enough, united at the next higher level to form a monophylum Neodermata. If all this is done, then the required information concerning the phylogenetic relationships of these taxa is stored perfectly and, at the same time, the question already put is answered clearly. For the arrangement of names states, unmistakeably, that the Monogenea and Cestoda are postulated to be adelphotaxa within the closed descent community of the Cercomeromorpha. Likewise, the sister-group relationship between the Trematoda and the Cercomeromopha can equally be read off from the phylogenetic system. Furthermore, by uniting all these parasitic taxa under their own name of Neodermata a final piece of information is stored—namely, that all of them together belong to a closed descent community in Nature. In other words, as a totality they represent the descendants of a stem species shared by them alone.

Features

Although all Man's attempts at ordering operate with agreements and differences in the pattern of features of different organisms, none of the competing order-structures allows any information on the features or characteristics of the classified taxa to be read off directly (Farris 1977, 1979c; Wiley 1981). In Farris' view, any illusions to the contrary depend tacitly on a biologist's already-acquired knowledge of the characteristic features of the group—features which the biologist connects inseparably with the group name. Thus he connects hair and the feeding of the young on the secretions of milk glands with the symbol Mammalia. And he associates the name Cestoda with endoparasitism and the taking-up of food through the body surface.

Farris has said (1979c, p.489) that: "Specific information about the properties of organisms is contained in a classification, however, in an indirect sense. The names of taxa form a system that refers to diagnoses and descriptions of those taxa". Nevertheless, whether or not specific information about the feature patterns of supraspecific taxa can be retrieved from an order-structure on the basis of diagnoses or descriptions depends solely on the construction of the order-structure itself. I refer back to the difference in principle between the diagnostic features of the supraspecific units of traditional classifications, on the one hand, and the constitutive features (autapomorphies) of the monophyletic taxa of the phylogenetic system, on the other (p.153).

Traditional classifications produce diagnoses in which groupings of species are delimited from "neighbouring" taxa by way of the combination of a particular set of shared features. However, such diagnoses can be used to characterize not only the equivalents of real unities in Nature but also the non-monophyletic groupings artificially created by Man. As examples of this, I have already given the diagnoses of the "Reptilia" and "Turbellaria" in traditional classifications. Phylogenetic systematics requires, on the other hand, that the monophyly of each supraspecific taxon be validated by constitutive features, i.e. on the basis of evolutionary novelties (autapomorphies) which evolved in the stem lineage of the taxon.

In a phylogenetic system, in other words, such unique features show that the taxa are replicas of closed descent communities in Nature. Only with a consistently phylogenetic system will it be both possible and obligatory to find data about these unique features for all the names of supraspecific taxa in the literature. If, on the other hand, the symbols stand for artificial products not equivalent to real unities in Nature, then the data in the literature will be correspondingly worthless.

2. Predictive power

Opinions differ widely concerning the predictive power of the phylogenetic system, or, indeed, of any other method of subdividing organisms. Before proceeding with the argument, therefore, I shall try to find a consensus about what we actually mean when we speak of a prediction made from an order-structure.

The word "**prediction**", in what is probably its generally accepted usage, means a statement, based on objective knowledge, concerning what will happen in the future. Given this meaning of the word, neither the phylogenetic system, nor any other order-structure created by Man, allows an exact prediction to be made. For physics, on the basis of natural laws, can formulate reliable statements on the course of natural phenomena. But no comparable statements about future phylogenetic developments can be made on the basis of the phylogenetic system. Any hypothesis about phylogenetic relationships created in Nature in the past is made entirely in the framework of the theory of evolution, "... which is acknowledged as a scientific theory although it can make almost no predictions" (Vollmer 1975, p.111).

If the word "**prediction**" is applied with **a dubious and very different meaning**, it is widely asserted that the phylogenetic system has a special "predictive power", claimed to be one of the most astounding achievements in biology. This is the ability to "predict" certain new discoveries in organisms which live now or existed in the past. Thus these predictions relate not to future, but to past, phylogenesis. To be more exact, they are of two sorts: 1) those that predict the discovery of already known features in new, still insufficiently analysed, organisms; and 2) those that predict the discovery of still unknown features in species, or species groups, which have already been well studied. Statements of both sorts will show that this distinction is not mere casuistical word-splitting.

As to the "exactitude of relevant predictions" of the first sort, I shall take Remane's well known *Dytiscus* example. Suppose that a new, still undescribed water beetle is placed in the taxon *Dytiscus* on the basis of external features in dry specimens. In that case, according to Remane (1952, p.5): "By inserting it into the system, it is now possible to make a great number of relevant predictions: on the structure of the nervous system (which will be of the "rope-ladder" type), on the slit-like ostia of the heart, the histology of the brain, the structure and function of the tracheal system, egg laying, development (with superficial cleavage etc), the structure of the larva, feeding in the larva (external digestion), and so forth, extending down to thousands of details."

This sort of "prediction" as to the future discoveries of Man concerning Nature does not concern anything new. It deals with nothing but self-evident trivialities.

A new representative of the genus *Dytiscus* consists, like any single species of any other closed descent community in Nature, of thousands of features. We know that these were not thrown together pell-mell, but were evolved, one after the other, in different organisms in a long historical process and were joined together successively. If, on the basis of a few features, I can place a new species in a system of internested supraspecific taxa, then this species—ignoring any possible evolutionary losses—must necessarily and self-evidently show the totality of all features, alongside each other, which existed in the corresponding ground patterns of the closed descent communities. Thus if the new species, on the basis of preliminary results, is placed in the taxon *Dytiscus*, then it automatically belongs in a whole series of supraspecific taxa of increasing superordination. As a brief selection, I cite the Dytiscidae, Coleoptera, Insecta, Arthropoda and Articulata. By way of example, I choose a few characteristic features which the species must possess by successive inheritance from the ground patterns of these taxa.

— Larvae with sickle-shaped sucking mandibles and with external digestion, for these features arose in the Mesozoic as evolutionary novelties in the stem lineage of a closed descent community whose equivalent in the system is given the name Dytiscidae.
— Fore-wings in the form of hard elytra and hind-wings consisting of delicate, foldable alae, since wings of this condition were evolved in the Palaeozoic, before the Upper Carboniferous, in the stem lineage of the Coleoptera and were already present in the latest common stem species of the living beetles which number about 350,000 species.
— A thorax of three segments and with three pairs of limbs, because this morphology was an autapomorphy in the ground pattern of the closed descent community Insecta and had already arisen as an evolutionary novelty still farther back in time, i.e. before the Devonian.
— A heart with ostia, a pericardium with a pericardial septum, a mixocoel and an "open" blood vascular system, because a basic change in the structure of the coelom and circulatory system had occurred in the stem lineage of the Arthropoda in the Pre-Cambrian.
— A "rope-ladder" nervous system on the ventral side of the body, because this remarkable manifestation of the nervous system can be postulated with reasonable certainty for the latest common stem species of all recent Articulata—and was thus evolved in the stem lineage of the Articulata, still farther back in the remotest past.

This is only a modest selection of the correlated and co-existing features in the feature mosaic of the new *Dytiscus* species. Moreover, if only one of the thousands of features were absent, and its absence was not the result of evolutionary loss, then either the species had been wrongly placed in the system, or else the system itself is an incorrect replica of the order in living Nature.

Concerning the second point—the "prediction" of previously unknown features—I quote Platnick (1978a, p.365): "To the extent that the classification has already been tested, the names can function in a storage and retrieval system for already known information about characters; to the extent that it corresponds to a real order in nature, the classification can function like any scientific theory, as a valuable basis for predicting unknown information (about characters)".

The first sentence deals with the question which I have just discussed, i.e. the indirect information about constitutive features (autapomorphies) which must exist in the relevant literature if the uninominal terms of the system are to represent monophyletic taxa. The second sentence is then an extremely vague extrapolation, without much meaning, from this situation. Of course, it is at any time possible that an obstinate comparative analysis of any adelphotaxa whatsoever may reveal synapomorphies beyond those already known. This is so, because we start from the expectation that several evolutionary novelties will usually have arisen in the stem lineage of any closed descent community in Nature. This expectation, however, is much too vague to be called a prediction. For nobody can say beforehand what new apomorphies will, perhaps, be found, nor will the expectation necessarily be fulfilled. If, for example, in the stem lineage of a closed descent community in Nature, only one evolutionary novelty had originated, then there *is* only one synapomorphy between the highest ranking adelphotaxa within the descent community. There is, therefore, no further unknown information which could be of use for phylogenetic research.

Those who have a high opinion of the "predictive power" of the phylogenetic system finish by formulating a classic vicious circle. Thus Platnick (1978b, p.472) has said that: "... the prediction embodied in any taxon is that we will find characters unique to that taxon ...". As to this, I repeat that unique features which can be interpreted as autapomorphies must be known for every supraspecific taxon if it is to be identified as a replica of a closed descent community in Nature. In other words, no prediction is needed that features of this sort will be found. They do not emerge as a result of erecting taxa—on the exact contrary, they are the precondition for validating the existence of supraspecific taxa in the phylogenetic system.

And so I sum up. Unlike the natural laws of physics, the hypotheses of the phylogenetic system allow no predictions about future happenings in Nature. If the word "prediction" is stretched beyond its true meaning and taken to refer to possible future discoveries by Man about Nature, then such predictions derived from the phylogenetic system are either trivial or vague. But even this sort of prediction, poor in meaning though it is, is rigorously connected with a system made up of monophyletic groups of species. For the sake of good order, I illustrate this with a fictitious example. On the bottom of the sea a flatworm is discovered which, because of its unique colour pattern, obviously represents a new species. A few, superficial observations on the ciliation, lack of an anus and hermaphroditism, however, are not enough to put it in any of the monophyletic taxa of free-living Plathelminthes. However, if this new species, because it combines a free mode of life with a ciliated epidermis, is provisionally classified as a member of the "Turbellaria", then no prediction could be made on any unknown feature. Human artefacts such as the

"Turbellaria" have simply no unique constitutive features which could be discovered among the species conventionally included in them.

III. Practicability

Having examined the scientific explanatory power of the phylogenetic system, the next question to answer is how far this system is suited for carrying out particular **"service functions"** in biology. Systematists are expected to produce a **robust and unified order-structure, stable in time and clear in structure**, so that all objects of any one scientific investigation can be recorded quickly and expounded in academic teaching without undue complication.

I shall therefore compare the phylogenetic system, in these respects, with traditional classifications. As Hennig said (1984, p.30): "A simple glance at various textbooks and handbooks which have appeared at roughly the same time shows that they do not at all agree in the classifications that they present". In this respect, I cite the widely used German textbooks of Kükenthal & Renner (1980), Kaestner (1969, 1980, 1982) or Remane, Storch & Welsch (1986). Equally, it is impossible to speak of the stability of traditional classifications in time. This becomes obvious by following the deep changes in classification which have taken place even as between Hertwig's famous textbook of zoology (1912), through the standard work of Claus, Grobben & Kühn (1932) to the present day. For an impressive number of taxa have been eliminated, together with their names, and an equally impressive number of new systematic units have been created.

These simple facts are, of course, in no way remarkable. For there is no unifying, objectifiable *principium divisionis* in traditional classifications (p.6) and this lack gives endless discretionary powers to subjective decisions on how the taxa should be delimited. It is worth asking, however, how it is that some particular subdivisions of the traditional classification have, nevertheless, shown remarkable stability over long periods of time. I see two different possibilities here, and these can be illustrated, once again, by the Mammalia and the Plathelminthes.

The subdivision of the mammals, into Monotremata, Marsupialia and Placentalia, stable since the last century, has persisted because all three taxa, as is now known, are in fact equivalent to closed descent communities in Nature. In this case, "correct" insights were reached very early. And this occurred although a rational basis, using derived features, was for a long time lacking, at least for the Monotremata, and also the appropriate grouping of Marsupialia and Placentalia under Parker & Haswell's name of Theria (1897) was not always felt to be obligatory.

The subdivision of the flatworms into the Turbellaria, Trematoda and Cestoda (Minot 1876) has equally remained stable for more than one hundred years. Their persistence, however, is due to nothing more than the thoughtless acceptance of an artificial classification created by "authorities". It does not correspond to the order in living Nature. It is one of a multitude of cases of suspect "classificatory peace" which are aptly described by Gaffney's epigram (1979a, p.103): "Stability is ignorance".

These examples lead to a result that is scarcely surprising. If, on Earth, only one phylogenesis has occurred in time, then on the basis of this sole objectifiable parameter, there can only be one order-structure which gives the uniformity and stability sought—and this is the consistently phylogenetic system. This is the only order-structure which in theory fulfils the preconditions for displaying the truth about the phylogenetic relationships which exist in living Nature. Of course, it is entirely another matter how far phylogenetic systematics will in practice lead to uniformity and stability and how long it will take to do so. The requirement that all the non-monophyletic groupings of traditional classifications should consistently be eliminated will necessarily cause great hubbub if it is to be translated into fact. But even when, for particular organisms, a well based phylogenetic systematization exists based on apomorphous agreements, then the hypotheses of relationship which are at present favoured may always come to be rejected. In so far as the progress of scientific knowledge demands it, they must be replaced by new, better validated hypotheses, and these would logically result in corresponding changes in the phylogenetic system. This form of instability is simply the price which has to be payed for the advance of research, just as in every other science. The crucial insight is that unity and stability can, in principle, be reached through phylogenetic systematization but not in the framework of traditional classifications.

This brings me to the problem of the **clarity** of the competing order-structures. There is probably nobody who now objects to the binominal nomenclature of every single evolutionary species, i.e. everyone is prepared to put up with the use of millions of species names in designating these real unities in Nature. Nevertheless, with wearisome regularity the big guns are fired off against phylogenetic systematics, aimed at the huge number of new supraspecific taxa, each with its own name, which are needed to catch the equivalents of closed descent communities in Nature. I have discussed this accusation earlier (p.244) and so it should now suffice to put one simple argument against it. Photosynthesis can be described by a very clear equation: $6\,CO_2 + 6\,H_2O \longrightarrow C_6H_{12}O_6 + 6\,O_2$. Nevertheless, no physiologist or biochemist would be prepared to relinquish the detailed step-by-step analysis of the complicated chemistry behind this equation nor the necessity to describe the complications appropriately. He would certainly not hesitate to alter the presentation of photosynthesis at any time as a result of new knowledge of the process. As a science, phylogenetic systematics is in a comparable position. For science must present its results in a form adequate to its knowledge. It cannot be deterred from this by subordinate questions of clarity. In other words, I insist, on principle, that all closed descent communities in Nature must, without question, be given proper names. I expressly mention the possibility of some compromise, however, especially as concerns taxa of relatively low rank (p.244).

But what is the **didactic value of a consistently phylogenetic system**? Traditionally the land-living vertebrates of the taxon Tetrapoda are divided into the four "classes" of Amphibia, Reptilia, Aves and Mammalia and these are arranged into an "ascending series". For teaching purposes this may seem clearer and easier than first to emphasize the sister-group relationship between Amphibia and Amniota and then to discuss the phylogenetic systematization of the Amniota in which the Aves, as adelphotaxon of the Crocodilia, take a relatively low rank (p.189). Likewise it may seem simpler to continue to teach the traditional tripartition of the Plathelminthes into the three "classes" of Turbellaria, Trematoda and Cestoda than to take as starting point the phylogenetic systematization of the group that is now available. The price for this form of "clarity", however, is definitely too high. No scientist or teacher could seriously advocate that the required simplifications should prevail, for teaching purposes, over the demands of scientific knowledge. The problem of how to present the best possible simplified version of a complex factual situation is shared by phylogenetic systematics with every other science. But, as Hennig wrote (1984, p.30): "Simplification and incompleteness, however, must not imply distortion." There are no patent recipes here. Each case must be judged on its merits.

In the already cited portions of the system for the Insecta, Amniota and also the Arachnida (p.188,189,240), for example, the additional burden of teaching and learning a few extra taxa and their proper names seems to me negligible in view of the increased factual content. For it must be compared with the meaningless arrangement, side by side, of four equal-ranking taxa of primitively wingless insects, of four "orders" of the artefact "Reptilia", and indeed ten "orders" of terrestrial arachnids. Explaining the whole structure of a complete segment of the phylogenetic system by means of these examples, and expounding the information content by presenting all the adelphotaxon relationships, is of priceless educational value.

In the case of the very extensive phylogenetic systematization of the Plathelminthes (p.283) a different conclusion may be possible. Here it might seem reasonable to choose particular groups of free-living Plathelminthes and to validate them clearly as monophyla but consciously to refrain from explaining the adelphotaxon relationships between them. But there is a danger, in this connection, that the position of the parasitic taxa will be presented wrongly in teaching. To avoid this danger, it is necessary, at least, to explain how the parasites are joined with those free-living plathelminths which have a forward-directed pharynx doliiformis to form the taxon Doliopharyngiophora. It is also necessary to validate the monophyly of all the parasitic taxa together by characterizing the Neodermata as equivalent to a closed descent community in Nature. Obviously there may be different opinions here and better solutions may be sought. But, in any case, this approach is more honest, from the pupil's point of view, than a comfortable exposition of the conventional classification into Turbellaria, Trematoda and Cestoda. For such an exposition, by perpetuating the artefact of the "Turbellaria", flies in the face of scientific knowledge and lacks any trace of explanatory value.

Simplifications in teaching must be governed by objective knowledge. In systematics they must pay attention to the currently favoured objectively testable hypotheses on the existence of closed descent communities in Nature. All attempts to simplify the system of organisms for teaching purposes will be grave mistakes if they run counter to well founded insights into phylogenetic relationships. They must be rejected as indefensible lapses.

There is a **proposal** which I hear more and more often in discussions on phylogenetic systematics. It is intended as a way of saving the traditional classification, at least in part, for teaching purposes. It completely accepts the "tree of descent"— what I have called the diagram of phylogenetic relationships—as a means of showing all the adelphotaxon relationships of a particular part of the system. It holds, however, that it should be allowable, when tabulating the same part of the system, to express aspects totally other than phylogeny and, among other things, to persist in the use of paraphyletic groupings. As Hennig has said (1984, p.31): "This demand is based on illusions". Apart from the scarcely soluble practical problem of explaining two versions of each part of the system, each with the taxa differently delimited, and expounding the differences point by point, the proposal makes, in actual fact, no sense at all. For paraphyletic taxa give no objectifiable information — neither on phylogenetic relationships, nor on any other evolutionary "parameter". All attempts to combine the objective knowledge given by phylogenetic systematics with the arbitrary and subjective notions of traditional classification lead to nothing but intolerable confusion. As a science, phylogenetic systematics rejects from its system all forms of subjectivity in the subdivision of organisms. In logic and practice there is no alternative to the requirement that the diagram of phylogenetic relationships shall be completely congruent with the hierarchical tabulation of the corresponding part of the system. Only when everyone agrees that every group of species named in the system — whether in the diagram of phylogenetic relationships or the hierarchical tabulation — must be validated as equivalent to a closed descent community in Nature, only then, in a better future, will it be possible in Hennig's words (1984, p.31), by consulting "any zoological textbook or any encyclopaedia to find out, quickly and reliably, what is known about the phylogenetic relationships, and thus the relationships by descent".

IV. Closing words

Of all order-structures intended to display the phylogenetic relationships in Nature, the phylogenetic system is the only one which cannot be faulted in its epistemology. Moreover, its explanatory value, as explained above, and its ability to solve practical problems fully justify its claim to be the general reference system for biology. The justification that I have given for this view is discussed by Mickevich (1978) and Farris (1979c) from the viewpoint of information theory. In Farris' words (1979c, p.518): "A single system, the phylogenetic system, best represents both genealogical and 'phenetic' information".

A consistent phylogenetic systematization is the basis of modern vicariance biogeography (Nelson & Platnick 1981). It is the starting point for new perspectives on the origin of the "macro-evolutionary" pattern in closed descent communities in Nature (Eldredge & Cracraft 1980). The characterization of the closed descent communities in Nature by revealing their means of evolutionary novelties (autapomorphies) is, in principle, independent of questions about the mechanisms of evolution. This characterization describes, "most accurately the adaptive history of the origin of each phylogenetic group" (Farris 1979c, p.518). These points may be enough to suggest the many-sided significance of a consistently phylogenetic system. "At some point most biologists want to make comparisons among different kinds of organisms, and it is at such times that a classification becomes useful" (Cracraft 1983a, p.281; cf. 1983b). Then, at least, every biologist should be able to judge the power of a consistently phylogenetic system in answering the questions that he is putting to Nature.

In advocating the phylogenetic system of organisms, I invest in the future. It is my intent to demonstrate, to the open-minded biologists of the rising generation, the rationality and efficiency of the arguments of the phylogenetic system. In my judgement, it is only by this rationality and efficiency that systematics can, within biology, make itself legitimate as a science.

Literature

Andersen, N. M. (1978) Some principles and methods of cladistic analysis with notes on the uses of cladistics in classification and biogeography. *Z. zool. Syst. Evolut.-forsch.* **16**, 242–255.

Anderson, D. T. (1973) *Embryology and phylogeny in annelids and arthropods.* Pergamon Press, Oxford, New York, Toronto, Sydney, Braunschweig.

Anderson, D. T. (1979) Embryos, fate maps, and the phylogeny of arthropods. In Gupta, A. P. (Ed.): *Arthropod phylogeny.* Van Nostrand Reinhold Company, New York.

Apelt, G. (1969) Fortpflanzungsbiologie, Entwicklungszyklen und vergleichende Frühentwicklung acoeler Turbellarien. *Marine Biology* **4**, 267–325.

Arnold, E. N. (1981) Estimating phylogenies at low taxonomic level. *Z. zool. Syst. Evolut.-forsch.* **19**, 1–35.

Ashlock, P. D. (1971) Monophyly and associated terms. *Syst. Zool.* **20**, 63–69.

Ashlock, P. D. (1972) Monophyly again. *Syst. Zool.* **21**, 430–438.

Ashlock, P. D. (1974) The uses of cladistics. *Ann. Rev. Syst. Ecol.* **5**, 81–99.

Ashlock, P. D. (1979) An evolutionary systematist's view of classification. *Syst. Zool.* **28**, 441–450.

Ax, P. (1956a) Monographie der Otoplanidae (Turbellaria). Morphologie und Systematik. *Akad. Wiss. Lit. Mainz, Abhandl. Math.-Nat. Kl. Jg. 1955, Nr. 13,* 1–298.

Ax, P. (1956b) Die Gnathostomulida, eine rätselhafte Wurmgruppe aus dem Meeressand. *Akad. Wiss. Lit. Mainz, Abhandl. Math.-Nat. Kl. Jg. 1956. Nr. 8,* 1–32.

Ax, P. (1961) Verwandtschaftsbeziehungen und Phylogenie der Turbellarien. *Ergebn. Biol.* **24**, 1–68.

Ax, P. (1963a) Relationships and phylogeny of the Turbellaria. In Dougherty, E. C. (Ed.): *The lower Metazoa.Comparative biology and phylogeny.* University of California Press, Berkeley, Los Angeles.

Ax, P. (1963b) Die Ausbildung eines Schwanzfadens in der interstitiellen Sandfauna und die Verwertbarkeit von Lebensformmerkmalen für die Verwandtschaftsforschung. *Zool. Anz.* **171**, 51–76.

Ax, P. (1964a) Das Hautgeißelepithel der Gnathostomulida. *Verhandl. Dtsch. Zool. Ges.* (München 1963), 452–461.

Ax, P. (1964b) Die Kieferapparatur von *Gnathostomaria lutheri* Ax (Gnathostomulida). *Zool. Anz.* **173**, 174–181.

Ax, P. (1965) Zur Morphologie und Systematik der Gnathostomulida. Untersuchungen an *Gnathostomula paradoxa.* Ax. *Z. zool. Syst. Evolut.-forsch.* **3**, 259–276.

Ax, P. (1966) Die Bedeutung der interstitiellen Sandfauna für allgemeine Probleme der Systematik, Ökologie und Biologie. *Veröffentl. Inst. Meeresf. Bremerhaven,* Sonderb. II, 15–66.

Ax, P. (1977) Problems of speciation in the interstitial fauna of the Galapagos. *Mikrofauna Meeresboden* **61**, 29–43.

Ax, P. (1985a) Stem species and the stem lineage concept. *Cladistics* **1**, 279–287.

Ax, P. (1985b) The position of the Gnathostomulida and Platyhelminthes in the phylogenetic system of the Bilateria. In Conway Morris, S., George, J. D., Gibson, R. & Platt, H. M. (Eds.): *The origins and relationships of lower invertebrates.* The Systematics Association. Special Volume **28**, 168–180. Clarendon Press, Oxford.

Ax, P. (1985c) Die stammesgeschichtliche Ordnung in der Natur. *Akad. Wiss. Lit. Mainz. Abhandl. Math.-Nat. Kl. Jg. 1955, Nr. 4,* 1–32.

Ax, P. & Ax, R. (1974) Interstitielle Fauna von Galapagos. V. Otoplanidae (Turbellaria, Proseriata). *Mikrofauna Meeresboden* **27**, 1–28.

Ax, P. & Ax, R. (1977) Interstitielle Fauna von Galapagos. XIX. Monocelididae (Turbellaria, Proseriata). *Mikrofauna Meeresboden* **64**, 1–44.

Ax, P. & Borkott, H. (1969) Organisation und Fortpflanzung von *Macrostomum romanicum* (Turbellaria, Macrostomida). *Verhandl. Dtsch. Zool. Ges.* (Innsbruck 1968), 344–347.

310

Ax, P. & Dörjes, J. (1966) *Oligochoerus limnophilus* nov. spec., ein kaspisches Faunenelement als erster Süßwasservertreter der Turbellaria Acoela in Flüssen Mitteleuropas. *Int. Rev. ges. Hydrobiol.* **57**, 15–144.

Ax, P. & Sopott-Ehlers, B. (1985) Monocelididae (Plathelminthes, Proseriata) von Bermuda. *Microfauna Marina* **2**, 371–382.

Ball, I. R. (1981) The order of life—towards a comparative biology. *Nature* **294**, 675–676.

Beatty, I. (1982) Classes and cladists. *Syst. Zool.* **31**, 25–34.

Beklemischew, W. N. (1958) *Grundlagen der vergleichenden Anatomie der Wirbellosen I.* VEB Deutscher Verlag der Wissenschaften, Berlin.

Beklemischew, W. N. (1960) *Grundlagen der vergleichenden Anatomie der Wirbellosen II.* VEB Deutscher Verlag der Wissenschaften, Berlin.

Beklemischev, V. N. (1963) On the relationship of the Turbellaria to other groups of the animal kingdom. In Dougherty, E. C. (Ed.): *The lower Metazoa. Comparative biology and phylogeny.* University of California Press, Berkeley, Los Angeles.

Bigelow, R. S. (1956) Monophyletic classification and evolution. *Syst. Zool.* **5**, 145–146.

Bigelow, R. S. (1958) Classification and phylogeny. *Syst. Zool.* **7**, 49–59.

Blackwelder, R. E. (1967) *Taxonomy. A text and reference book.* J. Wiley & Sons, New York, London, Sydney.

Bock, W. J. (1963) Evolution and phylogeny in morphologically uniform groups. *Am. Nat.* **97**, 265–285.

Bock, W. J. (1969) Comparative morphology in systematics. *Syst. Biology. Nat. Acad. Sci. publ. 1962.* 411–458.

Bock, W. J. (1973) Philosophical foundations of classical evolutionary classification. *Syst. Zool.* **22**, 375–392.

Bock, W. J. (1977a) Adaptation and the comparative method. In Hecht, M. K., Goody, P. C. & Hecht, B. M. (Eds.): *Major patterns in vertebrate evolution.* Plenum Press, New York, London.

Bock, W. J. (1977b) Foundations and methods of evolutionary classification. In Hecht, M. K., Goody, P. C. & Hecht, B. M. (Eds.): *Major patterns in vertebrate evolution.* Plenum Press, New York, London.

Bock, W. J. (1979) The synthetic explanation of macroevolutionary change—a reductionistic approach. *Bull. Carnegie Mus. Nat. Hist.* **13**, 20–69.

Bock, W. J. (1980) The definition and recognition of biological adaption. *Amer. Zool.* **20**, 217–227.

Bock, W. J. (1981) Functional-adaptive analysis in evolutionary classification. *Amer. Zool.* **21**, 5–20.

Bock, W. J. & von Wahlert, G. (1965) Adaptation and the form-function complex. *Evolution* **19**, 269–299.

Bonde, N. (1975) Origin of "higher groups". Viewpoints of phylogenetic systematics. *Coll. Int. C. N. R. S.* **218**, 293–324.

Bonde, N. (1977) Cladistic classification as applied to vertebrates. In Hecht, M. K., Goody, P. C. & Hecht, B. M. (Eds.): *Major patterns in vertebrate evolution.* Plenum Press, New York, London.

Bogomolov, S. I. (1960) |The development of *Convoluta* in connection with the morphology of the turbellarians.| *Trudy Obshch. Estest. Kazan. Gos. Univ.* **63**, 155–208 (in Russian).

Borkott, H. (1970) Geschlechtliche Organisation, Fortpflanzungsverhalten und Ursachen der sexuellen Vermehrung von *Stenostomum sthenum* nov. spec. *Z. Morph. Tiere.* **67**, 183–262.

Boudreaux, H. B. (1979) *Arthropod phylogeny with special reference to insects.* J. Wiley & Sons, New York, Chichester, Brisbane, Toronto.

Boyden, A. (1973) *Perspectives in zoology.* Pergamon Press, Oxford, New York, Toronto, Sydney, Braunschweig.

Brady, R. H. (1979) Natural selection and the criteria by which a theory is judged. *Syst. Zool.* **28**, 600–621.

Brady, R. H. (1982) Theoretical issues and "pattern cladistics". *Syst. Zool.* **31**, 286–291.

Brady, R. H. (1983) Parsimony, hierarchy, and biological implications. In Platnick, N. I. &

Funk, V. A. (Eds.): *Advances in cladistics.* Volume **2.** *Proceedings of the second meeting of the Willi Hennig Society.* Columbia University Press, New York.

Brandenburg, J. (1962) Elektronenmikroskopische Untersuchung des Terminalapparates von *Chaetonotus* sp. (Gastrotricha) als erstes Beispiel einer Cyrtocyte bei Askelminthen. *Z. Zellforsch.* **17,** 136–144.

Brandenburg, J. (1966) Die Reusenformen der Cyrtocyten. *Zool. Beitr.* (N.S.) **12,** 345–417.

Brandenburg, J. (1975) The morphology of protonephridia. *Fortschr. Zool.* **23,** 1–16.

Brauer, F. (1885) Systematisch-zoologische Studien. *Sitzungsber. Math-Naturw. Kl. Akad. Wiss. Wien* **91,** 237–413.

Bremer, K. & Wanntorp, H. E. (1981) The cladistic approach to plant classification. In Funk, V. A., & Brooks D. R. (Eds.): *Advances in cladistics. Proceedings of the first meeting of the Willi Hennig Society.* The New York Botanical Garden. Bronx, New York. Allen Press, Lawrence, Kansas.

Briggs, D. E. G., Clarkson, E. N. K. & Aldridge, R. J. (1983) The conodont animal. *Lethaia* **16,** 1–14.

Bruce, E. J. & Ayala, F. J. (1979) Phylogenetic relationships between man and the apes. Electrophoretic evidence. *Evolution* **33,** 1040–1050.

Brundin, L. (1966) Transantarctic relationships and their significance, as evidenced by Chironomid mites. *Kungl. Svenska. Vetenskap. Handl.* **11,** 1–472.

Brundin, L. (1968) Application of phylogenetic principles in systematics and evolutionary theory. In Ørvig, T. (Ed.): *Current problems of lower vertebrate phylogeny.* Almquist & Wiksell, Stockholm.

Brundin, L. (1972) Evolution, causal biology, and classification. *Zool. Scr.* **1,** 107–120.

Buck, R. C. & Hull, D. L. (1966) The logical structure of the Linnean hierarchy. *Syst. Zool.* **15,** 97–111.

Bush, G. L. (1975) Models of animal speciation. *Ann. Rev. Ecol. Syst.* **6,** 339–364.

Butler, P. M., Green, E. & Krebs, B. (1973) A pantotherian milk dentition. *Paläont. Z.* **47,** 256–258.

Camin, J. H. & Sokal, R. R. (1965) A method for deducing branching sequences in phylogeny. *Evolution* **19,** 311–326.

Cartmill, M. (1981) Hypothesis testing and phylogenetic reconstruction. *Z. zool. Syst. Evolut.-forsch.* **19,** 73–96.

Cavalier-Smith, T. (1981) The origin and evolution of the eucaryotic cell. In Carlile, M. J., Collins, J. F. & Moseley, B. E. B. (Eds.): *Molecular and cellular aspects of microbial evolution.* Cambridge University Press, Cambridge, London, New York, New Rochelle, Melbourne, Sydney.

Charig, A. J. (1982) Systematics in Biology: A fundamental comparison of some major schools of thought. In Joysey, K. A. & Friday, A. E. (Eds.): *Problems of phylogenetic reconstruction.* Academic Press, London, New York.

Ciochon, R. L. (1983) Hominoid cladistics and the ancestry of modern apes and humans. A summary statement. In Ciochon, R. L. & Corruccini, R. S. (Eds.): *New interpretations of ape and human ancestry.* Plenum Press, New York, London.

Ciochon, R. L. & Corruccini, R. S. (Eds.) (1983) *New interpretations of ape and human ancestry.* Plenum Press, New York, London.

Claus, C., Grobben, K. & Kühn, A. (1932) *Lehrbuch der Zoologie.* J. Springer, Berlin, Vienna. 10th. Edn.

Costello, D. P. & Henley, C. (1976) Spiralian development: A perspective. *Amer. Zool.* **16,** 277–291.

Cracraft, J. (1974) Phylogenetic models and classification. *Syst. Zool.* **23,** 71–90.

Cracraft, J. (1979) Phylogenetic analysis, evolutionary models and palaeontology. In Cracraft, J. & Eldredge, N. (Eds.): *Phylogenetic analysis and paleontology.* Columbia University Press, New York.

Cracraft, J. (1981a) The use of functional and adaptive criteria in phylogenetic systematics. *Amer. Zool.* **21,** 21–36.

Cracraft, J. (1981b) Toward a phylogenetic classification of the recent birds of the world (Class Aves). *The Auk* **98,** 681–714.

312

Cracraft, J. (1982) Phylogenetic relationships and monophyly of loons, grebes, and hesperornithiform birds, with comments on the early history of birds. *Syst. Zool.* **31**, 35–56.

Cracraft, J. (1983a) Cladistic analysis and vicariance biogeography. *Am. Scientist* **71**, 273–281.

Cracraft, J. (1983b) The significance of phylogenetic classification for systematic and evolutionary biology. In Felsenstein, J. (Ed.): *Numerical taxonomy.* NATO ASI Series, Vol. G I. Springer-Verlag, Berlin, Heidelberg, New York, Tokyo.

Cracraft, J. (1984) The terminology of allopatric speciation. *Syst. Zool.* **33**, 115–116.

Cracraft, J. & Eldredge, N. (Eds.) (1979) *Phylogenetic analysis and paleontology.* Columbia University Press.

Cronin, J. E. (1983) Apes, humans, and molecular clocks. A reappraisal. In Ciochon, R. L. & Corruccini, R. S. (Eds.): *New interpretations of ape and human ancestry.* Plenum Press, New York, London.

Crowson, R. A. (1970) *Classification and biology.* Heinemann, Educational Books Ltd, London.

Darwin, C. (1859) *On the origin of species by means of natural selection, or the preservation of favoured races in the struggle for life.* John Murray, London.

Dayhoff, M. O. (1972) *Atlas of protein sequence and structure,* Vol. **5**. National Biomedical Research Foundation, Silver Spring.

De Beer, G. R. (1937) *The development of the vertebrate skull.* Clarendon, Oxford.

De Beer, G. R. (1954a) *Archaeopteryx lithographica. A study based upon the British Museum specimen.* London.

De Beer, G. R. (1954b) *Archaeopteryx* and evolution. *Advanc. Sci. London* **11**, 160–170.

De Bonis, L. (1983) Phyletic relationships of miocene hominoids and higher primate classification. In Ciochon, R. L. & Corruccini, R. S. (Eds.): *New interpretations of ape and human ancestry.* Plenum Press, New York, London.

Delson, E., Eldredge, N. & Tattersall, J. (1977) Reconstruction of hominid phylogeny: A testable framework based on cladistic analysis. *J. Hum. Evol.* **6**, 263–278.

De Saint-Aubain, M. L. (1980) Amphibian limb ontogeny and its bearing on the phylogeny of the group. *Z. zool. Syst. Evolut.-forsch.* **19**, 175–194.

Diemer, A. (1968) *System und Klassifikation in Wissenschaft und Dokumentation.* Verlag Anton Glan, Meisenheim am Glan.

Dobzhansky, T., Ayala, F. J., Stebbins, G. L. & Valentine, J. W. (1977) *Evolution.* W. H. Freeman and Company, San Francisco.

Doe, D. A. (1976) The proboscis hooks in Karkinorhynchidae and Gnathorhynchidae (Turbellaria, Kalyptorhynchia) as basement membrane or intercellular specialisation. *Zool. Scr.* **5**, 105–115.

Doe, D. A. (1981) Comparative ultrastructure of the pharynx simplex in Turbellaria. *Zoomorphology* **97**, 133–193.

Doe, D. A. (1982) Ultrastructure of copulatory organs in Turbellaria. I. *Macrostomum* sp. and *Microstomum* sp. (Macrostomida). *Zoomorphology* **101**, 39–59.

Dohle, W. (1976) Zur Frage des Nachweises von Homologien durch die komplexen Zell- und Teilungsmuster in der embryonalen Entwicklung höherer Krebse (Crustacea, Malacostraca, Peracarida). *Sitzungsber. Ges. Naturf. Freunde Berlin* (N. S.) **16**, 125–144.

Dohle, W. (1980) Sind die Myriapoden eine monophyletische Gruppe? Eine Diskussion der Verwandtschaftsbeziehungen der Antennaten. *Abh. naturwiss. Ver. Hamburg* (N. S.) **23**, 45–104.

Doolittle, R. F., Wooding, G. L., Lin, Y. & Riley, M. (1971) Hominoid evolution as judged by fibrinopeptid structures. *J. Molec. Evolution* **1**, 74–83.

Dupuis, C. (1978) Permanence et actualité de la Systématique: La "Systématique phylogénétique" de W. Hennig (Historique, discussion, choix de références). *Cahiers des Naturalistes* **34**, 1–69.

Dutrillaux, B. (1975) Sur la nature et l'origine des chromosomes humains. *Monogr. Ann. Génét.* **3**, 1–104.

Ehlers, U. (1977) Vergleichende Untersuchungen über Collar-Rezeptoren bei Turbellarien. *Acta. Zool. Fenn.* **154**, 137–148.

Ehlers, U. (1981) Fine structure of the giant aflagellate spermatozoon in *Pseudostomum quadrioculatum* (Leuckart) (Plathelminthes, Prolecithophora). *Hydrobiologica* **84**, 287–300.

Ehlers, U. (1984) Phylogenetisches System der Plathelminthes. *Verh. natur. wiss. Ver. Hamburg* (N. S.) **27**, 291–294.

Ehlers, U. (1985a) Phylogenetic relationships within the Platyhelminthes. In Conway Morris, S., George, J. D., Gibson, R. & Platt, H. M. (Eds.): The origins and relationships of lower invertebrates. *The Systematics Association Special Volume* **28**. 143–158. Clarendon Press.

Ehlers, U. (1985b) *Das phylogenetische System der Plathelminthes.* G. Fischer, Stuttgart, New York.

Ehlers, B. & Ehlers U. (1980a) Struktur und Differenzierung penialer Hartgebilde von *Carenscoilia bidentata* Sopott (Turbellaria, Proseriata). *Zoomorphologie* **95**, 159–167.

Ehlers, B. & Ehlers, U. (1980b) Zur Systematik und geographischen Verbreitung interstitieller Turbellarien der Kanarischen Inseln. *Mikrofauna Meeresboden* **80**, 1–23.

Eisenberg, J. F. (1981) *The mammalian radiation. An analysis of trends in evolution, adaptation, and behavior.* The Athlone Press, London.

Eldredge, N. (1979) Cladism and common sense. In Cracraft, J. & Eldredge, N. (Eds.): *Phylogenetic analysis and paleontology.* Columbia University Press, New York.

Eldredge, N. & Cracraft, J. (1980) *Phylogenetic patterns and the evolutionary process. Methods and theory in comparative biology.* Columbia University Press, New York.

Eldredge. N. & Gould. S. J. (1972) Punctuated equilibria: An alternative to phyletic gradualism. In Schopf, T. J. M. (Ed.): *Models in paleontology.* Freeman, Cooper & Co., San Francisco.

Eldredge, N. & Tattersall, J. (1975) Evolutionary models, phylogenetic reconstruction and another look on hominid phylogeny. In Szalay, F. S. (Ed.): *Approaches in primate paleobiology. Contrib. Primat.* **5**, 218–242.

Endler, J. A. (1977) *Geographic variation, speciation, and clines.* Princeton University Press, Princeton, New Jersey.

Engelmann, G. F. & Wiley, E. O. (1977) The place of ancestor-descendant relationships in phylogeny reconstruction. *Syst. Zool.* **26**. 1–11.

Erben, H. K. (1981) *Leben heißt Sterben. Der Tod des einzelnen und das Aussterben der Arten.* Hoffmann und Campe, Hamburg.

Evdonin, L. A. (1977) |The proboscis-bearing turbellarians Kalyptorhynchia in the fauna of the U.S.S.R. and neighbouring regions.| *Fauna SSSR Turbellaria.* Vol. I, I. Akademia Nauk. Zool. Inst. Leningrad (in Russian).

Farris, J. S. (1974) Formal definitions of paraphyly and polyphyly. *Syst. Zool.* **23**, 548–554.

Farris, J. S. (1976) Phylogenetic classification of fossils with recent species. *Syst. Zool.* **25**, 271–282.

Farris, J. S. (1977) On the phenetic approach to vertebrate classification. In Hecht, M. K., Goody, P. C. & Hecht, B. M. (Eds.): *Major patterns in vertebrate evolution.* Plenum Press, New York.

Farris, J. S. (1979a) On the naturalness of phylogenetic classification. *Syst. Zool.* **28**, 200–214.

Farris, J. S. (1979b) The Willi Hennig Memorial Symposium. Willi Hennig and the development of modern systematics—an introduction. *Syst. Zool.* **28**, 415.

Farris, J. S. (1979c) The information content of the phylogenetic system. *Syst. Zool.* **28**, 483–519.

Farris, J. S. (1983) The logical basis of phylogenetic analysis. In Platnick, N. I. & Funk, V. A. (Eds.): *Advances in cladistics,* Volume 2. *Proceedings of the second meeting of the Willi Hennig Society.* Columbia University Press, New York.

Ferguson, A. (1980) *Biochemical systematics and evolution.* Blackie & Son. Glasgow.

Fioroni, P. (1980) Zur Signifikanz des Blastoporus-Verhaltens in evolutiver Hinsicht. *Rev. Suisse Zool.* **87**, 261–272.

Franzén, A. (1956) On spermiogenesis, morphology of the spermatozoon, and biology of fertilization among invertebrates. *Zool. Bidrag. Uppsala.* **31**, 355–482.

Franzén, A. (1977) Sperm structure with regard to fertilization biology and phylogenetics. *Verh. Dtsch. Zool. Ges.* (Erlangen). 123–138.

314

Franzen, J. L. (1972) Wie kam es zum aufrechten Gang des Menschen? *Natur und Museum* **102**, 161–172.

Funk, V. A. & Brook, D. R. (1981) (Eds.): *Advances in cladistics. Proceedings of the first meeting of the Willi Hennig Society.*The New York Botanical Garden, Bronx, New York. Allen Press, Lawrence, Kansas.

Gaffney, E. S. (1979a) An introduction to the logic of phylogeny reconstruction. In Cracraft, J. & Eldredge. N. (Eds.): *Phylogenetic analysis and paleontology.* Columbia University Press, New York.

Gaffney, E. S. (1979b) Tetrapod monophyly: a phylogenetic analysis. *Bull. Carnegie Mus. Nat. Hist.* **13**, 92–105.

Gaffney, E. S. (1980) Phylogenetic relationships of the major groups of Amniotes. In Panchen, A. L. (Ed.): *The terrestrial environment and the origin of the land vertebrates.* Academic Press, London, New York.

Gardiner, B. G. (1982) Tetrapod classification. *Zool. J. Linn. Soc. London* **74**, 202–232.

Ghiselin, M. T. (1966a) An application of the theory of definitions to systematic principles. *Syst. Zool.* **15**, 127–130.

Ghiselin, M. T. (1966b) On psychologism in the logic of taxonomic controversies. *Syst. Zool.* **15**, 207–215.

Ghiselin, M. T. (1974) A radical solution of the species problem. *Syst. Zool.* **23**, 536–544.

Ghiselin, M. T. (1980a) Biogeographical units: more on radical solutions. *Syst. Zool.* **29**, 80–85.

Ghiselin, M. T. (1980b) Natural kinds and literary accomplishments. *The Michigan Quart.* **19**, 73–88.

Ghiselin, M. T. (1981) Categories. life and thinking. *The behavioral and brain sciences* **4**, 269–313.

Goodman, M. (1975) Protein sequences and immunological specificity: their role in phylogenetic studies of primates. In Luckett, W. P. & Szalay, F. S. (Eds.): *Phylogeny of the primates.* Plenum Press, New York, London.

Goodman, M. (1977) Protein sequences in phylogeny. In Ayala, A. (Ed.): *Molecular evolution.* Sinauer Associates, Inc. Publishers, Sunderland, Massachusetts.

Goodman, M., Baba, M. L. & Darga, L. L. (1983) The bearing of molecular data on the cladogenesis and times of divergence of hominoid lineages. In Ciochon, R. L. & Corruccini, R. S. (Eds.): *New interpretations of ape and human ancestry.* Plenum Press, New York, London.

Goodrich, E. S. (1916) On the classification of the Reptilia. *Proc. R. Soc. London* **89**, 216–276.

Goodrich, E. S. (1930) *Studies on the structure and development of vertebrates.* Macmillan, London.

Gould, S. J. (1977) *Ontogeny and phylogeny.* The Belknap Press, Harvard University Press, Cambridge. Massachusetts.

Gould, S. J. & Eldredge, N. (1977) Punctuated equilibria: the tempo and mode of evolution reconsidered. *Paleobiology* **3**, 115–151.

Graebner, J. (1968) Erste Befunde über die Feinstruktur der Exkretionszellen der Gnathostomulidae (*Gnathostomula paradoxa,* Ax 1956 und *Austrognathia riedli,* Sterrer 1965). *Mikroskopie* **23**, 277–292.

Graebner, J. (1969a) Vergleichende elektronenmikroskopische Untersuchung der Spermienmorphologie und Spermiogenese einiger Gnathostomula-Arten: *Gnathostomula paradoxa* (Ax 1956), *Gnathostomula axi* (Kirsteuer 1964). *Gnathostomula jenneri* (Riedl 1969). *Mikroskopie* **24**, 131–160.

Graebner, J. (1969b) Ergebnisse einer elektronenmikroskopischen Untersuchung von Gnathostomuliden. *Verhandl. Dtsch. Zool. Ges.* (Innsbruck 1968), 580–599.

Grant, V. (1977) *Organismic evolution.* W. H. Freeman & Co., San Francisco.

Grant, V. (1982) Punctuated equilibria: A critique. *Biol. Zbl.* **101**, 175–184.

Green, H. L. H. H. (1937) The development and morphology of the teeth of *Ornithorhynchus. Phil. Trans. R. Soc. Lond.* (B) **228**, 367–420.

Gregory, W. K. (1947) The monotremes and the palimpsest theory. *Bull. Am. Mus. Nat. Hist.* **88**, 1–52.

Griffiths, G. C. D. (1974a) On the foundation of biological systematics. *Acta Biotheoretica* **23**, 85–131.

Griffiths, G. C. D. (1974b) Some fundamental problems in biological classification. *Syst. Zool.* **22**, 338–343.

Griffiths, G. C. D. (1976) The future of Linnaean nomenclature. *Syst. Zool.* **25**, 168–173.

Griffiths, M. (1978) *The biology of the monotremes.* Academic Press, New York, San Francisco, London.

Günther, K. (1971) Natürliches oder phylogenetisches System. *Erlanger Forschungen* (B)**4**, 29–41.

Guibé, J. (1970) Classe des Reptiles. Le squelette du tronc et des membres. In Grassé, P. P. (Ed.): *Traité de Zoologie* XIV, fasc. 2, Paris.

Gutmann, W. F. (1977) Phylogenetic reconstruction: Theory, methodology, and application to chordate evolution. In Hecht, M. K., Goody, P. C. & Hecht, B. M. (Eds.): *Major patterns in vertebrate evolution.* Plenum Press, New York, London.

Gutmann, W. F. (1981) Relationships between invertebrate phyla based on functional-mechanical analysis of the hydrostatic skeleton. *Amer. Zool.* **21**, 63–81.

Gutmann, W. F. & Bonik, K. (1981a) *Kritische Evolutionstheorie. Ein Beitrag zur* Überwindung altdarwinistischer Dogmen. Gerstenberg Verlag, Hildesheim.

Gutmann, W. F. & Bonik, K. (1981b) Hennigs Theorem und die Strategie des stammes-geschichtlichen Rekonstruierens. Die Agnathen-Gnathostomen-Beziehung als Beispiel. *Paläont. Z.* **55**, 51–70.

Gutmann, W. F. & Peters, D. S. (1973) Konstruktion und Selektion: Argumente gegen einen morphologisch verkürzten Selektionismus. *Acta Biotheoretica* **22**, 151–180.

Haeckel, E. (1866) *Generelle Morphologie der Organismen.* Vols. I and II. G. Reiner, Berlin.

Hadži, J. (1963) *The evolution of the Metazoa.* Pergamon Press, New York.

Hanson, E. D. (1963) Homologies and the ciliate origin of the Eumetazoa. In Dougherty, E. C. (Ed.): *The lower Metazoa. Comparative biology and phylogeny.* University of California Press, Berkeley, Los Angeles.

Hanson, E. D. (1977) *The origin and early evolution of animals.* Wesleyan University Press, Middletown, Connecticut.

Hartman, W. D. (1963) A critique of the enterocoel theory. In Dougherty, E. C. (Ed.): *The lower Metazoa. Comparative biology and phylogeny.* University of California Press, Berkeley, Los Angeles.

Hecht, M. K. (1976) Phylogenetic inference and methodology as applied to the vertebrate record. *Evol. Biol.* **9**, 335–363.

Hecht, M. K. & Edwards, J. L. (1977) The methodology of phylogenetic inference above the species level. In Hecht, M. K., Goody, P. C. & Hecht, B. M. (Eds.): *Major patterns in vertebrate evolution.* Plenum Press, New York, London.

Hendelberg, J. (1974) Spermiogenesis, sperm morphology, and biology of fertilization in the Turbellaria. In Riser, N. W. & Morse, M. P. (Eds.): *Biology of the Turbellaria.* McGraw-Hill, New York.

Hendelberg, J. (1975) Functional aspects of flatworm sperm morphology. In Afzelius, B. A. (Ed.): *The functional anatomy of the spermatozoon.* Pergamon Press, Oxford, New York.

Hendelberg, J. (1977) Comparative morphology of turbellarian spermatozoa studied by electron microscopy. *Acta Zool. Fenn.* **154**, 149–162.

Hendelberg, J. & Hedlund, K. O. (1974) On the morphology of the epidermal ciliary rootlet system of the acoelous turbellarian *Childia groenlandica. Zoon* **2**, 13–24.

Hennig, W. (1950) *Grundzüge einer Theorie der phylogenetischen Systematik.* Deutscher Zentralverlag, Berlin.

Hennig, W. (1953) Kritische Bermerkungen zum phylogenetischen System der Insekten. *Beitr. Ent.* **3**, 1–85.

Hennig, W. (1955) Meinungsverschiedenheiten über das System der niederen Insekten. *Zool. Anz.* **155**, 21–30.

316

Hennig, W. (1957) Systematik und Phylogenese. *Bericht. Hundertjahrfeier Deutsch. Ent. Ges.* (Berlin 1956), 50–71.

Hennig, W. (1960) Die Dipterenfauna von Neuseeland als systematisches und tiergeographisches Problem. *Beitr. Ent.* **10**, 221–329.

Hennig, W. (1965) Phylogenetic systematics. *Ann. Rev. Ent.* **10**. 97–116.

Hennig, W. (1966) *Phylogenetic systematics.* University of Illinois Press, Urbana, Chicago, London.

Hennig, W. (1969) *Die Stammesgeschichte der Insekten.* E. Kramer. Frankfurt/Main.

Hennig, W. (1971) Zur Situation der biologischen Systematik. *Erlanger Forschungen,* (B)**4**, 7–15.

Hennig, W. (1972) *Wirbellose II. Gliedertiere. Taschenbuch der Speziellen Zoologie, Teil 2.,* 3rd. Edn., H. Deutsch, Frankfurt/Main, Zürich.

Hennig, W. (1974) Kritische Bemerkungen zur Frage "Cladistic analysis or cladistic classification". *Z. zool. Syst. Evolut.-forsch.* **12**, 279–294.

Hennig, W. (1975) "Cladistic analysis or cladistic classification?": A reply to Ernst Mayr. *Syst. Zool.* **24**, 244–256.

Hennig, W. (1980) *Wirbellose I. Ausgenommen Gliedertiere. Taschenbuch der Speziellen Zoologie Teil 1.,* 4th. Edn. VEB G. Fischer Verlag, Jena 1979.—Verlag H. Deutsch, Thun, Frankfurt/Main.

Hennig, W. (1981) *Insect phylogeny* (Translated and edited by A. C. Pont. Revisionary notes by D. Schlee). J. Wiley & Sons. Chichester, New York, Brisbane, Toronto.

Hennig, W. (1982) *Phylogenetische Systematik* (Ed. Wolfgang Hennig). *Pareys Studientexte* **34**. P. Parey, Berlin, Hamburg.

Hennig, W. (1983) Stammesgeschichte der Chordaten (Ed. Wolfgang Hennig). *Fortschr. zool. Syst. Evolution.-forsch.* **2**. P. Parey, Hamburg, Berlin.

Hennig, W. (1984) Aufgaben und Probleme stammesgeschichtlicher Forschung (Ed. Wolfgang Hennig). *Pareys Studientexte* **35**. P. Parey, Berlin, Hamburg.

Hennig, W. & Schlee, D. (1978) Abriß der phylogenetischen Systematik. *Stuttgarter Beitr. Naturk.* (A)**319**. 1–11.

Hertwig, R. (1912) *Lehrbuch der Zoologie.* G. Fischer. Jena. 10th. Edn.

Higgins, A. (1983) The conodont animal. *Nature* **302**. 107.

Hill, C. R. & Crane, P. R. (1982) Evolutionary cladistics and the origin of angiosperms In Joysey, K. A. & Friday, A. E. (Eds.): *Problems of phylogenetic reconstruction.* The Systematics Association Special Volume **21**. Academic Press, London, New York.

Hofsten, N. V. (1941) On the phylogeny of the Reptilia. *Zool. Bidr. Uppsala* **20**. 501–520.

Holmes, E. B. (1975) A reconsideration of the phylogeny of the tetrapod heart. *J. Morph.* **147**. 209–228.

Holmes, E. B. (1980) Reconsideration of some systematic concepts and terms. *Evolut. Theory* **5**. 35–87.

Hopson, J. A. (1966) The origin of the mammalian middle ear. *Amer. Zool.* **6**, 437–450.

House, M. R. (Ed.) 1979 *The origin of major invertebrate groups.* Academic Press, London, New York.

Hull, D. L. (1976) Are species really individuals? *Syst. Zool.* **25**. 174–191.

Hull, D. L. (1979) The limits of cladism. *Syst. Zool.* **28**, 416–440.

Hull, D. L. (1983) Karl Popper and Plato's metaphor. In Platnick, N. I. & Funk, V. A. (Eds.): *Advances in Cladistics.* Volume **2**. *Proceedings of the second meeting of the Willi Hennig Society.* Columbia University Press, New York.

Huxley, T. H. (1963) *Evidences as to man's place in nature* (1863). Translation by G. Heberer: *Zeugnisse für die Stellung des Menschen in der Natur.* G. Fischer-Verlag, Stuttgart.

Hyman, L. H. (1951) *The invertebrates: Platyhelminthes and Rhynchocoela.* Vol. II. McGraw-Hill Book Company, New York.

Hyman, L. H. (1959) *The invertebrates: Smaller coelomate groups.* Vol. V. McGraw-Hill Book Company, New York.

Ihle, J. E. W., Van Kampen, P. N., Nierstrasz, H. F. & Versluys, J. (1971) *Vergleichende Anatomie der Wirbeltiere.* Springer Verlag, Berlin. 1927 (Reprint Springer-Verlag, Berlin, Heidelberg, New York.)

Immelmann, K. (1979) *Einführung in die Verhaltensforschung.* 2nd. Edn. Parey, Berlin, Hamburg.

Ivanov, A. V. (1968) [The origin of the Metazoa.] Verlag Nauka, Leningrad. (in Russian).

Ivanov, A. V. (1970) Verwandtschaft und Evolution der Pogonophoren. *Z. zool. Syst. Evolut.-forsch.* **8**, 109–119.

Ivanov, A. V. & Mamkaev, Y. V. (1973) [The turbellarians—their origin and evolution.] Nauka, Leningrad (in Russian).

Ivanova-Kasas, O. M. (1959) Entstehung und Evolution der Spiralfurchung. *Naturwiss. Beiträge (Berlin)* **12**, 1267–1279.

Ivanova-Kasas, O. M. (1981) *Phylogenetic significance of spiral cleavage.* Evolutionary embryology. Plenum Publishing Corporation. 275–283 (1982). (Translated from *Biologiya Morya* **5**, 3–14.)

Jägersten, C. (1972) *Evolution of the metazoan life cycle. A comprehensive theory.* Academic Press, London, New York.

Jarvik, E. (1980) *Basic structure and evolution of vertebrates.* Volume 1. Academic Press Inc, London.

Jarvik, E. (1980) *Basic structure and evolution of vertebrates.* Volume 2. Academic Press, London.

Jefferies, R. P. S. (1979) The origin of chordates—a methodological essay. In House, M. R. (Ed.): *The origin of major invertebrate groups.* Academic Press, London, New York.

Jefferies, R. P. S. (1980) Zur Fossilgeschichte des Ursprungs der Chordaten und der Echinodermen. *Zool. Jb. Anat.* **103**, 285–353.

Jeffrey, C. (1973) *Biological nomenclature.* E. Arnold, London.

Jong, R. De. (1980) Some tools for evolutionary and phylogenetic studies. *Z. zool. Syst. Evolut.-forsch.* **18**, 1–23.

Joysey, K. A. & Friday, A. E. (1982) *Problems of phylogenetic reconstruction. Systematic Association Special Volume 21.* Academic Press, London.

Jürgens, H. W. & Vogel, C. (1965) *Beiträge zur menschlichen Typenkunde.* Enke-Verlag, Stuttgart.

Kaestner, A. (1969) *Lehrbuch der Speziellen Zoologie. Band 1: Wirbellose Tiere. 1. Teil.* 3rd. Edn., G. Fischer, Stuttgart.

Kaestner, A. (1980) (Ed. H. E. Gruner): *Lehrbuch der Speziellen Zoologie. Band 1. Wirbellose Tiere 1. Teil.* 4th. Edn., G. Fischer, Stuttgart.

Kaestner, A. (1982) (Ed. H. E. Gruner): *Lehrbuch der Speziellen Zoologie. Band 1. Wirbellose Tiere 3. Teil* 4th. Edn., G. Fischer, Stuttgart.

Karling, T. G. (1961) Zur Morphologie, Entstehungsweise und Funktion des Spaltrüssels der Turbellaria Schizorhynchia. *Ark. Zool.*, (2)**13**, 253–286.

Karling, T. G. (1965) *Haplopharynx rostratus* Meixner (Turbellaria) mit den Nemertinen verglichen. *Z. zool. Syst. Evolut.-forsch.* **3**, 1–18.

Karling, T. G. (1966) On the defecation apparatus in the genus *Archimonocelis* (Turbellaria, Monocelididae). *Sarsia* **24**, 37–44.

Karling, T. G. (1974) On the anatomy and affinities of the turbellarian orders. In Riser, N. W. & Morse, M. P. (Eds.): *Biology of the Turbellaria.* McGraw-Hill Book Company, New York.

Karling, T. G., Mack-Fira, V. & Dörjes, J. (1972) First report of marine microturbellarians from Hawaii. *Zool. Scr.* **1**, 251–269.

Kato, K. (1940) On the development of some Japanese polyclads. *Japan. J. Zool.* **8**, 537–573.

Keast, A. (1977) Historical biogeography of the marsupials. In Stonehouse, B. & Gilmore, D. (Eds.): *The biology of marsupials.* Macmillan Press, London, Basingstoke.

Kemp, T. S. (1969) The atlas-axis complex of the mammal-like reptiles. *J. Zool. Soc. Lond.* **159**, 223–248.

Kemp, T. S. (1982) *Mammal-like reptiles and the origin of mammals.* Academic Press, New York, London.

Kielan-Jaworowska, Z. (1979) Pelvic structure and nature of reproduction in Multituberculata. *Nature* **277**, 402–403.

King, M. C. & Wilson, A. C. (1975) Evolution at two levels in humans and chimpanzees. *Science* **188**, 107–116.

Kirsch, J. A. W. (1977a) The classification of marsupials. In Husaker, D. (Ed.): *The biology of marsupials.* Academic Press, New York. San Francisco, London.

Kirsch. J. A. W. (1977b) The comparative serology of Marsupialia, and a classification of marsupials. *Austral. J. Zool., Suppl. Series* **52**, 1–152.

Kluge, A. G. (1983) Cladistics and the classification of the great apes. In Ciochon, R. L. & Corruccini, R. S. (Eds.): *New interpretations of ape and human ancestry.* Plenum Press, New York, London.

Kluge, A. G. (1984) The relevance of parsimony to phylogenetic inference. In Duncan, T. & Stuessy, T. E. (Eds.): *Cladistics: Perspectives on the reconstruction of evolutionary history.* Columbia University Press, New York.

Kluge, A. G. & Farris, J. S. (1969) Quantitative phyletics and the evolution of anurans. *Syst. Zool* **18**, 1–32.

Knauss, E. B. (1979) Indication of an anal pore in Gnathostomulida. *Zool. Scr.* **8**, 181–186.

Königsmann, E. (1975) Termini der phylogenetischen Systematik. *Biol. Rdsch.* **13**, 99–115.

Kraus, O. (1976) Phylogenetische Systematik und evolutionäre Klassifikation. *Verh. Dtsch. Zool. Ges.*, (Hamburg), 84–99.

Krebs, B. (1974) Die Archosaurier. *Naturwissenschaften* **61**, 17–24.

Kristensen, N. P. (1975) The phylogeny of hexapod "orders". A critical review of recent accounts. *Z. zool. Syst. Evolut.-forsch.* **13**, 1–44.

Kristensen, N. P. (1981) Phylogeny of insect orders. *Ann. Rev. Entomol.* **26**, 135–157.

Kristensen, R. M. & Norrevang, A. (1977) On the fine structure of *Rastrognathia macrostoma* gen. et sp. n. placed in Rastrognathiidae fam. n. (Gnathostomulida). *Zool. Scr.* **6**, 27–41.

Kuhn, H. J. (1971) Die Entwicklung und Morphologie des Schädels von *Tachyglossus aculeatus. Abh. senckenberg. naturforsch. Ges.* **528**, 1–192.

Kühne, W. G. (1973) The systematic position of monotremes reconsidered (Mammalia). *Z. Morph. Tiere* **75**, 59–64.

Kühne, W. G. (1975) Marsupium and marsupial bone in Mesozoic mammals and in the Marsupionta. *Coll. int. C. N. R. S.* **218**, 585–590.

Kükenthal, W. & Renner, M. (1980) *Leitfaden für das Zoologische Praktikum.* G. Fischer, Stuttgart, New York, 18th Edn.

Kümmel, G. (1962) Zwei neue Formen von Cyrtocyten. Vergleich der bisher bekannten Cyrtocyten und Erörterung des Begriffes "Zelltyp". *Z. Zellforsch.* **57**, 172–201.

Lakatos, I. & Musgrave, A. (Eds.) (1970) *Criticism and the growth of knowledge.* Cambridge University Press, London, New York.

Lakatos, I. & Musgrave, A. (Eds.) (1974) *Kritik und Erkenntnisfortschritt.* Vieweg, Braunschweig.

Lambert, A. (1980) Oncomiracidiums et phylogenèse des Monogenea (Plathelminthes). *Ann. Parasit.* **55**, 281–325.

Lammert, V. (1985) The fine structure of protonephridia in Gnathostomulida and their comparison within Bilateria. *Zoomorphologie* **105**, 308–316.

Lang, A. (1881) Der Bau von *Gunda segmentata* und die Verwandtschaft der Plathelminthen mit Coelenteraten und Hirudineen. *Mitt. Zool. Stat. Neapel* **3**, 187–251.

Lang, A. (1884) Die Polycladen (Seeplanarien) des Golfes von Neapel und der angrenzenden Meeresabschnitte. *Fauna Flora Neapel* **11**.

Lauterbach, K. E. (1972a) Über die sogenannte Ganzbein-Mandibel der Tracheata, insbesondere der Myriapoden. *Zool. Anz.* **188**, 145–154.

Lauterbach, K. E. (1972b) Die morphologischen Grundlagen für die Entstehung der Entognathie bei den apterygoten Insekten in phylogenetischer Sicht. *Zool. Beitr.* (N. S.) **18**, 25–69.

Lauterbach, K. E. (1973) Schlüsselereignisse in der Evolution der Stammgruppe der Euarthropoda. *Zool. Beitr.* (N. S). **19**, 251–299.

Lauterbach, K. E. (1974) Über die Herkunft des Carapax der Crustacea. *Zool. Beitr.* (N. S.) **20**, 273–327.

Lauterbach. K. E. (1978) Gedanken zur Evolution der Euarthropoden-Extremität. *Zool. Jb. Anat.* **99**, 64–92.

Lauterbach. K. E. (1980a) Schlüsselereignisse in der Evolution des Grundplans der Mandibulata (Arthropoda). *Abh. naturw. Ver. Hamburg* (N. S.) **23**, 105–161.

Lauterbach. K. E. (1980b) Schlüsselereignisse in der Evolution des Grundplanes der Arachnata (Arthropoda). *Abh. naturw. Ver. Hamburg* (N. S.) **23**, 163–327.

Lauterbach. K. E. (1983a) Zum Problem der Monophylie der Crustacea. *Verh. Hamburg naturw. Ver.* (N. S.) **26**, 293–320.

Lauterbach. K. E. (1983b) Synapomorphien zwischen Trilobiten- und Cheliceratenzweig der Arachnata. *Zool. Anz.* **210**, 213–238.

Lewontin, R. C. (1978) Adaptation. *Scientific American* **239**, 212–230.

Lillegraven, J. A., Kielan-Jaworowska, A. Z., & Clements, W. A. (Eds.) (1979) *Mesozoic mammals. The first two-thirds of mammalian history*. University of California Press, Berkeley, Los Angeles, London.

Linnaeus, C. (1735) Systema naturae, sive regna tria naturae systematice proposita per classes, ordines, genera et species. *Lugduni Batavorum*.

Linnaeus, C. (1758) Systema naturae per regna tria naturae, secundum classes, ordines, genera, species cum caracteribus, differentiis, synonymis, locis. Editio decima reformata. Tom I. *Laurentii Salvii, Holmiae*.

Löther. R. (1972) *Die Beherrschung der Mannigfaltigkeit. Philosophische Grundlagen der Taxonomie*. VEB G. Fischer Verlag, Jena.

Lorenzen. S. (1976) Zur Theorie der phylogenetischen Systematik. *Verh. Dtsch. Zool. Ges.* (Hamburg). 229.

Lorenzen. S. (1981) Entwurf eines phylogenetischen Systems der freilebenden Nematoden. *Veröffentl. Inst. Meeresforsch. Bremerhaven*, Suppl. **7**, 1–472.

Lorenzen. S. (1983) Phylogenetic Systematics: Problems, achievements and its application to the Nematoda. In Stone, A. R., Platt, H. M. & Khalil, I. F. (Eds.): *Concepts in nematode systematics*. Systematics Association Special Volume **22**. Academic Press, London and New York.

Løvtrup. S. (1977) *The phylogeny of Vertebrata*. J. Wiley & Sons, London, New York, Sydney, Toronto.

Lyons, K. M. (1973) Epidermal fine structure and development in the oncomiracidium larva of *Entobdella soleae* (Monogenea). *Parasitology* **66**, 321–333.

Mai, L. L. (1983) A model of chromosome evolution and its bearing on cladogenesis in the hominoidea. In Ciochon, R. L. & Corruccini, R. S. (Eds.): *New interpretations of ape and human ancestry*. Plenum Press, New York, London.

Mainitz, M. (1977) The fine structure of the stylet apparatus in Gnathostomulida Scleroperalia and its relationship to turbellarian stylets. *Acta Zool. Fenn.* **154**, 163–174.

Mainitz, M. (1979) The fine structure of gnathostomulid reproductive organs. *Zoomorphologie* **92**, 241–272.

Manton, S. M. (1977) *The Arthropoda. Habits, functional morphology, and evolution*. Clarendon Press, Oxford.

Manton, S. M. (1979) Functional morphology and the evolution of the hexapod classes. In Gupta, A. P. (Ed.): *Arthropod phylogeny*. Van Nostrand Reinhold Company, New York.

Manton, S. M. & Anderson, D. T. (1979) Polyphyly and the evolution of arthropods. In House, M. R. (Ed.): *The origin of major invertebrate groups*. Systematics Association Special Volume **12**. Academic Press, London, New York.

Marcus, E. (1950) Turbellaria Brasileiros (8). *Bol. Fac. Fil. Ciênc. Letr. Univ. São Paulo, Zoologia* **15**, 5–192.

Marshall, L. (1979) Evolution of metatherian and eutherian (mammalian) characters: a review based on cladistic methodology. *Zool. Journ. Linn. Soc. Lond.* **66**, 369–410.

Martens, P. M. (1983) Three new species of Minoninae (Turbellaria, Proseriata, Monocelididae) from the North Sea, with remarks on the taxonomy of the subfamily. *Zool. Scr.* **12**, 153–160.

Maslin, T. P. (1979) Morphological criteria of phyletic relationships. *Syst. Zool.* **1** 49–70.

Mayr, E. (1963) *Animal species and evolution*. Harvard University Press, Cambridge, Massachusetts.

Mayr, E. (1965) Numerical phenetics and taxonomic theory. *Syst. Zool.* **14**, 73–97.

Mayr, E. (1967) *Artbegriff und Evolution*. P. Parey, Hamburg, Berlin.

Mayr, E. (1969) *Principles of systematic zoology*. McGraw-Hill Book Company, New York.

Mayr, E. (1974) Cladistic analysis or cladistic classification? *Z. zool. Syst. Evolut.-forsch.* **12**, 94–128.

Mayr, E. (1975) *Grundlagen der zoologischen Systematik*. P. Parey, Hamburg, Berlin.

Mayr, E. (1976) Is the species a class or an individual? *Syst. Zool.* **25**, 192.

Mayr, E. (1981) Biological classification: Towards a synthesis of opposing methodologies. *Science* **214**, 510–516.

Mayr, E. (1982a) Speciation and macroevolution. *Evolution* **36**, 1119–1132.

Mayr, E. (1982b) Adaptation and selection. *Biol. Zbl.* **101**, 161–174.

Mayr, E. (1982c) *The growth of biological thought. Diversity, evolution, and inheritance*. The Belknap Press, Harvard University Press, Cambridge, Massachusetts.

Maze, J. & Bradfield, G. E. (1982) Neo-darwinian evolution—panacea or popgun. *Syst. Zool.* **31**, 92–95.

McKenna, M. C. (1975) Towards a phylogenetic classification of the Mammalia. In Luckett, W. P. & Szalay, F. S. (Eds.): *Phylogeny of the primates*. Plenum Press, New York, London.

Meissner, K. (1976) *Homologieforschung in der Ethologie*. VEB Fischer Verlag, Jena.

Mickevich, M. F. (1978) Taxonomic congruence. *Syst. Zool.* **27**, 143–158.

Miller, D. A. (1977) Evolution of primate chromosomes. Man's closest relative may be the gorilla, not the chimpanzee. *Science* **198**, 1116–1124.

Minot, C. S. (1876) On the classification of some of the lower worms. *Proc. Boston Soc. Natur. Hist.* **19**.

Mishler, B. D. & Donoghue, M. J. (1982) Species concepts: A case for pluralism. *Syst. Zool.* **31**, 491–503.

Moraczewski, J. (1981) Fine structure of some Catenulida (Turbellaria, Archoophora). *Zool. Polonia* **28**, 368–415.

Müller, U. & Ax, P. (1971) Gnathostomulida von der Nordseeinsel Sylt mit Betrachtungen zur Lebensweise und Entwicklung von *Gnathostomula paradoxa* Ax. *Mikrofauna Meeresboden* **9**, 1–41.

Napier, J. (1962) The evolution of the hand. *Scientific American* **140**, 56–62.

Nelson, G. J. (1969) Gill arches and the phylogeny of fishes, with notes on the classification of vertebrates. *Bull. Amer. Mus. Nat. Hist.* **141**, 475–552.

Nelson, G. (1972) Phylogenetic relationships and classification. *Syst. Zool.* **21**, 227–231.

Nelson, G. J. (1973) Classification as an expression of phylogenetic relationships. *Syst. Zool.* **22**, 344–359.

Nelson, G. J. (1978) Ontogeny, phylogeny, paleontology, and the biogenetic law. *Syst. Zool.* **27**, 324–345.

Nelson, G. & Platnick, N. (1981) *Systematics and biogeography. Cladistics and vicariance.* Columbia University Press, New York.

Nelson, G. & Rosen, D. E. (Eds.) (1981) *Vicariance biogeography. A critique*. Columbia University Press, New York.

Norrevang, A. & Wingstrand, K. G. (1979) On the occurrence and structure of choanocyte-like cells in some echinoderms. *Acta Zool.* **51**, 249–270.

Olson, E. C. (1959) The evolution of mammalian characters. *Evolution* **13**, 344–353.

Osche, G. (1961) Aufgaben und Probleme der Systematik am Beispiel der Nematoden. *Verhandl. Dtsch. Zool. Ges.* (Bonn 1960), 329–384.

Osche, G. (1973) Das Homologisieren als eine grundlegende Methode der Phylogenetik. *Aufsätze Red. senckenb. naturf. Ges.* **24**, 155–165.

Osche, G. (1975) Die Vergleichende Biologie und die Beherrschung der Mannigfaltigkeit. *Biologie in unserer Zeit* **5**, 139–146.

Osche, G. (1982) Rekapitulationsentwicklung und ihre Bedeutung für die Phylogenetik— Wann gilt die "Biogenetische Grundregel"? *Verh. naturw. Ver. Hamburg* (N. S.) **25**, 5–31.

321

Ostrom, J. H. (1972) Description of the *Archaeopteryx* specimen in the Teyler Museum, Haarlem. *Proc. Koninkl. Nederl. Akad. Wet., (B)* **75**, 289–305.

Ostrom, J. H. (1976) *Archaeopteryx* and the origin of birds. *Biol. J. Linn. Soc. Lond.* **8**, 91–182.

Owen, R. (1848) Report on the archetype and homologies of the vertebrate skeleton. *Reports 16th Meeting, British Assoc. Adv. Sci.* 169–340.

Panchen, A. L. (1982) The uses of parsimony in testing phylogenetic hypothesis. *Zool. J. Linn. Soc. Lond.* **74**, 305–328.

Parsons, T. S. (1970) The nose and Jacobson's organ. In Gans, C. & Parsons, T.S. (Eds.): *Biology of the Reptilia.* Vol. 2. Academic Press, London, New York.

Parsons, T. S. & Williams, E. E. (1962) The teeth of Amphibia and their relation to amphibian phylogeny. *J. Morph.* **110**, 375–383.

Parsons, T. S. & Williams, E. E. (1963) The relationships of the modern Amphibia. *Quart. Rev. Biol.* **38**, 26–53.

Patterson, C. (1978a) Verifiability in systematics. *Syst. Zool.* **27**, 218–222.

Patterson, C. (1978b) Arthropods and ancestors. *Antenna, Bull. Roy. Ent. Soc. London* **2**, 99–103.

Patterson, C. (1980) Cladistics. *The Biologist* **27**, 234–240.

Patterson, C. (1981a) Significance of fossils in determining evolutionary relationships. *Ann. Rev. Ecol. Syst.* **12**, 195–223.

Patterson, C. (1981b) Methods of palaeobiogeography. In Nelson, G. J. & Rosen, D. E. (Eds.): *Vicariance biogeography, a critique.* Columbia University Press, New York.

Patterson, C. (1982a) Morphological characters and homology. In Joysey, K. A. & Friday, A. E. (Eds.): *Problems of phylogenetic reconstruction. Systematics Association Special Volume* **21**. Academic Press, London.

Patterson, C. (1982b) Classes and cladists or individuals and evolution. *Syst. Zool.* **31**, 284–286.

Patterson, C. & Rosen, D. E. (1977) Review of ichthyodectiform and other Mesozoic teleost fishes and the theory and practice of classifying fossils. *Bull. Am. Mus. Natur. Hist.* **158**, 81–172.

Paulus, H. F. (1979) Eye structure and the monophyly of Arthropoda. In Gupta, A. P. (Ed.): *Arthropod phylogeny.* Van Nostrand Reinhold Company, New York.

Peters, D. S. (1970) Über den Zusammenhang von biologischem Artbegriff und phylogenetischer Systematik. *Aufsätze Red. Senckenberg naturforsch. Ges.* **18**, 1–35.

Peters, D. S. (1972) Das Problem konvergent entstandener Strukturen in der anagetischen und genealogischen Systematik. *Z. zool. Syst. Evolut.-forsch.* **10**, 161–173.

Peters, D. S. & Gutmann, W. F. (1971) Über die Lesrichtung von Merkmals- und Konstruktions-Reihen. *Z. zool. Syst. Evolut.-forsch.* **9**, 237–263.

Peters, G. (1978) Die Taxonomie auf dem Weg zur Analyse der Stammverwandten. In Böhme, H., Hagemann, R. & Löther, R. (Eds.): *Beiträge zur Genetik und Abstammungslehre.* 2nd. Edn., Volk und Wissen, Volkseigener Verlag, Berlin.

Peters, G. & Klausnitzer, B. (1978) Phylogenetische Systematik als Methode zur Erforschung der Stammesgeschichte der Tiere. *Biol. Rdsch.* **16**, 88–98.

Platnick, N. I. (1977a) Paraphyletic and polyphyletic groups. *Syst. Zool.* **26**, 195–200.

Platnick, N. I. (1977b) The hypochiloid spiders: A cladistic analysis, with notes on the Atypoidea (Arachnida, Araneae). *Amer. Mus, Novitates* **2627**, 1–23.

Platnick, N. I. (1978a) Classifications, historical narratives, and hypothesis. *Syst. Zool.* **27**, 365–369.

Platnick, N. I. (1978b) Gaps and prediction in classification. *Syst. Zool.* **27**, 472–474.

Platnick, N. I. (1979) Philosophy and the transformation of cladistics. *Syst. Zool.* **28**, 537–546.

Platnick, N. I. (1982) Defining characters and evolutionary groups. *Syst. Zool.* **31**, 282–284.

Platnick, N. I. & Funk, V. A. (Eds.) (1983) *Advances in cladistics.* Volume **2**. *Proceedings of the Willi Hennig Society.* Columbia University Press, New York.

Platnick, N. I. & Gaffney, E. S. (1977) Systematics: A Popperian perspective. *Syst. Zool.* **26**, 360–365.

322

Platnick, N. I. & Gaffney, E. S. (1978) Systematics and the Popperian paradigma. *Syst. Zool.* **27**, 381–388.

Platnick, N. I. & Gertsch, W. J. (1976) The suborders of spiders. A cladistic analysis (Arachnida, Araneae). *Amer. Mus. Novitates.* **2607**, 1–15.

Popper, K. R. (1974) *Objektive Erkenntnis. Ein evolutionärer Entwurf.* 2nd. Edn. Hoffmann & Campe, Hamburg.

Popper, K. R. (1976) *Logik der Forschung.* 6. Edn. Mohr (Paul Siebeck), Tübingen.

Popper, K. R. (1979) *Ausgangspunkte. Meine intellektuelle Entwicklung.* Hoffmann und Campe.

Prothero, D. R. & Lazarus, D. B. (1980) Planktonic microfossils and the recognition of ancestors. *Syst. Zool.* **28**, 119–129.

Rähr, H. (1981) The ultrastructure of the blood vessels of *Branchiostoma lanceolatum* (Pallas) (Cephalochordata). I. Relations between blood vessels, epithelia, basal lamina and "connective tissue". *Zoomorphology* **97**, 53–74.

Reed, C. A. (1960) Polyphyletic or monophyletic ancestry of mammals, or: what is a class? *Evolution* **14**, 314–322.

Reisinger, E. (1970) Zur Problematik der Evolution der Coelomaten. *Z. zool. Syst. Evolut.-forsch.* **8**, 81–109.

Reisinger, E. (1972) Die Evolution des Orthogons der Spiralia und das Archicoelomaten-problem. *Z. zool. Syst. Evolut.-forsch.* **10**, 1–43.

Reisinger, E. & Kelbetz, S. (1964) Feinbau und Entladungsmechanismus der Rhabditen. *Z. wiss. Mikrosk.* **65**, 472–508.

Reisz, R. R. (1980) The Pelycosauria: A review of phylogenetic relationships. In Panchen, A. L. (Ed.): *The terrestrial environment and the origin of land vertebrates.* Academic Press, London, New York.

Remane, A. (1952) *Die Grundlagen des natürlichen Systems, der vergleichenden Anatomie und der Phylogenetik.* Akademische Verlagsgesellschaft Geest & Portig, K.-G., Leipzig.

Remane, A. (1955) Morphologie als Homologienforschung. *Verh. Dtsch. Zool. Ges.* (Tübingen 1954), 159–183.

Remane, A. (1963) The enterocelic origin of the celom. In Dougherty, E. C. (Ed.): *The lower Metazoa Comparative biology and phylogeny.* University of California Press, Berkeley, Los Angeles.

Remane, A. (1967) Die Geschichte der Tiere. In Heberer, G. (Ed.): *Die Evolution der Organismen.* Vol. 1. 3rd Edn., G. Fischer Verlag, Stuttgart.

Remane, A., Storch, V. & Welsch, U. (1972) *Evolution.* Deutscher Taschenbuch Verlag, München.

Remane, A., Storch, V. & Welsch, U. (1986) *Systematische Zoologie.* G. Fischer, Stuttgart, New York, 3rd Edn.

Rensch, B. (1947) *Neuere Probleme der Abstammungslehre. Die transspezifische Evolution.* Enke, Stuttgart.

Riedl, R. (1969) Gnathostomulida from America. First record of the new phylum from North America. *Science* **163**, 445–452.

Riedl, R. (1971) On the genus *Gnathostomula* (Gnathostomulida). *Int. Rev. ges. Hydrobiol.* **56**, 385–496.

Riedl, R. (1975) *Die Ordnung des Lebendigen. Systembedingungen der Evolution.* P. Parey, Hamburg, Berlin.

Riedl, R. (1977) A system-analytical approach to macroevolutionary phenomena. *Quart. Rev. Biol.* **52**, 352–370.

Riedl, R. (1980) Homologien: ihre Gründe und Erkenntnisgründe. *Verhandl. Dtsch. Zool. Ges.* (Berlin), 164–176.

Rieger, R. M. (1976) Monociliated epidermal cells in Gastrotricha: Significance for concepts of early metazoan evolution. *Z. zool. Syst. Evolut.-forsch.* **14**, 198–226.

Rieger, R. M. (1978) Multiple ciliary structures in developing spermatozoa of marine Catenulida (Turbellaria). *Zoomorphologie* **89**, 229–236.

Rieger, R. M. (1980) A new group of interstial worms, Lobatocerebridae nov. fam. (Annelida) and its significance for metazoan phylogeny. *Zoomorphologie* **95**, 41–84.

Rieger, R. M. (1981) Morphology of the Turbellaria at the ultrastructural level. *Hydrobiologia* **84**, 287–300.

Rieger, R. M., & Doe, D. (1975) The proboscis armature of the Turbellaria Kalyptorhynchia. A derivative of the basement lamina. *Zool. Scr.* **4**, 25–32.

Rieger, R. M., & Mainitz, M. (1977) Comparative fine structure study of the body wall in Gnathostomulida and their phylogenetic position between Plathyhelminthes und Aschelminthes. *Z. zool. Syst. Evolut.-forsch* **15**, 9–35.

Rieppel, O. (1979) Ontogeny and the recognition of primitive characters. *Z. zool. Syst. Evolut.-forsch.* **17**, 57–61.

Rieppel, O. (1980a) Why to be a cladist. *Z. zool. Syst. Evolut.-forsch.* **18**, 81–90.

Rieppel, O. (1980b) Homology, a deductive concept? *Z. zool. Syst. Evolut.-forsch.* **18**, 315–319.

Rieppel, O. (1983) The "tertium comparationis" of competing evolutionary theories. *Z. zool. Syst. Evolut.-forsch.* **21**, 1–6.

Rieppel, O. (1984) Können Fossilien die Evolution beweisen? Das Problem fossiler Zwischenformen. *Natur und Museum* **114**, 69–74.

Romer, A. S. (1957) *Osteology of the reptiles.* Chicago Univ. Press.

Rosen, D. E. & Buth, D. G. (1980) Empirical evolutionary research versus neo-Darwinian speculation. *Syst. Zool.* **29**, 300–308.

Rosen, D. E., Forey, P. L., Gardiner, B. G. & Patterson, C. (1981) Lungfishes, tetrapods, palaeontology, and plesiomorphy. *Bull. Am. Mus. Nat. Hist.* **167**, (4), 159–276.

Rosen, D. E., Nelson, G. & Patterson, C. (1979) Foreword in Hennig, W.: *Phylogenetic systematics.* University of Illinois Press, Urbana, Chicago, London (unchanged impression of the 1966 edition).

Ross, H. H. (1974) *Biological systematics.* Addison Wesley Publishing Company, Inc. Reading, Massachusetts.

Ruppert, E. E. (1978) A review of metamorphosis of turbellarian larvae. In Chia, F. S. & Rice, M. (Eds.): *Settlement and metamorphosis of marine invertebrate larvae.* Elsevier, North-Holland, Biomedical Press.

Ruppert, E. E. & Carle, K. J. (1983) Morphology of metazoan circulatory systems. *Zoomorphology* **103**, 193–208.

Salvini-Plawen, L. v. (1972) Zur Morphologie und Phylogenie der Mollusken: Die Beziehungen der Caudofoveata und der Solenogastres als Aculifera, als Mollusca und als Spiralia. *Z. wiss. Zool.* **184**, 205–394.

Salvini-Plawen, L. v. (1978) On the origin and evolution of the lower Metazoa. *Z. zool. Syst. Evolut.-forsch.* **16**, 40–88.

Salvini-Plawen, L. v. (1980a) Phylogenetischer Status und Bedeutung der mesenchymaten Bilateria. *Zool. Jb. Anat.* **103**, 354–373.

Salvini-Plawen, L. v. (1980b) Was ist eine Trochophora? Eine Analyse der Larventypen mariner Protostomier. *Zool. Jb. Anat.* **103**, 389–423.

Salvini-Plawen, L. v. (1981) On the origin and evolution of the Mollusca. In: Origine dei grandi phyla dei Metazoi. *Atti dei Convegni Lincei* **49**, Acad. Naz. dei Lincei, Roma.

Sarich, V. M. (1983) Retrospective on hominoid macromolecular systematics. Appendix to Cronin, J. E.: Apes, humans and molecular clocks. A reappraisal. In Ciochon, R. L., & Corruccini, R. S. (Eds.): *New interpretations of ape and human ancestry.* Plenum Press, New York, London.

Sarich, V. M. & Cronin, J. E. (1976) Molecular systematics of the primates. In Goodman, M. & Tashian, R. E. (Eds.): *Molecular anthropology* Plenum Press, New York, London.

Schaeffer, B., Hecht, M. K. & Eldredge, N. (1972) Paleontology and phylogeny. *Evol. Biol.* **6**, 31–44.

Scheibe, E. (1983) Kriterien zur Beurteilung der Naturwissenschaften. In Funke, G. & Scheibe, E.: Wissenschaft und Wissenschaftsbegriff. *Akad. Wiss. Lit. Mainz. Abhandl. Math.-Nat. Kl. Jg.* 1983, No. 3, 25–41.

Schilke, K. (1969) Zwei neuartige Konstruktionsytpen des Rüsselapparates der Kalyptorhynchia (Turbellaria). *Z. Morph. Tiere* **65**, 287–314.

324

Schilke. K. (1970) Zur Morphologie und Phylogenie der Schizorhynchia (Turbellaria, Kalyptorhynchia). *Z. Morph. Tiere.* **67**, 118–171.

Schindewolf, O. H. (1969) Über den "Typus" in morphologischer und phylogenetischer Biologie. *Akad. Wiss. Lit. Mainz. Abhandl. Math.-Naturw. Kl. Jg. 1969. Nr. 4*, 57–131.

Schlee. D. (1969) Hennig's principle of phylogenetic systematics, an "Intuitive statistico-phenetic taxonomy"? *Syst. Zool.* **18**, 127–134.

Schlee, D. (1971) Die Rekonstruktion der Phylogenese mit Hennig's Prinzip. *Aufsätze Red. senckenberg. naturforsch. Ges.* **20**, W. Kramer, Frankfurt/M.

Schlee. D. (1975a) An analysis of numerical phenetics. *Ent. scand.* **6**, 1–9.

Schlee, D. (1975b) Numerical phenetics: An analysis from the viewpoint of phylogenetic systematics. *Ent. scand.* **6**, 193–208.

Schlee. D. (1976) Structures and functions, their general significance for phylogenetic reconstruction in recent and fossil taxa. *Zool. Scr.* **5**, 181–184.

Schlee, D. (1978a) Anmerkungen zur phylogenetischen Systematik: Stellungnahme zu einigen Mißverständnissen. *Stuttgarter Beitr. Naturk.*, (A) **320** 1–14.

Schlee, D. (1978b) In Memoriam Willi Hennig 1913–1976. Eine biographische Skizze. *Ent. Germ.* **4**, 377–391.

Schlee, D. (1981) Grundsätze der phylogenetischen Systematik (Eine praxisorientierte Übersicht). *Paläont. Z.* **55**, 11–30.

Schmidt. G. A. (1966) *Evolutionäre Ontogenie der Tiere.* Akademie-Verlag, Berlin.

Schuh, R. T. (1976) Pretarsal structure in the Miridae (Hemiptera) with a cladistic analysis of the relationships within the family. *Americ. Mus. Novitates* **2601**, 1–36.

Schultz, A. H. (1936) Characters common to higher primates and characters specific for man. *Quart. Rev. Biol.* **11**, 259–283 & 425–455.

Schultze. H. P. (1970) Folded teeth and the monophyletic origin of tetrapods. *Americ. Mus. Novitates* **2408**, 1–10.

Schwartz, J. H., Tattersall, I. & Eldredge, N. (1978) Phylogeny and classification of the primates revisited. *Yearbook Phys. Anthro.* **21**, 95–133.

Seuanez, H. N. (1979) *The phylogeny of human chromosomes.* Springer-Verlag, Berlin, Heidelberg, New York.

Siewing. R. (1960) Zum Problem der Polyphylie der Arthropoda. *Z. wiss. Zool.* **164**, 238–270.

Siewing. R. (1969) *Lehrbuch der vergleichenden Entwicklungsgeschichte der Tiere.* P. Parey, Hamburg, Berlin.

Siewing, R. (1972) Zur Deszendenz der Chordaten—Erwiderung und Versuch einer Geschichte der Coelomaten. *Z. zool. Syst. Evolut.-forsch.* **10**, 267–291.

Siewing. R. (1976) Probleme und neuere Erkenntnisse in der Großsystematik der Wirbellosen. *Verh. Dtsch. Zool. Ges.* (Hamburg), 59–83.

Siewing, R. (1980a) Körpergliederung und phylogenetisches System. *Zool. Jb. Anat.* **103**, 196–210.

Siewing, R. (1980b) Das Archicoelomatenkonzept. *Zool. Jb. Anat.* **103**, 439–482.

Siewing. R. (1981) Problems and results of research on the phylogenetic origin of Coelomata. In: Origine dei grandi phyla dei Metazoi. *Atti dei Convegni Lincei* **49**. Acad. Naz. dei Lincei. Rome.

Simpson, G. G. (1959) Mesozoic mammals and the polyphyletic origin of mammals. *Evolution* **13**, 405–414.

Simpson, G. G. (1961) *Principles of animal taxonomy.* Columbia University Press. New York.

Smith, J. P. S. (1981) Fine-structural observations on the central parenchym in *Convoluta* spec. *Hydrobiologia* **84**, 259–264.

Smith, J., Tyler, S., Thomas, M. B. & Rieger, R. M. (1982) The morphology of turbellarian rhabdites: phylogenetic implications. *Trans. Amer. Microsc. Soc.* **101**, 209–228.

Smyth, J. D. & Clegg, J. A. (1959) Egg-shell formation in Trematodes and Cestodes. *Exp. Parasitol.* **8**, 286–323.

Sneath, P. H. A. & Sokal, R. R. (1973) *Numerical taxonomy. The principles and practice of numerical classification.* W. H. Freeman and Company, San Francisco.

Sober, E. R. (1983) Parsimony methods in systematics. In Platnick, N. I. & Funk, V. A.

(Eds.): *Advances in cladistics*. Volume 2. *Proceedings of the second meeting of the Willi Hennig Society*. Columbia University Press, New York.

Sokal, R. R. & Sneath, P. H. A. (1963) *Principles of numerical taxonomy*. W. H. Freeman and Company, San Francisco, London.

Sopott-Ehlers, B. & Ax, P. (1985) Proseriata (Plathelminthes) von der nord-amerikanischen Pazifikküste III. Monocelididae. *Microfauna Marina* **2**, 331–346.

Spies, T. (1981) Structure and phylogenetic interpretation of diplopod eyes (Diplopoda). *Zoomorphology* **98**, 241–260.

Stackebrandt, E. & Woese, C. R. (1981) The evolution of procaryotes. In Carlile, M. J., Collins, J. F. & Moseley, B. E. B. (Eds.): *Molecular and cellular aspects of microbial evolution*. Cambridge University Press, Cambridge, London, New York, New Rochelle, Melbourne, Sydney.

Stanley, S. M. (1979) *Macroevolution. Pattern and Process*. W. H. Freeman and Company, San Francisco.

Starck D. (1974) Die Stellung der Hominiden im Rahmen der Säugetiere. In Heberer, G. (Ed.): *Die Evolution der Organismen*. Vol III. C. Fischer Verlag, Stuttgart.

Starck, D. (1978a) Das evolutive Plateau Säugetier. Eine Übersicht unter besonderer Berücksichtigung der stammesgeschichtlichen und systematischen Stellung der Monotremata. *Sonderbd. naturwiss. Verein Hamburg*. **3**. 7–33.

Starck, D. (1978b) *Vergleichende Anatomie der Wirbeltiere 1*. Springer-Verlag, Berlin, Heidelberg, New York.

Starck, D. (1979) *Vergleichende Anatomie der Wirbeltiere 2*. Springer-Verlag, Berlin, Heidelberg, New York.

Starck, D. (1982) *Vergleichende Anatomie der Wirbeltiere 3*. Springer-Verlag, Berlin, Heidelberg, New York.

Starck, D. & Siewing, R. (1980) Zur Diskussion der Begriffe Mesenchym und Mesoderm. *Zool. Jb. Anat.* **103**. 374–388.

Steinböck, O. (1958) Zur Phylogenie der Gastrotrichen. *Verhandl. Dtsch. Zool. Ges.*, (Graz 1957), 196–218.

Steinböck, O. (1963) Origin and affinities of the lower Metazoa: The "aceloid" ancestry of the Eumetazoa. In Dougherty, E. C. (Ed.): *The lower Metazoa. Comparative biology and phylogeny*. University of California Press, Berkeley, Los Angeles.

Stern, J. T. (1970) The meaning of "adaptation" and its relation to the phenomenon of natural selection. *Evol. Biol.* **4**. 39–66.

Sterrer, W. (1968) Beitrag zur Kenntnis der Gnathostomulida. I. Anatomie und Morphologie des Genus *Pterognathia* Sterrer. *Ark. Zool.* (2)**22**. 1–125.

Sterrer, W. (1971) On the biology of Gnathostomulida. *Vie et Milieu, Suppl.* **22**, 493–508.

Sterrer, W. (1972) Systematics and evolution within the Gnathostomulida. *Syst. Zool.* **21**, 151–173.

Sterrer, W. (1974) Gnathostomulida. In Giese, A. C. & Pearse, J. S. (Eds.): *Reproduction of marine invertebrates. I. Acoelomate and pseudocoelomate metazoans*. Academic Press, Inc., New York.

Sterrer, W., Mainitz, M. & Rieger, R. M. (1985) Gnathostomulida: enigmatic as ever. In Conway Morris, S., George, J. D., Gibson, R. & Platt, H. M. (Eds.): *The origins and relationships of lower invertebrates. The Systematic Association*. Special Volume **28**, 181–199. Clarendon Press, Oxford.

Sterrer, W. & Rieger, R. (1974) Retronectidae—a new cosmopolitan marine family of Catenulida (Turbellaria). In Riser, W. & Morse, M. P. (Eds.): *The biology of Turbellaria*. McGraw-Hill Book Company, New York.

Stevens, P. F. (1980) Evolutionary polarity of character states. *Ann. Rev. Ecol. Syst.* **11**, 333–358.

Straus, W. L. (1949) The riddle of man's ancestry. *Quart. Rev. Biol.* **24**. 200–223.

Szalay, F. S. & Delson, E. (1979) *Evolutionary history of the primates*. Academic Press, New York.

Szarski, H. (1977) Sarcopterygii and the origin of tetrapods. In Hecht, M. K., Goody, P. C.

& Hecht, B. M. (Eds.): *Major patterns in vertebrate evolution*. Plenum Press, New York, London.

Takhtajan, A. (1959) *Die Evolution der Angiospermen*. VEB G. Fischer-Verlag, Jena.

Tarsitano, S. & Hecht, M. K. (1980) A reconsideration of the reptilian relationship of *Archaeopteryx*. *Zool. Journ. Linn. Soc. London* **69**, 149–182.

Tattersall, J. & Eldredge, N. (1977) Fact, theory and fantasy in human paleobiology. *Amer. Sci.* **65**, 204–211.

Teuchert, G. (1973) Die Feinstruktur des Protonephridialsystems von *Turbanella cornuta* Remane, einem marinen Gastrotrich der Ordnung Macrodasyoidea. *Z. Zellforsch.* **136**, 277–289.

Teuchert, G. (1977a) The ultrastructure of the marine gastrotrich *Turbanella cornuta* Remane (Macrodasyoidea) and its functional and phylogenetical importance. *Zoomorphologie* **88**, 189–246.

Teuchert, G. (1977b) Leibeshöhlenverhältnisse von dem marinen Gastrotrich *Turbanella cornuta* Remane (Macrodasyoidea) und eine phylogenetische Bewertung. *Zool. Jb. Anat.* **97**, 586–596.

Teuchert, G. & Lappe, A. (1980) Zum sogenannten "Pseudocoel" der Nemathelminthes.— Ein Vergleich der Leibeshöhlen von mehreren Gastrotrichen. *Zool. Jb. Anat.* **103**, 424–438.

Thenius, E. (1979) *Die Evolution der Säugetiere. Eine Übersicht über Ergebnisse und Probleme*. G. Fischer Verlag, Stuttgart, New York.

Thomas, M. B. (1975) The structure of the 9 + 1 axonemal core as revealed by treatment with trypsin. *J. Ultrastructure. Research* **52**, 409–422.

Tschulok, S. (1922) *Deszendenzlehre (Entwicklungslehre). Ein Lehrbuch auf historisch-kritischer Grundlage*. G. Fischer, Jena.

Tuttle, R. H. (1969) Knuckle-walking and the problem of human origin. *Science* **166**, 953–961.

Tuttle, R. (1975) Parallelism, brachiation and hominoid phylogeny. In Luckett, W. P. & Szalay, F. S. (Eds.): *Phylogeny of the primates*. Plenum Press, New York, London.

Turbeville, J. M. & Ruppert, E. E. (1981) Ultrastructure of the nemertine circulatory system: Blood vessels or coelomic channels. *Am. Zool.* **21**, 989.

Tyler, S. (1976) Comparative ultrastructure of adhesive systems in the Turbellaria. *Zoomorphologie* **84**, 1–76.

Tyler, S. (1979) Distinctive features of cilia in metazoans and their significance for systematics. *Tissue & Cell* **11**, 385–400.

Tyler, S. & Rieger, R. M. (1975) Uniflagellate spermatozoa in Nemertoderma and their phylognetic significance. *Science* **188**, 730–732.

Tyler, S. & Rieger, R. M. (1977) Ultrastructural evidence for the systematic position of the Nemertodermatida (Turbellaria). *Acta Zool. Fenn.* **154**, 193–207.

Van Valen, L. (1978) Why not to be a cladist. *Evol. Theory* **3**, 285–299.

Vollmer, G. (1975) *Evolutionäre Erkenntnistheorie*. S. Hirzel Verlag, Stuttgart.

Voigt, W. (1973) *Homologie und Typus in der Biologie*. VEB G. Fischer Verlag, Jena.

Wake, M. H. (Ed.) (1979) *Hyman's comparative vertebrate anatomy*. Third edition. The University of Chicago Press, Chicago, London.

Washburn, S. L. (1968) Behaviour and the origin of man. *Proc. Royal Anthropol. Inst. Great Britain and Ireland* 1966, 21–27.

Watrous, L. E. & Wheeler, Q. D. (1981) The out-group comparison method of character analysis. *Syst. Zool.* **30**, 1–11.

Watson, D. M. S. (1954) On *Bolosaurus* and the origin and classification of the reptiles. *Bull. Mus. Comp. Zool. Harv.* **111**, 295–449.

Weinert, H. (1944) *Ursprung der Menschheit. Über den engeren Anschluß des Menschengeschlechts an die Menschenaffen*. Enke, Stuttgart.

Wellnhofer, P. (1974) Das fünfte Skelettexemplar von *Archaeopteryx*. *Palaeontographica Abt. A*, **147**, 169–216.

Werner, F. C. (1970) *Die Benennung der Organismen und Organe nach Größe, Form, Farbe und anderen Merkmalen*. VEB Max Niemeyer Verlag, Halle (Saale).

327

Westblad, E. (1949) *Xenoturbella bocki* n. g. n. sp. a peculiar primitive turbellarian type. *Ark. Zool.*, Ser. 2, **1**, 11–29.

Westgaard, B. (1983) A new detailed model for mammalian dentitional evolution. *Z. zool. Syst. Evolut.-forsch.* **21**, 68–78.

Weygoldt, P. (1979) Significance of later embryonic stages and head development in arthropod phylogeny. In Gupta, A. P. (Ed.): *Arthropod phylogeny.* Van Nostrand Reinhold Company, New York.

Weygoldt, P. & Paulus, H. P. (1979a) Untersuchungen zur Morphologie, Taxonomie und Phylogenie der Chelicerata. I. Morphologische Untersuchungen. *Z. zool. Syst. Evolut.-forsch.* **17**, 85–116.

Weygoldt, P. & Paulus, H. P.(1979b) Untersuchungen zur Morphologie, Taxonomie und Phylogenie der Chelicerata. II. Cladogramme und die Entfaltung der Chelicerata. *Z. zool. Syst. Evolut.-forsch.* **17**, 177–200.

White, M. J. D. (1978) *Modes of speciation.* W. H. Freeman & Company, San Francisco.

Wickler, W. (1965) Die Evolution von Mustern der Zeichnung und des Verhaltens. *Naturwiss.* **52**, 335–341.

Wickler, W. (1967) Vergleichende Verhaltensforschung und Phylogenetik. In Heberer, G. (Ed.): *Die Evolution der Organismen.* 3. Edn. Vol. 1. C. Fischer-Verlag, Stuttgart.

Wiley, E. O. (1975) & Karl. R. Popper, systematics, and classification—a reply to Walter Bock and other evolutionary taxonomists. *Syst. Zool.* **24**, 233–243.

Wiley, E. O. (1976) The phylogeny and biogeography of fossil and recent gars (Actinopterygii: Lepisosteidae). *Misc. Publ. Mus. Natur. Hist. Univ. Kansas* **64**, 1–111.

Wiley, E. O. (1978) The evolutionary species concept reconsidered. *Syst. Zool.* **27**, 17–26.

Wiley, E. O. (1979a) Cladograms and phylogenetic trees. *Syst. Zool.* **18**, 88–92.

Wiley, E. O. (1979b) Ancestors, species and cladograms—remarks on the symposium. In: Cracraft, J. & Eldredge, N. (Eds.): *Phylogenetic analysis and paleontology.* Columbia University Press, New York.

Wiley, E. O. (1979c) An annotated Linnean hierarchy, with comments on natural taxa and competing systems. *Syst. Zool.* **28**, 308–337.

Wiley, E. O. (1979d) Ventral gill arch muscles and the interrelationships of gnathostomes, with a new classification of the Vertebrata. *Zool. J. Linn. Soc. Lond.* **67**, 149–180.

Wiley, E. O. (1980) Is the evolutionary species fiction?—A consideration of classes, individuals, and historical entities. *Syst. Zool.* **29**, 76–80.

Wiley, E. O. (1981) *Phylogenetics. The theory and practice of phylogenetic systematics.* J. Wiley and Sons, New York, Chichester, Brisbane, Toronto.

Wiley, E. O. & Brooks, D. R. (1982) Victims of history—a nonequilibrium approach to evolution. *Syst. Zool.* **31**, 1–24.

Wilhelmi, J. (1913) Platodaria, Plattiere. *Handbuch der Morphologie der wirbellosen Tiere* **3**, 1–148.

Willmann, R. (1983) Biospecies und phylogenetische Systematik. *Z. zool. Syst. Evolut.-forsch.* **21**, 241–249.

Willmann, R. (1985) *Die Art in Raum und Zeit. Das Artkonzept in der Biologie und Paläontologie.* P. Parey, Berlin, Hamburg.

Wilson, R. A., & Webster, L. A. (1974) Protonephridia. *Biol. Rev.* **49**, 127–160.

Wuketits, F. M. (1978) *Wissenschaftstheoretische Probleme der modernen Biologie.* Dunker und Humblodt, Berlin.

Wuketits, F. (1981) *Biologie und Kausalität. Biologische Ansätze zur Kausalität, Determination und Freiheit.* P. Parey, Berlin, Hamburg.

Wuketits, F. M. (1982) *Grundriß der Evolutionstheorie.* Wissenschaftliche Buchgesellschaft, Darmstadt.

Wuketits, F. M. (1983) *Biologische Erkenntnis: Grundlagen und Probleme.* G. Fischer Verlag, Stuttgart.

Yalden, D. W. (1985) Feeding mechanisms as evidence for cyclostome monophyly. *Zool. Journ. Linnean Soc.* **84**, 201–300.

Zilch, R. (1979) Cell lineage in arthropods? *Fortschr. zool. Syst. Evolut.-forsch* **1**, 19–41.

Subject Index

constructive and functional alteration of, 125–126
definition, 105
distribution of, 115
heterobathmy of, 109
homologous relationship between, 164–174
homology of, 162
in ontogeny and in the adult, 122–134
information content of multiple series of, 134–142
mosaic of, 109–113
requirement to be separable in an organism, 107
species-specific or group-specific pattern, 106, 112
Features as sources of information for phylogenetic research, 105–157
Fertilization mode, 267–268
Flagellum, 125, 126
Follicular gonads, 293
Fore limbs, 91, 172
Fore-wings, 303
Fossil record, phylogenetic research, 229–231
Fossils, 43, 121, 161, 193, 201–231, 233, 241, 299
Frontal organ, 274, 285

Gametes, 267
Gastro vascular system, 254, 263, 293
Genealogical relationship, 31, 192
Generic name, 237
Genetic variation, 127
Genus, 238
Germanium, 269, 293
Gills, 124, 155
Gland cells, 285, 288
Graphical representation of phylogenetic relationships, 42
Grasping hand, 165, 166, 177
Grasping hooks, 138–141, 173, 178
Ground pattern, 147–153, 303
  closed descent community, 147–149
  Mammalia, 149
  Onychophora, 150
  Peathelminthomorpha, 252–277
Groups, 134–235
Gut, 253–254, 256, 257, 293

Hallux, 94, 230
Head pouch, 168
Heart, 196, 302, 303
Heart chambers, 93, 98
Hermaphroditism, 270, 277, 278
Heterobathmy of features, 109–113

Hierarchical structure of the phylogenetic system, 39–41
Holophyly, 28
Homoiology, 53
Homoiothermy, 88, 94, 104, 154
Homologous relationship between features, 164–174
Homology, 158–178
  antonym of, 159
  'criteria' of, 160, 161
  definition, 159
  independent hypotheses of, 161, 162
  meaning of term, 160
  of features, 162
  with agreement, 165–168
  with alteration in the same direction in two lineages, 172–174
  with alteration on one side only, 168–170
  with divergent alteration in two lineages, 170–171
Homoplasy, 53
Humerus, 176
Hybridization, 13, 15, 18, 32, 233

Incertae sedis, taxa, 233, 234
Indentation, 243–244
Indications, 9
Individuals, 10, 19, 22, 23, 31, 106, 202, 203
Information storage, 300
Information theory, 308
In-group comparison, 121–122
Inner ear, 88, 92, 118
International Rules of Nomenclature, 232, 300
Intervertebral discs, 92
Intraspecific level, 129

Jacobson's organ, 93, 97, 98, 101
Johnston's organ, 125, 168

Labium, 168
Labyrinth, 107, 108, 119
Lagena, 93
Lamellar rhabdites, 287, 288
Larvae, 273, 274, 275, 303
Lesrichtungs-Kriterium, 130
Level of reproductive output, 128
Life cycle, 273–275
Life span of species, 13
Lineage, 12
Linnaean hierarchy, 8, 19, 24, 26, 42, 153, 232, 237, 239, 243, 281
Lungs, 94

Male copulatory organ, 88, 93
Malpighian tubules, 162, 163, 178, 195

331

Index of Animals

Testicardines, 256
Tetrapoda, 32, 87, 90, 91, 103, 117, 145, 146,
    149, 154, 193, 176, 237, 244, 248, 306
Thaiodoclada, 242
Thecodontia, 33
Therapsida, 33, 103, 193, 194, 219–221, 224
Theria, 39, 40, 41, 44, 45, 47, 82, 84, 86–88,
    90, 108, 111–113, 121, 122, 126,
    145–147, 149, 154, 194, 208, 217, 218,
    220–223, 229, 239, 244, 248, 305
Theromorpha, 220
Therosauria, 219, 220
Thylacinus, 83
Thysanura, 187–189, 195
Toads, 124
Tortoises, 91
Tracheata, 114, 118, 152, 155, 162, 163, 169,
    170, 177, 178, 195, 196, 198, 199
Tracheospira, 242
Trematoda, 33, 156, 183, 185, 281, 282, 283,
    296, 297, 300, 301, 305–307
Trepaxonemata, 254, 269, 274, 282, 283,
    288, 292
Triassochelys, 102
Trichoplax adhaerens, 258, 275
Tricladida, 253, 254, 262, 263
Triconodonta, 219, 221–223
Trilobita, 22, 33, 43, 155, 203, 212–215, 223,
    249
Tritylodontia, 84, 120, 231
Tubulidentata, 24, 76, 239, 240
Tunicata, 108, 119, 183
Turbanella, 267

Turbellaria, 33, 156, 157, 183, 185, 281, 300,
    302, 304–307
Turdus merula, 2
Turtles, 91
Typhlomolge, 124
Typhloplanoida, 281, 282, 294

Uniramia, 195
Ungulata, 172
Urodela, 91, 104, 124
Uropygi, 240

Vermes, 183, 184, 193
Vertebrata, 26, 69, 70, 108, 117–119, 144,
    154, 161, 177, 178, 183–185, 189, 237,
    238, 239, 252, 306

Water beetles, 133
Whales, 123
Woodpecker, 131

Xenarthra, 89, 170
Xenotoplana tridentis, 288
Xenoturbella, 250
Xenoturbella bocki, 250
Xenoturbellida, 250
Xenusion auerswaldae, 203
Xiphosura, 163, 194, 197, 201, 212–214,
    216, 241

Zaglossus, 41, 83, 85, 121
Zaglossus bruijni, 107, 122, 223
Zygentoma, 122, 174, 177, 186, 187, 188,
    192